ADVANCED ENGINEERING AND TECHNOLOGY III

PROCEEDINGS OF THE 3RD ANNUAL CONGRESS ON ADVANCED ENGINEERING AND TECHNOLOGY (CAET 2016), HONG KONG, 22–23 OCTOBER 2016

Advanced Engineering and Technology III

Editor

Liquan Xie

Department of Hydraulic Engineering, Tongji University, China

CRC Press
Taylor & Francis Group
Boca Raton London New York Leiden

CRC Press is an imprint of the
Taylor & Francis Group, an **informa** business
A BALKEMA BOOK

CRC Press/Balkema is an imprint of the Taylor & Francis Group, an informa business

Typeset by V Publishing Solutions Pvt Ltd., Chennai, India
Printed and bound in Great Britain and USA

Published by: CRC Press/Balkema
P.O. Box 11320, 2301 EH Leiden, The Netherlands
e-mail: Pub.NL@taylorandfrancis.com
www.crcpress.com – www.taylorandfrancis.com

ISBN: 978-1-138-03275-0 (Hbk)
ISBN: 978-1-315-38722-2 (eBook PDF)

Advanced Engineering and Technology III – Xie (Ed.)
© 2017 Taylor & Francis Group, London, ISBN 978-1-138-03275-0

Table of contents

Water science and environmental engineering

Food engineering and technology

Advanced Engineering and Technology III – Xie (Ed.)
© 2017 Taylor & Francis Group, London, ISBN 978-1-138-03275-0

Preface

To create a sustainable world, engineers, academicians, and other practitioners are facing more difficult and unprecedented challenges in increasing demands in civil engineering, environmental engineering, water science and hydraulic engineering, food and biochemical engineering, and other related fields. To overcome the current engineering issues and challenges, some recent essential ideas and advanced techniques have been presented. The 3rd Annual Congress on Advanced Engineering and Technology (CAET 2016) will be held during 22–23 October 2016 in Hong Kong. The main objective of the congress is to promote technological progress and activities, technical transfer and cooperation, and opportunities for engineers and researchers to maintain and improve scientific and technical competence in their fields.

This book represents the CAET 2016 and published 35 papers from the congress. Each of the papers has been peer-reviewed by recognized specialists and revised prior to acceptance for publication. The papers embody a mix of theory and practice, planning and reflection participation, and observation to provide the rich diversity of perspectives represented at the congress.

The papers related to civil engineering mainly focus on advanced theories and technologies of soil mechanics and foundation engineering, rock mechanics and rock engineering, geotechnical fabrics and applications, new materials and sensors in civil engineering, fiber-reinforced ultra-lightweight concrete, suspended floor damping system as an anti-seismic design, and modeling, computing, and data analysis. The papers dealing with water science and environmental engineering address new concepts, theories, methods, and techniques related to water science, water engineering, sediment transportation, groundwater, water resources and ecological environments, hydropower management, river management, South-to-North water diversion project (China), farmland ecosystem, marine functional zoning, ecological restoration, waste disposal, and sustainable utilization of water resources and environments. The papers related to food engineering and technology presents the most recent developments of advanced theories and technologies in the engineering field of food processing, packaging, ingredient manufacturing, and control.

Although these papers represent only modest advances toward overcoming major scientific problems in engineering, some of the technologies might be key factors in the success of future engineering advances. It is expected that this book will stimulate new ideas, methods and applications in ongoing engineering advances.

Last but not least, we would like to express our deep gratitude to all authors, reviewers for their excellent work, and Léon Bijnsdorp, Lukas Goosen and other editors from Taylor & Francis Group for their wonderful work.

Civil engineering

Advanced Engineering and Technology III – Xie (Ed.)
© 2017 Taylor & Francis Group, London, ISBN 978-1-138-03275-0

Optimal design of photonic cooling film for buildings based on improved genetic algorithm

Z. Li
Department of Geotechnical Engineering, Tongji University, Shanghai, China

L. Xie
Department of Hydraulic Engineering, Tongji University, Shanghai, China

ABSTRACT: Photonic crystal approach is a newly developed energy-free radiative cooling method for buildings with distinguished advantages in terms of both reflecting sunlight efficiently and emitting indoor thermal radiation selectively. However, the application on windows requires remarkable visible light transparency and complicated photonic band gap arrangement, which leads to immense difficulties in quantifying the optimal number and individual thickness of film layers. This paper presents the design progress of photonic cooling film under approximate direct sunlight (0–20° incidence) by genetic algorithm, which is modified during the process of selection, crossing, and mutation. After validation by sensitivity analysis, the computational results indicate an average transmittance of 54% for visible light and 86% for indoor thermal radiation, while showing an average reflectivity of 73% for ultraviolet radiation and 80% for near-infrared radiation. Moreover, the net cooling power of the optimal design was as high as 108.7 $W \cdot m^{-2}$.

1 INTRODUCTION

Windows and glass curtain walls, which are the frequently used external constituents for buildings, unfortunately serve as the most active heat-transferring parts, thus causing a series of energy conservation problems (Hong, T.Z. 2009). Several achievements have been made to block heat conduction and convection by applying thermal insulation materials and stringent sealing. However, in the thermal radiation process, energy-saving methods have always been a formidable task (Nia, H. et al. 2016).

In the 1960s, European researchers tried to diversify infrared-reflecting layers on windows and defined them as low-emissivity (low-E) glasses. Since then, tremendous radiative coatings have been attempted on glasses. In the late 1950s, Holland & Siddall (1958) conducted their studies on BiO_x and Au layers to examine distinguished heat radiation and maintain enough visible light transmittance. In the mid-1980s, low-E companies deposited Au, Ag, and Cu layers to achieve a better coating quality (Gläser, H.J. 1980). Thereafter, numerous interface layers were tested (Berning, P.H. 1983, Grosse, P. et al. 1997, Gläser, H.J. 2008). Clearly, the efficiency of low-E glasses has achieved magnificent advancement and universal applications. However, if there were not the restraints of maximum infrared reflectivity and unduly broad band gap which inevitably blocks indoor thermal radiation, the cooling performance of low-E glasses would be more prominent.

In recent years, the photonic crystal cooling approach, with two photonic materials of diverse refractive index periodically exposited on a substrate, has been shown to have an advantage on band gap management (Shen, W.C. 2014) and arrange thermal flows more desirably than low-E glasses. Raman, A.P. & Zhu, L.X. (2014, 2015) proposed a photonic crystalline device that combines sunlight radiation and indoor thermal radiation, reaching a cooling temperature of 4.9 degrees Celsius under direct sunlight on the rooftop. In contrast, when the application of photonic crystalline film transfers from roofs to windows, it is fairly challenging to fulfill the demands for visible light transparency while maintaining considerable ultraviolet and near-infrared radiation. Moreover, the manufacturing cost for photonic crystal burgeons with the increasing layers; however, a sufficient cooling efficiency requires an extensive band gap range with sufficient tiers. In this regard, an optimal design for the total number and the individual thickness of film layers is an important requirement.

This paper elaborately presents the design and program operation for photonic cooling film with genetic algorithm. To accelerate optimization and avoid the local optimum solution, huge efforts were made in the modifying process of selection, crossing, and mutation. Furthermore, the sensitivity analysis and the spectral results are presented to demonstrate the effectiveness of the theoretical design.

2 PRELIMINARY DESIGN

2.1 *Photonic band gap arrangement*

According to the theory of black body radiation, the wavelength for the peak of irradiance decreases gradually with the increasing surface temperature. The marked temperature variation between the sun and the indoor temperature renders disparate sunlight and indoor heat spectra (Fig. 1), enabling photonic crystals to alter thermal flows as purposed. To fulfill the bidirectional cooling target on windows, it is imperative to ensure high transmission for visible light and indoor thermal radiation while arranging the photonic band gap in the ultraviolet and near-infrared regions, as shown in Figure 1. Particularly, the layout of the photonic band is of 10–400 nanometers and 0.76–1.2 micrometers. Moreover, the peak indoor thermal radiation wavelengths (8–13 micrometers) coincide with the atmospheric transparency window, which enables the discharge of indoor thermal radiation to the cold atmosphere. Therefore, high transmittance should be precisely applied to the region because other indoor irradiance wavelengths (above 13 micrometers or below 8 micrometers) are likely to be absorbed and will cause secondary heating.

2.2 *Photonic material selection*

Owing to complex demands for photonic band gap arrangement under sunlight (Section 1.1), the refractive indices of the selected materials require both relatively stationary perfor-

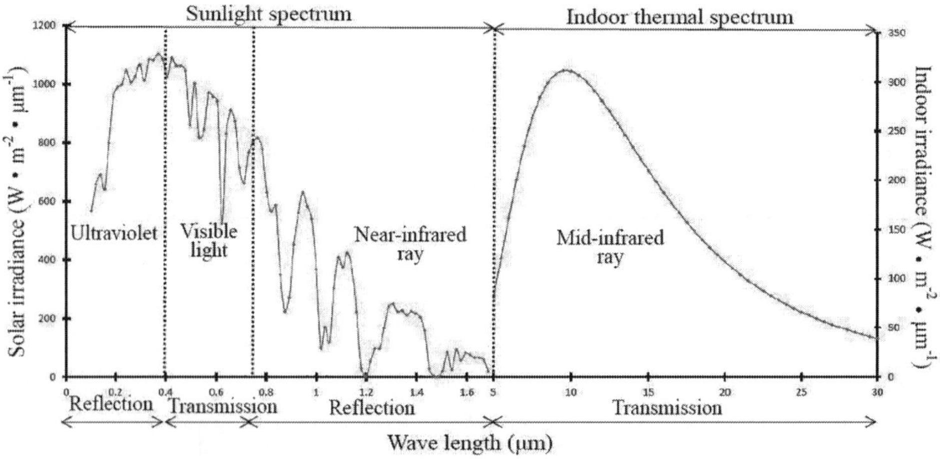

Figure 1. Indoor and outdoor spectra. Sunlight irradiance refers to the Chinese observed data (GBT 17683.1), while the indoor spectrum is calculated by the ideal black body formula at a temperature of 300 K.

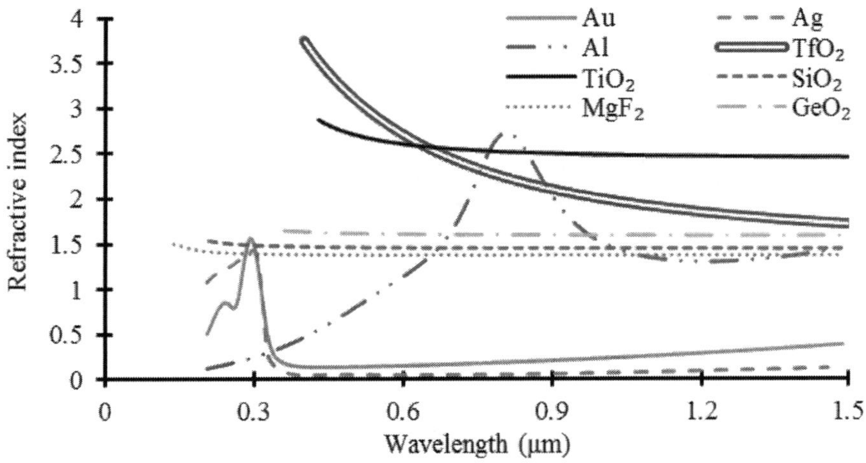

Figure 2. Refractive index of commonly used photonic materials in the short wave zone. The refractive index refers to the data in the literature (Rakic, A.D. et al. 1998, Zhou, W.W. et al. 2008, Jan, K. et al. 2008, Tero, P. et al. 2008).

mance in the short wave region and remarkable distinction between two materials. As shown in Figure 2, among the most commonly used photonic stuff, the curves of three metallic materials (Au, Ag, Al) fluctuate excessively. Also, TfO_2 eclipses when compared with TiO_2. In this regard, TiO_2 satisfied the criteria. Furthermore, SiO_2 is not only is commonly used and easily accessible, but also meets the requirement for the remarkable refractive index compared with TiO_2. In this respect, the next option was considered.

3 GENETIC ALGORITHM OPERATION

3.1 *Operation process*

Genetic algorithm, initially proposed by Holland, J.H. (1992), is an efficacious and efficient choice for global research by simulating heredity, natural selection, hybridization, and aberrance among organisms. The overall program operation presented in this paper is illustrated in Figure 3 and introduced in detail as follows.

3.1.1 *Definition of group and weight*

First, a group of X with a total of 100 units was generated. Then, the thickness of each layer D and the number of layers N were documented in X by real number compiling:

$$X = \left[D_1\ D_2\ D_3\ D_4 \ldots D_N\right] \tag{1}$$

The solar irradiance $I_{outdoor}$ (Fig. 1) was proposed by Lagrange's linear interpolation for calculation. Besides, the indoor irradiance I_{indoor} (Fig. 1) was computed by the black body formula with a temperature T of 300 K (27°C) as a reference statistical value of the average temperature in Shanghai, China (Lin, S. et al. 2016):

$$I_{indoor} = \frac{2hv^3}{c^2} \times \frac{1}{e^{\frac{hv}{kT}} - 1}, \quad T = 300\,K \tag{2}$$

where h is Planck's constant (= 6.626×10^{-34} J/s), ν is the wave frequency, c is the light velocity (= 3×10^8 m/s), k is the Boltzmann constant (= $1.3806505 \times 10^{-23}$ J/K).

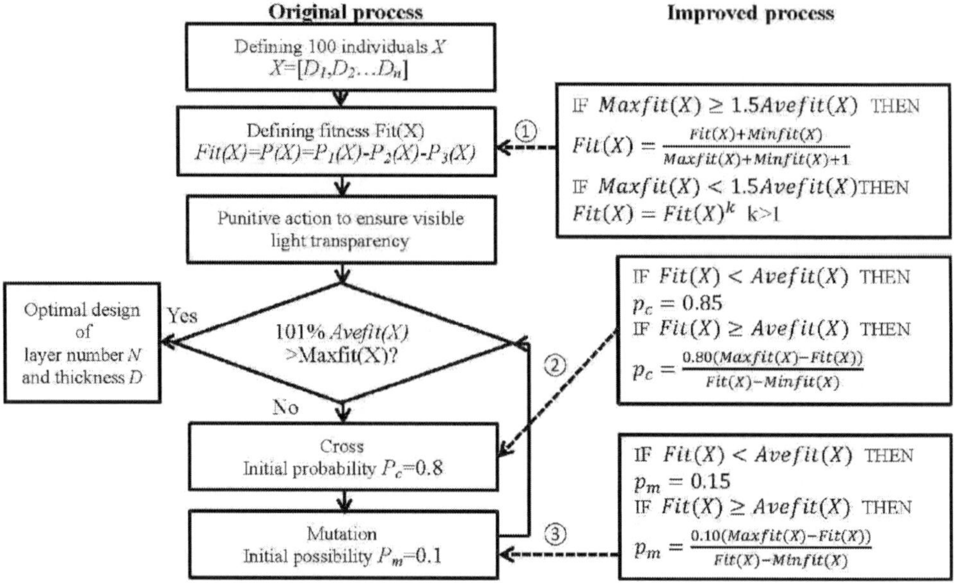

Figure 3. Algorithm operation.

In addition, the atmospheric irradiance was calculated in the same manner. The temperature *T* was considered as 310 K(37°C) because it is the lowest baseline value in Shanghai (Lin, S. et al. 2016) and an indoor and outdoor temperature difference of 10°C was maintained:

$$I_{atmosphere} = \frac{2hv^3}{c^2} \times \frac{1}{e^{\frac{hv}{kT}} - 1}, \quad T = 310\,K \tag{3}$$

3.1.2 *Target function management*

The transmittance $T(\lambda)$ for each wavelength λ was first calculated by the transmission matrix method (Guo, J. et al. 2009). Then, the indoor thermal power $P_1(X)$ emitted at angles θ (0–20°) was calculated by integral summation of all transmitted indoor irradiances I_{indoor}. Similarly, the entered sunlight power $P_2(X)$ and the entered atmospheric heat power $P_3(X)$ were calculated:

$$P_1(X) = \int_{\theta=0^\circ}^{\theta=20^\circ} \int_{\lambda=8\,\mu m}^{\lambda=13\,\mu m} [T(\lambda) \times I_{indoor}]\, d\lambda\, d\theta \tag{4}$$

$$P_2(X) = \int_{\theta=0^\circ}^{\theta=20^\circ} \int_{\lambda=0.01\,\mu m}^{\lambda=1.20\,\mu m} [T(\lambda) \times I_{outdoor}]\, d\lambda\, d\theta \tag{5}$$

$$P_3(X) = \int_{\theta=0^\circ}^{\theta=20^\circ} \int_{\lambda=0.01\,\mu m}^{\lambda=8.00\,\mu m} [T(\lambda) \times I_{atmosphere}]\, d\lambda\, d\theta \tag{6}$$

Assuming as absolutely heat-insulated and perfectly sealed in theory, the film was devoid of convection and conduction power losses. Hence, the cooling power $P(X)$, also considering fitness $Fit(X)$ which determines the possibility of survival during selection, can be calculated by outgoing energy minus all accessible power:

$$P(X) = P_1(X) - P_2(X) - P_3(X) \tag{7}$$

$$Fit(X) = P(X) \tag{8}$$

In addition, the highest, lowest and the average fitness *Fit*(*X*) for 100 individuals *X* were, respectively, denoted by *Maxfit* (*X*), *Minfix* (*X*), and *Avefit* (*X*). The program should be terminated when *Maxfit* (*X*) is identical to *Avefit* (*X*) (variation <1%), which implies that all individuals are optimal.

3.1.3 *Consideration for visible light transparency*
To maintain enough transmittance $T(\lambda)$ within the range of visible light, a punitive method was used in which units with an average visible light transparency below 60% would gradually die out:

$$Fit(X)\begin{cases} Fit(X), \int_{\theta=0^\circ}^{\theta=20^\circ}\int_{\lambda=0.40\mu m}^{\lambda=0.76\mu m} T(\lambda)d\theta d\lambda > 0.6\int_{\theta=0^\circ}^{\theta=20^\circ}\int_{\lambda=0.40\mu m}^{\lambda=0.76\mu m} d\theta d\lambda \\ \dfrac{Fit(X)}{2}, \int_{\theta=0^\circ}^{\theta=20^\circ}\int_{\lambda=0.40\mu m}^{\lambda=0.76\mu m} T(\lambda)d\theta d\lambda \le 0.6\int_{\theta=0^\circ}^{\theta=20^\circ}\int_{\lambda=0.40\mu m}^{\lambda=0.76\mu m} d\theta d\lambda \end{cases} \tag{9}$$

3.1.4 *Crossing and mutation*
It is well known that the crossing procedure reconstructs existing genes while the mutant process generates novel genes (Deep, K. 2007a, b). In this paper, the initial probability of crossing P_c was set as 0.8 and the incipient possibility of mutation P_m was considered as 0.1.

3.2 *Optimization accelerating*

To reduce time complexity of the algorithm and prevent the locally optimal solution, we modify the function of fitness *Fit*(*x*), crossing P_c, and mutation P_m.

3.2.1 *Fitness Fit(x) optimization*
Given that some individuals possess overly significant fitness, their reproduction should be confined; otherwise, they will exert an excessively overwhelming guidance for the final solution. In this regard, at the starting point of the program, amendments were made for *Fit*(*X*) to control their size on a rational scale:

$$Fit(X)=\frac{Fit(X)+Minfit(X)}{Maxfit(X)+Minfit(X)+1}, \quad Maxfit(X)\ge 1.5Avefit(X) \tag{10}$$

However, during the ultimate period, there are hardly any distinct variations in fitness *Fit*(*X*) among individuals, thus leading to poor efficiency in program determination. Therefore, the power of *Fit*(*X*) is recomposed under the condition that *Maxfit*(*X*) is 1.5 times less than *Avefit*(*X*):

$$Fit(X)=\begin{cases} Fit(X)^{1.1}, \ Maxfit(X)<1.5Avefit(X) \\ Fit(X)^{1.2}, \ Maxfit(X)<1.25Avefit(X) \end{cases} \tag{11}$$

3.2.2 *Crossing* P_c and mutation P_m *optimization*
Given that strong crossing P_c and mutation P_m abilities lead to the destruction of existing benign units while poor crossing P_c and mutation P_m abilities lead to hardship in introducing new genes, dialectically, limited P_c and P_m were adopted for well-behaved individuals (*Fit*(*X*) ≥ *Avefit*(*X*)) while intensive P_c *and* P_m were applied for others (*Fit*(*X*) < *Avefit*(*X*)):

$$p_c=\begin{cases} 0.85, Fit(X)<Avefit(X) \\ \dfrac{0.8((Maxfit(X)-Fit(X))}{Fit(X)-Minfit(X)}, Fit(X)\ge Avefit(X) \end{cases} \tag{12}$$

$$p_m = \begin{cases} 0.15, Fit(X) < Avefit(X) \\ \dfrac{0.10\,(Maxfit(X) - Fit(X))}{Fit(X) - Minfit(X)}, Fit(X) \geq Avefit(X) \end{cases} \quad (13)$$

4 RESULTS AND DISCUSSION

4.1 *Layer number selection*

Layer numbers 2, 4, 6, 8, 10, 12, 14, 16, 18, and 20 were arranged in sequence, as shown in Figure 4. The height of the gray bar indicates that the design of 14 layers serves as the optimal choice, and that the cooling power $P(X)$ of the optimal design is 108.7 W · m^{-2}.

As shown in Figure 4, all the three curves decline as layers n are increased, indicating that thick films can block the transmittance of radiation and lead to limited irradiance power. Initially, the increase in layers significantly facilitates the blockage effect on sunlight power P_1, hence influencing the cooling efficiency $P(X)$. In contrast, with increasing layers, the film suffers from formidable hardship in infrared emission so that the cooling efficiency $P(X)$ declines, although the film exhibits a better performance in blocking sunlight when compared previously.

4.2 *Layer thickness option*

Figure 5 illustrates the original and optimized convergence procedures, which demonstrate that the optimization accelerating is of remarkable effect due to approximately reducing steps for Max*fit*(*X*) and *AveFit*(*X*) aggregating from 280 to 90.

The layers are respectively indicated expressed by capital letters from A to N, and the final thickness choice in the design is presented in Table 1. Figures 6a, b and 7a, b show the simulated transmittance *T* for sunlight and indoor thermal radiation, respectively, under approxi-

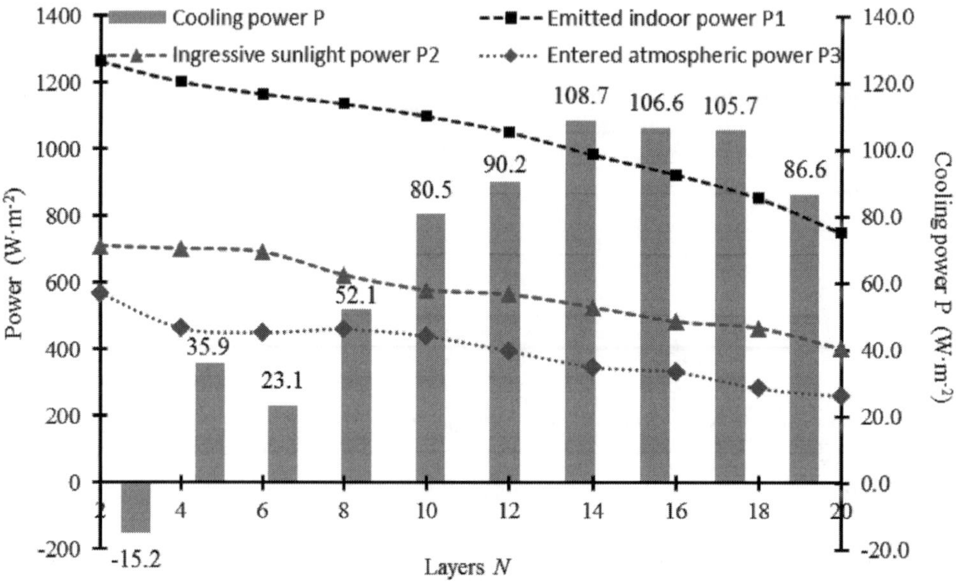

Figure 4. Cooling power under different layers. The data indicated by the broken lines represent the left y-axis while that indicated by bars represent the right y-axis.

mate direct incidence (0–20°). As shown in Figures 6a and 7a, the resolution in diminutive incidence angle (0–10°) shows a good performance of the desired transmittance averaged to 62% for visible light and 87% for mid-infrared light. Moreover, it presents forceful reflectivity for ultraviolet and near-infrared radiation, with mean radiolucent rates of only 28% and 22%. As shown in Figures 6b and 7b, the transmittance for ultraviolet, visible light, near-infrared, and mid-infrared radiation respectively reached 26%, 46%, 18%, and 85%. On the whole, the average transmittance respectively achieved 27%, 54%, 20%, and 86% for the region described above in sequence.

4.3 *Sensitivity analysis*

Several kinds of disturbance including variations in the numbers of individuals X and initial possibility of crossing P_c and mutation P_m were introduced to test the stability of the algorithm. The results of the sensitivity analysis are shown in Figure 8a,b. It indicates that the better the calculated cooling capacity $P(X)$ after disturbances approach the initial data ($108.7\ W \cdot m^{-2}$), the more stable the algorithm will be. As shown in the figure, the outcomes significantly get closer to the primary data on the premise that P_c and P_m are managed within a rational range.

Moreover, the figure indicates several characteristics of the genetic algorithm. First, the excessively strong effect of crossing P_c and mutation P_m facilitates the destruction of benign individuals while the reverse retards the selection process (Section 2.2.2), both of which show poor efficiency in achieving the optimal result. Secondly, although the program suffers from deviance owing to exceptional P_c and P_m, the side effect can be compensated with an increase

Table 1. Thickness result.

Structure	ABCDEFGHIJKLMN													
Material	SiO_2							TiO_2						
Layer	A	C	E	G	I	K	M	B	D	F	H	J	L	N
Thickness/nm	309	89	448	99	131	95	159	702	459	524	648	232	516	805

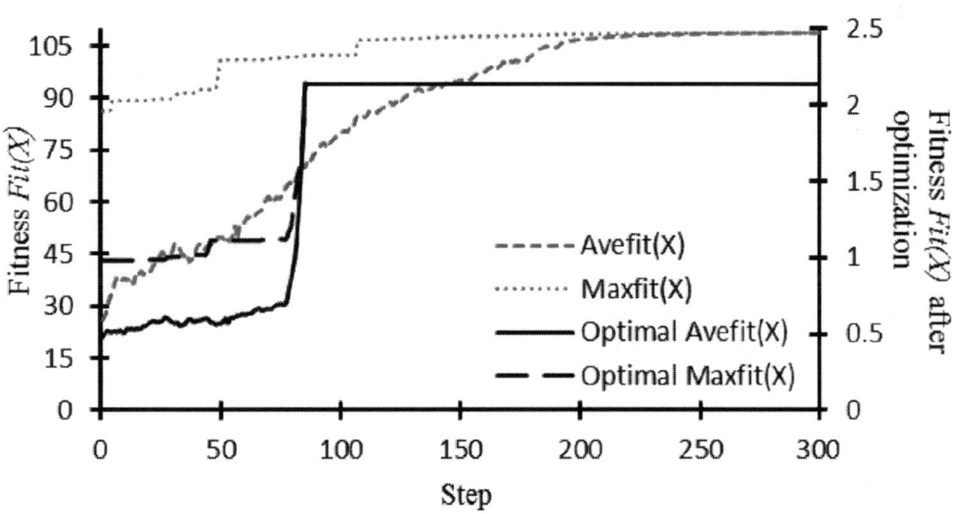

Figure 5. Original and optimized convergence procedure. The data indicated by the two dark black lines correspond to the left y-axis while that indicated by the two light gray curves refer to the right y-axis.

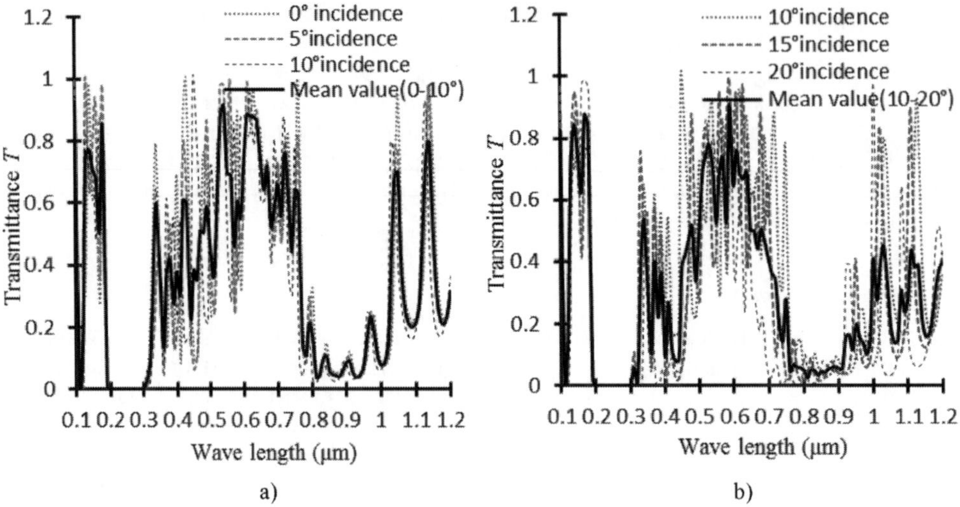

Figure 6. Transmittance in the short wave range with incidence of a) 0–10° and b) 10–20°.

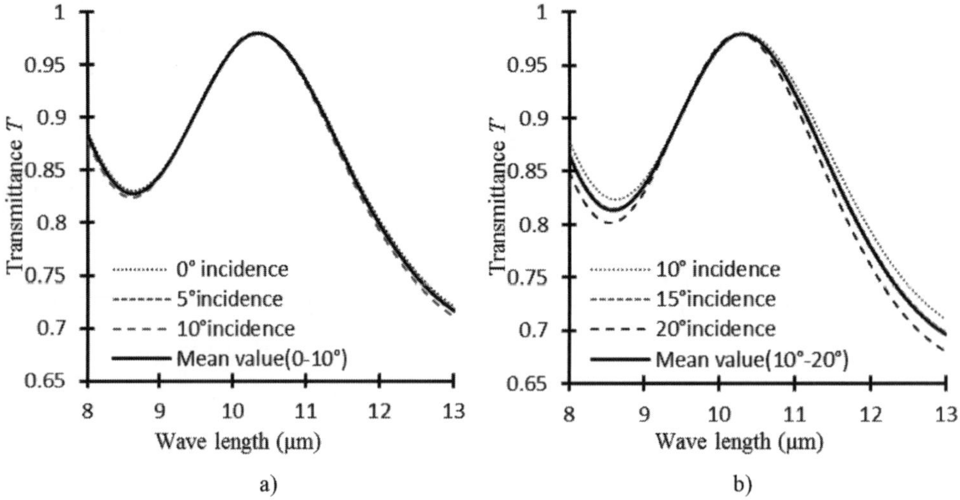

Figure 7. Transmittance the long wave range with incidence of a) 0–10° and b) 10–20°.

in group individuals N because the tolerance of the model gets expanded in this way. Thirdly, the increase in individual amounts N hardly affects the cooling efficiency $P(X)$ if P_c and P_m vary reasonably, which also proves the stability of the improved genetic algorithm.

5 CONCLUSION

This paper focuses on the design of photonic crystalline film that exerts a bidirectional radiative cooling effect for windows and glass walls in buildings. It has also proposed the optimal option for the total number N and the individual thickness D of layers by an optimized genetic algorithm. The conclusions are summarized as follows.

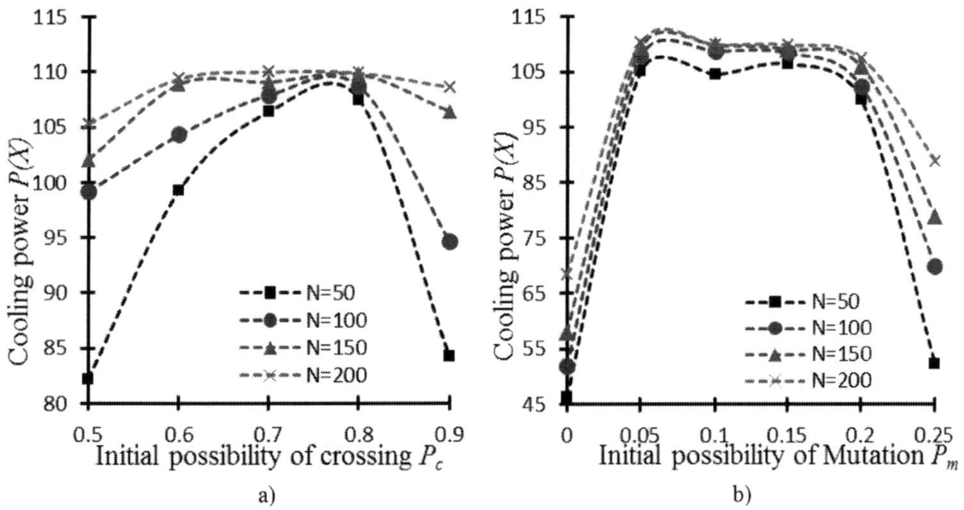

Figure 8. The sensitivity analysis results with variations in a) P_c and b) P_m.

1. The optimization accelerating in selection, crossing, and mutation significantly reinforces the efficiency of the algorithm. The sensitivity analysis proves the stability of the program.
2. The film design is comparatively transparent in visible light and mid-infrared range, with an average transmittance reaching 54% and 86%, respectively. Furthermore, it intensively rejects ultraviolet and near-infrared radiation, with a mean transmittance reaching 27% and 20%, respectively. In this regard, the film exerts a benign bidirectional cooling effect by both efficiently reflecting sunlight and selectively emitting indoor thermal radiation. The cooling power is calculated as 108.7 W $\cdot$ m^{-2}.
3. High probabilities of crossing P_c and mutation P_m lead to the destruction of existing benign units while poor crossing P_c and mutation P_m values result in hardship in introducing new genes. For the efficiency of the algorithm, P_c values should be in the range of 0.6 to 0.85 and P_m values from 0.05 to 0.2. Moreover, the increase in individual numbers X can, in part, compensate for the side effect caused by the deviance of the two values.

REFERENCES

Berning, P.H. 1983. Principles of design of architectural coatings. *Applied Optics* 22: 4127–4141.

Chinese reference solar spectral irradiance at the ground at different receiving conditions. 1999. *GBT 17683.1.*

Deep, K. & Thakur, M. 2007. A new crossover operator for real coded genetic algorithms. *Applied Mathematics and Computation* 188(1): 895–911.

Deep, K. & Thakur, M. 2007. A new mutation operator for real coded genetic algorithms. *Applied Mathematics and Computation* 193(1): 211–230.

Gläser, H.J. 1980. Improved insulating glass with low emissivity coatings on gold, silver, or copper films embedded in interference layers. *Glass Technology* 21: 254–261.

Gläser, H.J. 2008. History of the development and industrial production of low thermal emissivity coatings for high heat insulating glass units. *Applied Optics* 47: 193–199.

Grosse, P., Hertling, R. & Müggenburg, T. 1997. Design of low-E systems based on three layer coating. *Non-Crystal Solids* 218: 38–43.

Guo, S.H. 2011. Electrodynamics, Beijing: Higher Education Press.

Holland, J.H. 1992. *Adaptation in natural and artificial systems: an introductory analysis with applications to biology, control, and artificial intelligence*. Cambridge: MIT Press.

Holland, L. & Siddall, G. 1958. Heat-reflecting windows using gold and bismuth oxide films. *Applied Physics* 9: 359–361.
Hong, T.Z. 2009. A close look at the China design standard for energy efficiency of public buildings. *Energy and Buildings* 41: 426–435.
Kischkat, J., Peters, S., Gruska, B., Semtsiv, M., Chashnikova, M., Klinkmüller, M., Fedosenko, O., Machulik, S., Aleksandrova, A., Monastyrskyi, G., Flores, Y. & Masselink, W.T. 2012. Mid-infrared optical properties of thin films of aluminum oxide, titanium dioxide, silicon dioxide, aluminum nitride, and silicon nitride. *Applied Optics* 51(28): 6789–6798.
Lin, S., Feng, J.M. & Wang, J. 2016. Modeling the contribution of long-term urbanization to temperature increase in three extensive urban agglomerations in China. *Journal of Geophysical Research-Atmospheres* 121: 1683–1697.
Nia, H., Rezaei, M., Brown, R., Jang, S.J., Turay, A., Fathipour, V. & Mohseni, H. 2016. Efficient luminescence extraction strategies and anti-reflective coatings to enhance optical refrigeration of semiconductors. *Journal of Luminescence* 170: 841–854.
Rakic, A.D., Djurisic, A.B. & Elazar, J.M. 1998. Optical properties of metallic films for vertical-cavity optoelectronic devices. *Applied Optics* 37: 5271–5283.
Raman, A.P., Anoma, M.A., Zhu, L.X., Rephaeli, E. & Fan, S.H. 2014. Passive radiative cooling below ambient temperature under direct sunlight. *Nature* 11: 540–550.
Shen, Y.C., Ye, D.C., Celanovic, I., Johnson, S.G., Joannopoulos, J.D. & Soljačić, M. 2014. Optical Broadband Angular Selectivity. *Nature* 3: 1499–1501.
Tero, P., Puukilainen, E. & Ulrich, K. 2008. Atomic layer deposition of MgF_2 thin films using TaF_5 as a novel fluorine source. *Chemistry of Materials* 20(15): 5023–5028.
Zhu, L.X., Raman, A.P. & Fan, S.H. 2015. Radiative cooling of solar absorbers using a visibly transparent photonic crystal thermal blackbody. *Proceedings of the National Academy of Sciences* 112: 12282–12287.
Zou W.W., He Z.Y. & Hotate, K. 2008. Investigation of strain and temperature dependences of brillouin frequency shifts in GeO_2-Doped optical fibers. *Journal of Lightwave Technology* 26(13): 1854–1861.

Advanced Engineering and Technology III – Xie (Ed.)
 ISBN 978-1-138-03275-0

Information lifecycle management of cloud calculation information park based on BIM

Cheng Chen & Hailong Zhu
National Computer Network Emergency Response Technical Team, Coordination Center of China, China

Hao Lu & Xuguang Huang
Huaxin Consulting and Designing Institute Co. Ltd., Architectural Design and Research Institute, Hangzhou, Zhejiang, China

ABSTRACT: The existing cloud calculation information park model is incomplete and cannot be used for the full lifecycle management. Thus, it is necessary to study and establish a more comprehensive cloud calculation information park model. The building information model, the supporting production information model, and the supporting process information model are established by the research of cloud calculation information. In order to express these models accurately, the parametric concept is proposed. Therefore, by combining with the corresponding building information, the cloud calculation information park is realized in the whole lifecycle. The maximize return is obtained by the suitable operation and maintenance.

1 INTRODUCTION

With the rapid development of information industry, a large number of Internet companies set higher requirements on information technology and communication buildings. Thus, cloud calculation information park is proposed. The parks built recently are in their initial phase. The fully functional building, serviced as a vehicle of the information technology, is superimposed simply by a few communication buildings. During the real operation, only information data can be obtained at any time; however, the device data and member data are stored by the traditional method. When either electronic drawing or paper printing drawing cannot be obtained at any time, thus, the initial phase is incomplete. Therefore, it is necessary to propose a much more comprehensive model for cloud calculation information park.

BIM research group at Tsinghua University studied the implementation method of common building information data (BIM group of Tsinghua University, 2011; BIM group of Tsinghua University, 2013). Based on this study, Wang and Zhang (Wang Tingkwei, Zhang Ruiyi, 2014) analyzed the visual management of construction equipment, and Liu and Wang studied the BIM technology integration during the whole lifecycle (Liu Zhansheng, Wang Zeqiang, Zhang Tongrui, Xu Ruilong, 2013). The aforementioned methods help establish a perfect cloud calculation information park model. However, it could not set up the guidance for establishing the cloud calculation information park model. By analyzing the professional content and the above literature on the cloud calculation information park, it is found that the vehicle of different buildings is different, which, in turn, leads to the finding that research data of different buildings are different. For cloud calculation information park, it is necessary to study the visualization of not only the building elements, but also the production equipment and process. First, the appropriate parameters should be chosen and then the element and equipment should be established.

In addition to the information of the common building, the information of production equipment and production process is included based on the analysis of the cloud computing information park. The above-mentioned data contain elements and equipment information. Comprehensive information technology construction is used after the kickoff phase. The models of building information, production information, and process information are established by BIM and precise parametric technology. The processes of design, construction, operation, maintenance, and renewal realize informatization. The above information is stored, exchanged, operated, and renewed by the communication technology of cloud calculation information park. The established information model could provide timely and accurate visual information to participants, which could improve the whole productivity.

2 MAJOR ELEMENTS OF CLOUD CALCULATION INFORMATION PARK

2.1 *Building information model*

Building is the basis of the cloud calculation information park, which contains architecture, structure, electromechanics, ventilation, and fire protection. In the process of cloud calculation information park, the electromechanics, ventilation and fire protection are listed in the process information model, because the information is complex. In the process of design, both the general building information and the detailed professional information should be imported in the cloud calculation information park model. The complete model is available for construction, operation, maintenance, and renewal. Table 1 provides the aforementioned professional information.

Based on the realistic graphics technology, the building information model is shown in a real way. Moreover, it could renew between the building design and the user. Parameterized design is convenient to modify and satisfy the requirements of the building information model. The standard parts resource and the reference price resource are composed of parameterized material and price. These resources could provide a reference in the process of design, construction, operation, maintenance, and renewal. The BIM technology can analyze the energy conservation, comfort degree, sunlight, ventilation, voice, visibility, and security by the building information model. Fig. 1 shows a building information model of cloud calculation information park.

Table 1. Building information model.

Major elements	Details
General layout	Ground building, road, pipe line, greening, etc.
Building	Structural member, hole, equipment, fire zone, etc.
Structure	Foundation, pillar, beam, plate, wall, support, stair, special structure, etc.

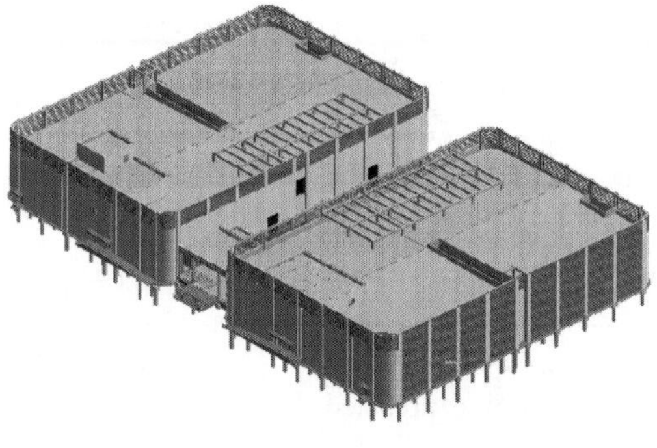

Figure 1. A building information model of cloud calculation information park.

2.2 *Supporting production information model*

The service object of the cloud calculation information park is communication, which is different from other buildings. Thus, the general requirements of ventilation, electric, fire protection, and water supply and drainage, as well as the process information design are necessary to satisfy the requirements of communication. It is also necessary to study the process information involved in the cloud calculation information park.

Table 2 presents the production information model. Fig. 2 shows a pipeline model of cloud calculation information park.

2.3 *Supporting process information model*

Supporting process information model is a key part of cloud calculation information park. In general, the model is submitted to the user by the table. During the maintenance and operation stage, the user would replace equipment directly when the requirement is changed, rather than writing the changed equipment in the file, which leads to confusion. Thus, the supporting process information model containing the building information and the supporting production information should be proposed. Table 3 presents the supporting process information model and Fig. 3 shows a history case.

Table 2. Supporting production information model.

Major elements	Content
Electric	Distribution box, lamps and lanterns, switch, socket, etc.
Fire protection	Automatic fire alarm, Monitoring, gas fire-extinguishing system, etc.
Water supply and drainage	Pump house, water treatment room, pipeline, valve, etc.
Ventilation	Boiler room, heat sink, pipeline, air-condition, etc.

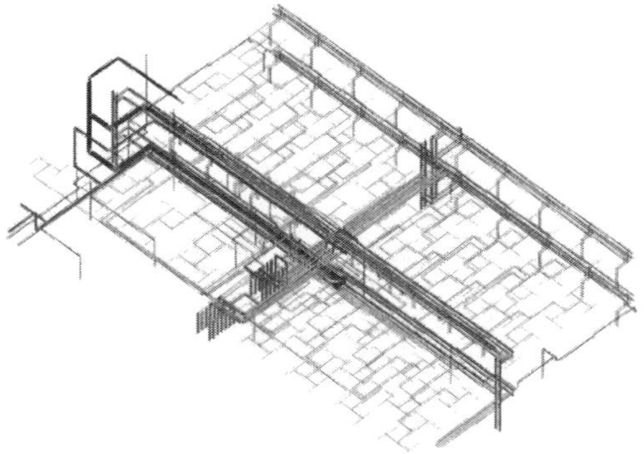

Figure 2. A pipeline model of cloud calculation information park.

Table 3. Supporting process information model.

Major elements	Content
Communication	Cabinet, equipment, sole piece, chamfer, etc.
Electrical source	Dynamo, switch box, mother line, lightning protection, etc.
Rotating seal ring	Monitoring, sensing equipment, alarm, etc.
Ventilation	Air conditioning terminal subsystem, etc.

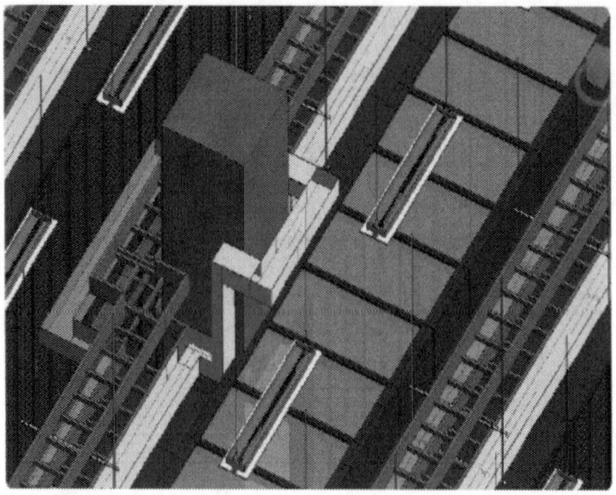

Figure 3. Supporting process information model.

Table 4. Coordinate system.

Coordinate system	Content
Overall coordinate system	Site layout, structure position, etc.
Global coordinate system	Single building location, structure position, etc.
Floor coordinate system	Machine room position, etc.
Machine coordinate system	Pipeline, process equipment position, etc.
Equipment coordinate system	Cabinet direction, detailed construction, etc.

Table 5. A part of physical parameter.

Physical parameter	Content
Size	Length, width, height, diameter, depth, angle, etc.
Material	Metal, concrete, aluminum, etc.
Structure	Component structure, equipment structure, etc.
Acousto-optic	Reflection coefficient, absorption coefficient, etc.

3 PARAMETERIZATION

3.1 *Coordinate parameterization*

The components and equipment should be established at the same coordinate system. Based on the cloud calculation information park, the coordinate system can be divided into five levels: overall coordinate system, global coordinate system, floor coordinate system, machine coordinate system, and equipment coordinate system. The details are provided in Table 4.

3.2 *Physical parameterization*

The properties and size of the equipment will influence on the buildings. Thus, the properties and size should be parametric. Table 5 presents a part of physical parameter.

3.3 *Function parameterization*

Every function of equipment should be parametric and listed in the file. During the operation, when a function is wrong, the corresponding equipment would find the problem, based on the automatic information feedback system. Table 6 presents a part of supporting production information function parameterization.

Table 6. A part of supporting production information function parameterization.

Equipment	Content
Electric	Various parameters, distribution box, lamp power, etc.
Fire control	Alarm system, fire-extinguishing system, etc.
Water supply and drainage	Running state, valve, etc.
Ventilation	Machine room, cabinet temperature, etc.

4 INFORMATION MANAGEMENT

4.1 *Equipment information*

Based on the parameterization, the global cloud calculation information park is obtained. It is easy to replace and check the equipment and the structure. The physical and functional parameterization make the information system close to the real situation. The software identifies physical parameterization easily. The software checks and renews the structure easily by the functional parameterization. The aforementioned information could improve the management level of cloud calculation information park.

4.2 *Information platform*

Management information system should establish the corresponding information platform. Information system could be stored in the cloud space. The system is managed by a numerical file. The client is assigned by the top management system that makes synchronous communication requirements among different major elements.

4.3 *Operation maintenance*

Cloud calculation information park runs in a safe and efficient state by reasonable operation and maintenance. All equipment could be performed by information technology. The efficiency is improved by optimized asset management of cloud calculation information park, thus achieving maximum revenue.

5 CONCLUSION

This paper analyzes information lifecycle management of cloud calculation information park based on BIM. Based on the analysis, the following conclusions are drawn:

1. The model of cloud calculation information park consists of the building information model, the supporting production information model, and the supporting process information model.
2. The model of structure and equipment is established by parameterization.
3. The structure and equipment can be maintained, removed, and renewed by operating in the information platform.

REFERENCES

BIM group of Tsinghua University. 2011. The standard framework research of China building information model [M]. Beijing: China Architecture & Building Press.

BIM group of Tsinghua University, 2013. BIM implementation standards in design [M]. Beijing: China Architecture & Building Press.

Liu Zhansheng, Wang Zeqiang, Zhang Tongrui, Xu Ruilong. 2013. Study on the application of whole life cycle of BIM technology [J]. Construction technology, 42(18): 91–95.

Wang Tingkwei, Zhang Ruiyi. 2014. Plant visualization management based on BIM [J]. Journal of Engineering Management, 28(3): 32–36.

Advanced Engineering and Technology III – Xie (Ed.)
© 2017 Taylor & Francis Group, London, ISBN 978-1-138-03275-0

Analysis of the influence of the weather on the driving speed on horizontal curves

R. Matuszkova, M. Radimsky, T. Apeltauer, J. Cepil & M. Heczko
Faculty of Civil Engineering, Brno University of Technology, Brno, Czech Republic

O. Budik
HBH Projekt spol. s r.o., Brno, Czech Republic

ABSTRACT: Weather may be an important factor when an accident occurs. This is because it influences not only the skid resistance of the pavement, but also the driving abilities of the driver and the vehicle. This paper presents the measurement of driving speeds in selected locations with significant bendiness. Measurements were made during different meteorological phenomena at different values of meteorological variables. During each measurement, the influence of weather on the driving speed was proved. Following the analysis of necessary braking distance, which directly depends on the driving speed and skid resistance, it was proved that drivers on wet pavement reduce the driving speed sufficiently to reduce the braking distance to values similar to or smaller than the braking distance during a clear weather.

1 INTRODUCTION

Weather is the state of the atmosphere at a selected moment of time or time period and at a selected place. It is characterized by meteorological variables and atmospheric phenomena. Meteorological variables we distinguished into air pressure, temperature, wind speed and direction, rain, etc. Atmospheric phenomena are divided into hydrometeors (e.g. rain, snowfall, fog, icing), lithometeors (e.g. sand storms), photometeors (e.g. rainbow, fata morgana), and electrometeors (e.g. storms, thunders, aurora).

As different atmospheric phenomena are related to different symptoms that influence the drivers' abilities, it is necessary for drivers to assess the situation correctly and react appropriately. For example, rain causes wet pavement and lower skid resistance. While dry asphalt has a skid coefficient of 0.6–0.9, wet asphalt has 0.3–0.8, and icy pavement can have a skid coefficient up to 0.15. The stopping distance at a driving speed of 50 km/h increases by 5 m when the pavement is wet and 50 m when the pavement is icy. Another issue connected with atmospheric phenomena can be the sight distance, which is reduced by fog or snowfall. A reduced sight distance influences the drivers' react time. These are the reasons why drivers should adapt their behavior to the actual weather.

The analysis of accident rates carried out in 2014 concluded that 74 567 accidents happened during clear weather conditions, while 11 283 accidents took place during bad weather conditions (fog, rain, snowfall, etc.). Accidents during clear weather conditions are six times more frequent than those during bad weather conditions. In addition, it is necessary to understood that clear weather conditions are more frequent than bad weather conditions. Whether accidents result in injury or death, the number of accidents that occur is similar. Every fourth accident results in injury or death, regardless of weather. The difference becomes evident when analyzing the number of fatalities: every 143th accident result in fatality during clear weather conditions; however, every 106th accident results in fatality during bad weather conditions.

Figure 1. Locations—Lisen, Ostrovacice, Brezina.

2 MEASURING AND EVALUATING THE METHODOLOGY

Locations with high bendiness were selected for analysis. We selected the locations (Fig. 1) where it was presumed that weather could affect the driving speed significantly, especially in terms of perception of skid resistance in horizontal curves. Each location was measured by statistical radars (Sierzega), which record the driving speeds of vehicles separately. The radars were placed on traffic equipment used to direct vehicles through curves. Locations were not influenced by any other factors such as intersections or railway crossings. The data collected were analyzed and the average speed of vehicles was determined in hourly intervals.

The Czech Hydrometeorological Institute provided the data of nearest meteorological stations for each measured location. The values of meteorological variables were compared to average driving speeds in hourly intervals. The following meteorological variables were selected for comparison:

- Average air temperature (°C);
- Maximal wind speed (m/s);
- Sunlight (min);
- Precipitation (mm);
- New snow depth (cm);
- Atmospheric phenomena.

Using different shades of gray, the graphs show hourly average speeds during measurement. The graphs also show maximal wind speed and the average temperature. Precipitation, fog, and snowfall are represented by dashed curves. Visibility is illustrated according to daytime. Locations were measured between 19th January 2016 and 27th January 2016 when the sunrise was between 7:33 and 7:41 and the sunset between 16:27 and 16:39. Based on these times, the sunset and sunrise hours were excluded from measurement to avoid the transient condition of dawn and dusk.

Based on the analysis of the graph, which shows the influence of weather on speed changes, locations will be described and further analyzed. Particularly, we determined how each meteorological phenomenon influenced drivers and how the stopping distance changed (which is important aspect of safe driving).

3 MEASUREMENT

3.1 *Road II/386 Ostrovacice*

In the area of interest, two horizontal curves were measured (Fig. 1). The whole section was situated in a field. The radars were placed in location from 19th January 2016 to 21th January 2016. The route went uphill/downhill throughout the section. The curve radii are 35 m and 40 m. Road marking was applied throughout the section in the form of the center line, and no edge marker posts were used.

The outside temperature was freezing cold during the measurement interval. On the second day of measurement, snowfall was recorded from 15:00 to 00:00. A few hours after that, the snow depth on the pavement was above 1 cm. On the third day, snowfall was recorded after 12:00, with lower intensity and only snow powder lying on the pavement.

The curves in the graph show that the average hourly driving speed fluctuates with regards to horizontal curves (Fig. 2). After excluding the influence of weather, differences between day and night were found to be minimal. A significant decrease in speed (compared to the standard speed) was evident during snowfall on the second day of measurement, when the driving speed decreased by 5 km/h. Throughout snowfall, the driving speed was lower, however, the minimal value (10 km/h lower than the standard value) was reached near the end of snowfall during early morning hours when the pavement was covered with snow. Interestingly, snowfall during the third day of measurement did not reduce the driving speed. Many factors may be attributed to this fact, snow intensity was lower, snowfall was during day hours, and throughout snowfall sunlight was detected, which might have given the drivers the needed confidence to maintain the speed.

3.2 *Road II/373 Brezina*

In the area of interest, six horizontal curves were measured (Fig. 1). The whole section was situated in a forest. The radars were placed in location from 23rd January 2016 to 24th January 2016. The route passed throughout the section with a flat slope. Road marking was applied throughout the section in form of side lines, and no edge marker posts are used.

During the selected interval of the first day of measurement, intense snowfall and fog was observed. Afterwards, throughout night until 7 AM, the snow depth on the pavement kept above 5 cm.

Reduced driving speeds can be observed at the selected location from the start of the measurement due to intense snowfall and fog (Fig. 3). Lower speeds can be observed till 7 AM, when the snow cover started to disappear. During the first part of snowfall, the speed decreased to 7 km/h, which continued to decrease further. Differences between snowfall and clear day were, on average, about 12 km/h, with maximum values reaching up to 16 km/h. The lower level of driving speed, comparable to the driving speed during snowfall, was measured until morning. Afterwards, the temperature rose, the snow melted, and the speed increased to the standard day values. Differences in speeds during snowfall and fog at daytime and night time were about 3 km/h.

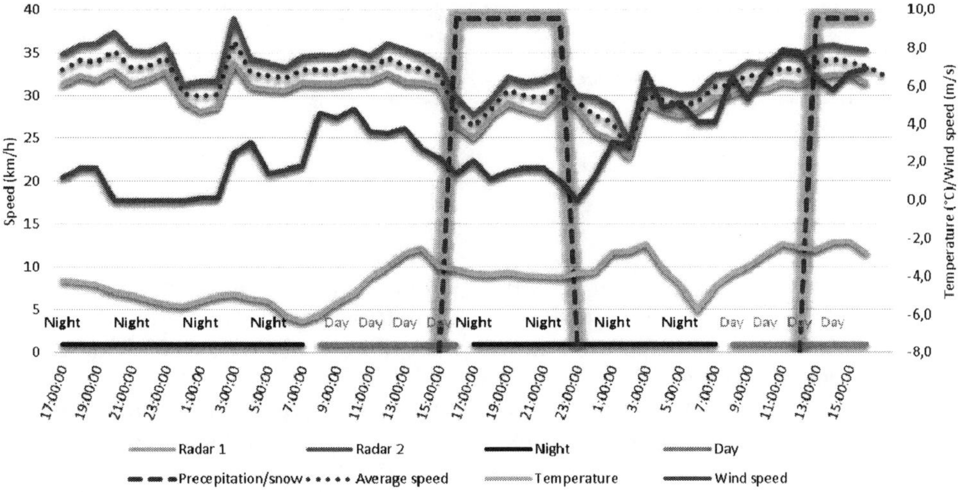

Figure 2. Road II/386 Ostrovacice.

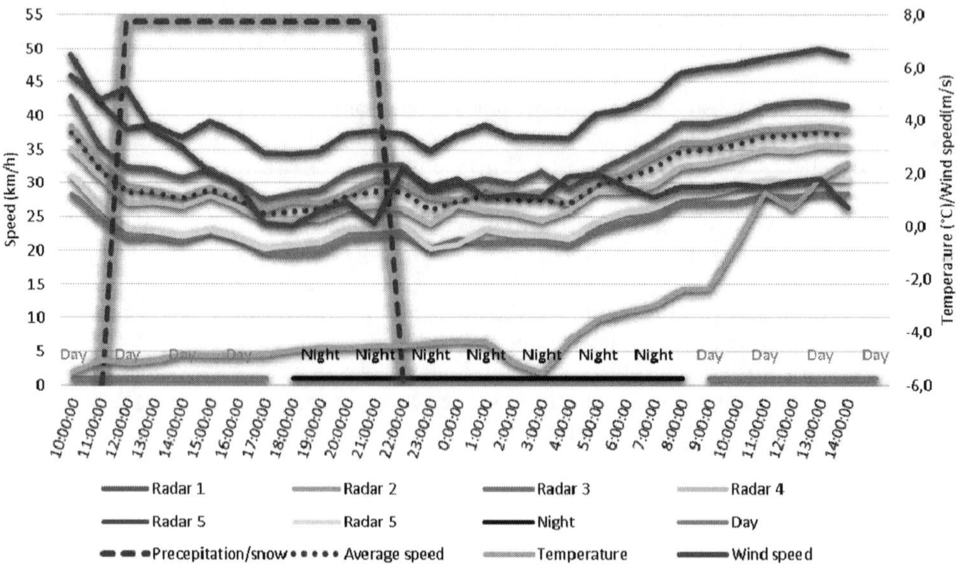

Figure 3. Road II/373 Brezina.

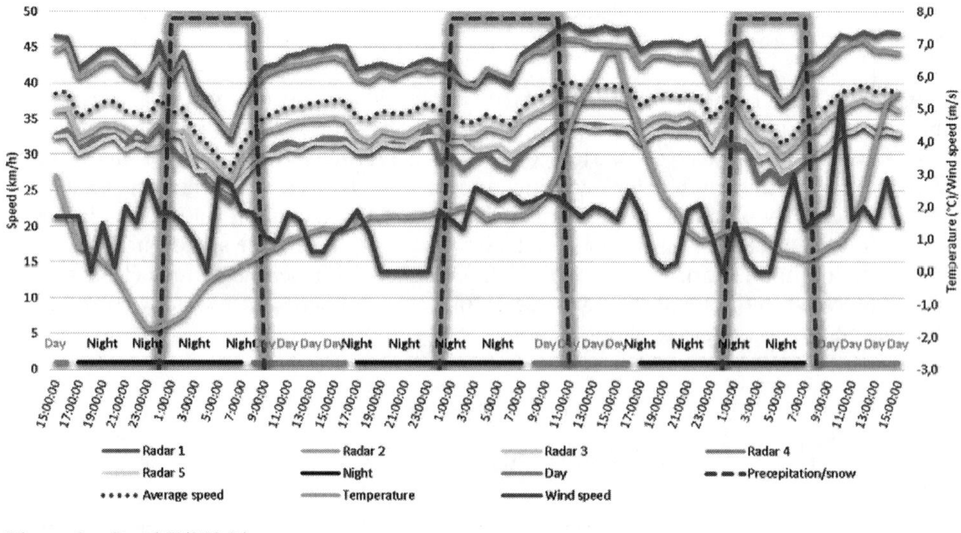

Figure 4. Road II/373 Lisen.

3.3 *Road II/373 Lisen*

In the area of interest, six horizontal curves were measured (Fig. 1). The whole section was situated in a forest. The radars were placed in location from 24th January 2016 to 27th January 2016. The route had a variable longitudinal slope. Road marking was applied throughout the section in the form of a center line and side lines, which were in a poor condition, and no edge marker posts were used.

The selected location was measured throughout four days. On the first night of measurement, freezing rain was observed, creating ice glaze over the pavement. The glaze was over the pavement from 0:00 to 9:30. On the next two nights, only fog was observed.

During the first night of measurement, a significant decrease in driving speeds was observed due to freezing rain (Fig. 4). Decrease was up to 10 km/h compared with the typical value of driving speed during night. The average daytime speed was comparable to the night time speed. However, during sunny day and higher temperatures, the typical daytime speed of 3 km/h higher was measured. The measurement also indicated that, on the selected section of the route, fog influenced only during night time (as shown in the graph during the second night). During the same meteorological phenomena, the driving speed during night time was 6 km/h lower than that during daytime. Driving speed values during foggy daytime was comparable to the typical speed. During the third night of measurement, the driving speed decreased, on average, by about 10 km/h.

4 CONCLUSION

At each location of interest, influence of meteorological phenomena on the driving speed was proved. Interestingly, the influence of fog was observed only during nighttime, and drivers did not react to fog by reducing the speed during daytime. During snowfall or freezing rain, the driving speed was reduced were from 5 km/h to 16 km/h, which in the most cases corresponded to about 30% of the current speed.

Considering average speeds across all radars, we can compare the differences of stopping distance (react distance and braking distance) for different driving speeds and skid resistances. For a standard pavement, the skid resistance coefficient is 0.8, while for a wet pavement, it is 0.5, and for icy pavement, it is 0.15. It is necessary to indicate that the comparison is very simplified, because we consider the total average speed of all drivers; moreover, appropriate measurements for determining the real skid resistance coefficient on a dry pavement or a wet pavement in selected road sections are missing.

According to the evaluation made in Ostrovacice (Fig. 2), the average speed (without the influence of weather) is 33 km/h, which corresponds to the stopping distance of 15 m. During snowfall, the speed was reduced to 28 km/h and then to 23 km/h, which corresponds to stopping distances of 14 m and 11 m. In Brezina (Fig. 3), the standard average speed was 37 km/h, which corresponds to a stopping distance of 17 m. When the speed was reduced to 30 km/h and then to 25 km/h, stopping distance got to 15 m and 17 m, respectively. It is clear that when there is snow and change in skid resistance, drivers reduce their driving speed sufficiently to achieve the stopping distance on the wet pavement lower than the stopping distance on the dry pavement. This fact does not apply to Lisen (Fig. 4), where the driving speed was reduced by 7 km/h; however, due to the icy pavement, which has a lower skid resistance coefficient, the stopping distance was 15 m longer than that on the dry pavement.

ACKNOWLEDGEMENTS

This study was performed under the project No. FAST-S-15-2911 Assessment of the real driving speed to horizontal curves' parameters and project No. LO1408 AdMaS UP—Advanced Materials, Structures and Technologies, supported by the Ministry of Education, Youth and Sports under the National Sustainability Programme I.

REFERENCES

Bradac, A. 1999. *Soudni inzenyrstvi*. Brno: Akademicke nakladatelstvi, s.r.o. CERM. 335–504.
Husak, J. 2008. URSUS.cz: *Atmosfericke jevy—zakladni rozdeleni*.
Policie CR. 2015. *Policie CR: Statistika nehodovosti*.

Advanced Engineering and Technology III – Xie (Ed.)
© 2017 Taylor & Francis Group, London, ISBN 978-1-138-03275-0

Research on the monitoring of cracks in concrete structures based on fiber Bragg grating sensors

Lina Zhang & Fengchen Li
Department of Civil and Architectural Engineering, East China University of Technology, Nanchang, China

Dapeng Wang
School of Civil Engineering, Suzhou University of Science and Technology, Suzhou, Jiangsu, China

Cide Peng
Department of Civil and Architectural Engineering, East China University of Technology, Nanchang, China

ABSTRACT: This paper presents the monitoring of cracks in reinforced concrete structures using fiber Bragg grating sensors. Quasi-distributed FBG sensors and non-encapsulated quasi-distributed FBG sensors are arranged at the bottom of a reinforced concrete simply supported beam equipped with Fiber Reinforced Plastic (FRP) package. The quasi-distributed FBG sensor is used to study the problems of effective perception and positioning of cracks in reinforced concrete members. The results indicate that the fiber Bragg grating sensor can locate the approximate position. In addition, strain sensitivity can be reduced using the FRP-packaged FBG sensor relative to the bare fiber grating.

1 INTRODUCTION

Reinforced concrete structure in the process of service may develop concrete cracks due to fatigue, overload, and other factors, which has become the safety hidden trouble (Mehta, 2001). At present, the conventional detection of concrete cracks mainly relies on concrete absorption, scattering of the ultrasonic, X-ray or R-ray, and other technologies (Wither, 2006 and Vikram, 2006). The devices used in these methods are numerous and complex, and the operation is inconvenient and difficult to detect in real time. Recently, some researchers have put forward the application of long-distance Fiber Bragg Grating (FBG) sensors and quasi-distributed FBG sensors in civil engineering. Grattan et al (Grattan, 2007) conducted an experiment in which fiber grating sensors and strain gauges were embedded into the concrete to verify the feasibility of FBG sensors for the corrosion measurement of internal steel reinforcement in concrete structures. Inaudi et al (Kirikera, 2011) measured the deflection of the structure under load by using the long-distance FBG sensor. The test results indicated that the sensor can provide very effective data, especially for concrete and wood structures. Li et al (Li, 2007) studied the sensitivity of long-distance FBG sensors, and found that with the increase in gauge length of the sensor, the sensitivity becomes stronger. Kamath et al (Kamath, 2010) proposed that the strain sensor can be used to measure the strain of the structure and identify damage by comparing different strains in the intact structure and the damage structure with the appropriate identification algorithm. In this paper, we applied the quasi-distributed FBG strain sensor to the reinforced concrete simply supported beam to study the problem of effective perception and positioning of cracks during the loading process.

2 OPTICAL FIBER BRAGG GRATING SENSOR

Fiber Bragg grating is an optical fiber sensor with the nonlinear function of wavelength modulation. When the broadband optical wave is transmitted in the grating, the incident light will be reflected back at the corresponding wavelength, and the rest of the transmitted light is not affected. Thus, the fiber grating plays an important role in the selection of the optical wave. The sensing principle is shown in Figure 1. According to the orthogonal relation of fiber grating transmission mode, the Bragg wavelength λ_B is associated with the effective refractive index n_{eff} of the fiber core and the fiber grating period Λ, whose expression is given by:

$$\lambda_B = 2n_{eff}\Lambda \tag{1}$$

The working principle of the sensor is based on the measurement of the Bragg grating center wavelength λ_B. Through the detection of λ_B drift caused by external disturbance, the pressure, stress, and temperature are obtained. The relationship between the central wavelength of the fiber grating and the temperature and strain is given by:

$$\Delta\lambda_B = (k_T \Delta T + k_\varepsilon \varepsilon)\lambda_B \tag{2}$$

where K_T is the temperature sensitivity coefficient of fiber Bragg grating and K_S is the strain sensitivity coefficient of fiber Bragg grating. Therefore, the linear relationship between the variation quantity of fiber Bragg grating center wavelength and the external disturbance is temperature or strain.

Figure 2 and Figure 3 show the strain response curves of fiber Bragg grating strain sensors at the center wavelength of 1532.000 nm and 1540.000 nm, respectively. As shown in

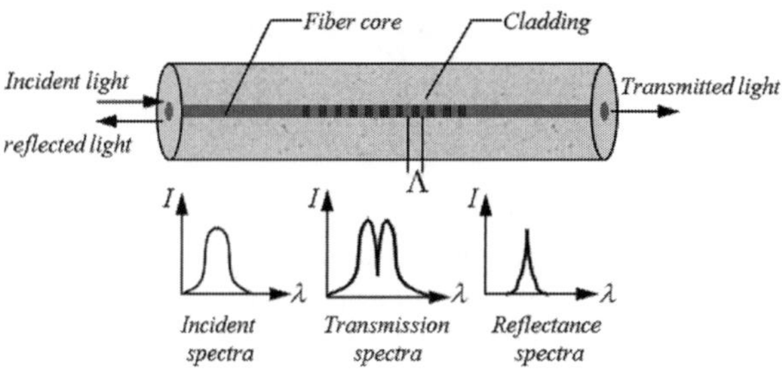

Figure 1. Schematic of the fiber Bragg grating sensor.

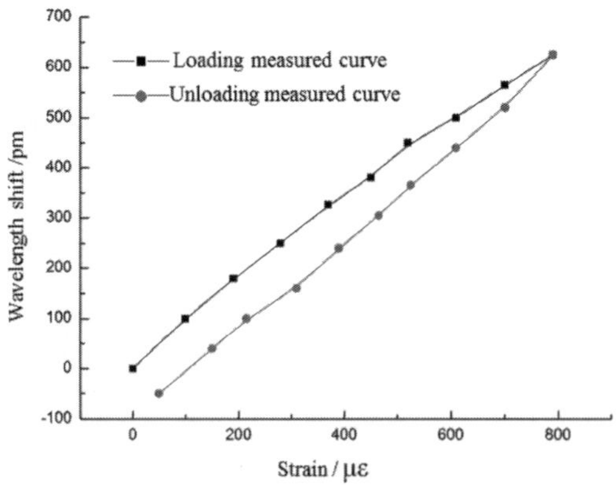

Figure 2. Strain response curve of strain sensors at the center wavelength of 1532.000 nm.

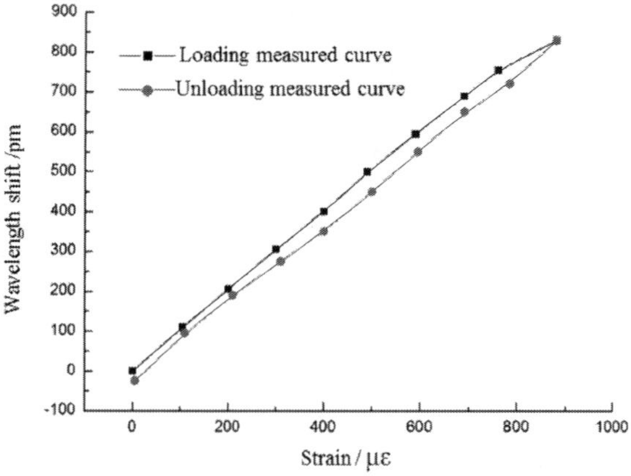

Figure 3. Strain response curve of strain sensors at the center wavelength of 1540.000 nm.

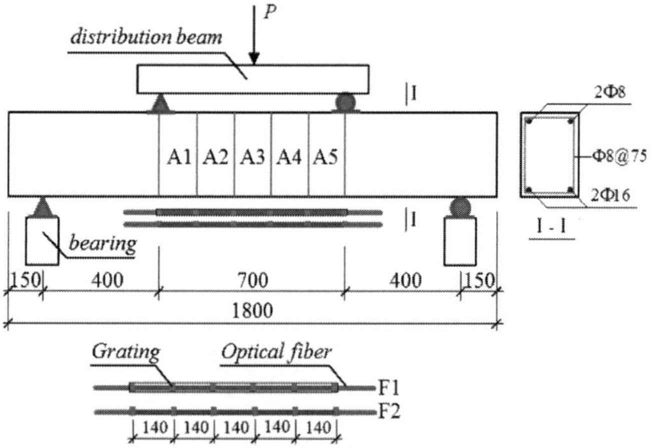

Figure 4. View of the structure and the load.

Table 1. Parameters of the concrete beam.

Concrete	Rebar	Beam length/mm	Beam width/mm	Beam height/mm	Protective layer thickness/mm
C30	HRB335	1800	120	160	25

the figures, the variation of strain and wavelength has a good linear relationship, and the correlation is more than 99%. Thus it can be seen that the wavelength shift value of the fiber Bragg grating strain sensor has a good linear relationship with the strain, and thus it is an ideal strain sensor.

3 EXPERIMENTAL PREPARATION

The experimental objects were a reinforced concrete simple supported beam, material and reinforcement parameters, and the load form, as shown in Figure 4 and Table 1.

The pure bending section of the reinforced concrete beam was selected as the experiment section. This section was divided into five units, denoted as A1, …, A5, respectively. The lower surface of concrete beams contained five elements on which two quasi-distributed FBG strain sensors (F1 and F2) were fixed. F1 was encapsulated by fiber reinforced plastic composite materials. The interior part contained five gratings, marked, respectively, as $G_{1,1}$, ..., $G_{1,5}$. The average spacing of the grating was 140 mm. F2 was a strain sensor with non-encapsulated bare FBG. Its interior part also contained five gratings, marked, respectively, as $G_{2,1}$, ..., $G_{2,5}$. The average spacing of the grating was also 140 mm.

4 EXPERIMENTAL RESULTS AND ANALYSIS

The first crack appeared in the A3 zone of the concrete beam when the load was increased to the third level (7.5 kN). Figure 5 shows the load-strain curve of each measurement point of the grating in the sensor F1.

Similarly, Figure 6 shows the load-strain curves of each measurement point of the grating in the sensor F2. As can be seen from Figs. 5 and 6 during the process of loading from 5 kN to 7.5 kN, the strain of each measurement point of the grating in F1 and F2 sensors had different degrees of mutation. It can be deduced that cracks appeared during this stage. Among the measurement points, G1,3 and G2,3 were found to be the maximum strain measurement points in the sensors, which belong to each other. Therefore, it can be concluded that the first crack appeared in the A3 zone, which is consistent with the actual location of the first crack. In this way, the approximate location of the first crack in the concrete beam could be realized.

Figure 7 shows the load-strain curves of $G_{1,3}$ and $G_{2,3}$. It can be seen that under the condition of the same load classification, the strain variation of the bare fiber grating strain sen-

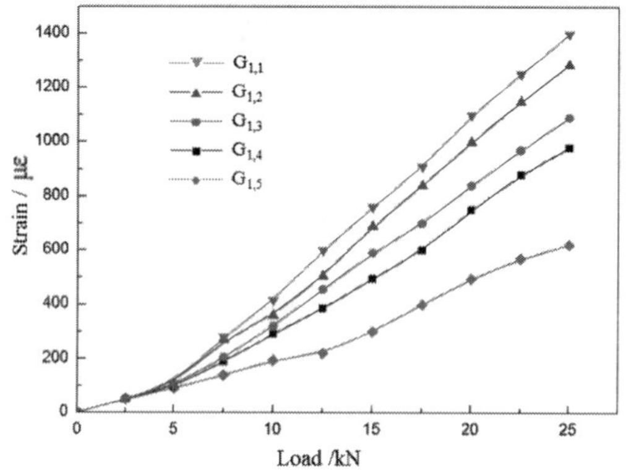

Figure 5. Load—strain curve of $G_{1,1}$–$G_{1,5}$.

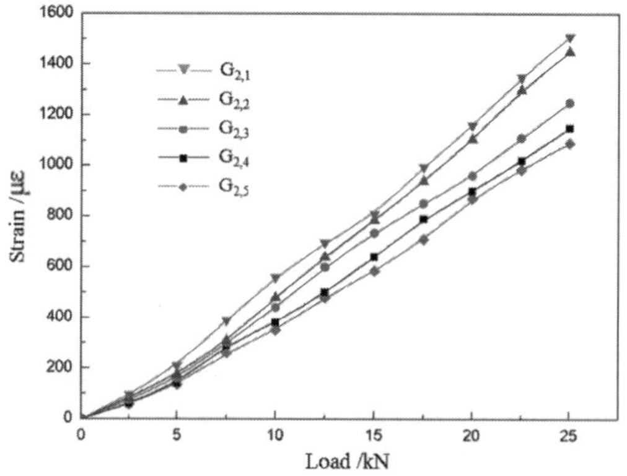

Figure 6. Load—strain curve of $G_{2,1}$–$G_{2,5}$.

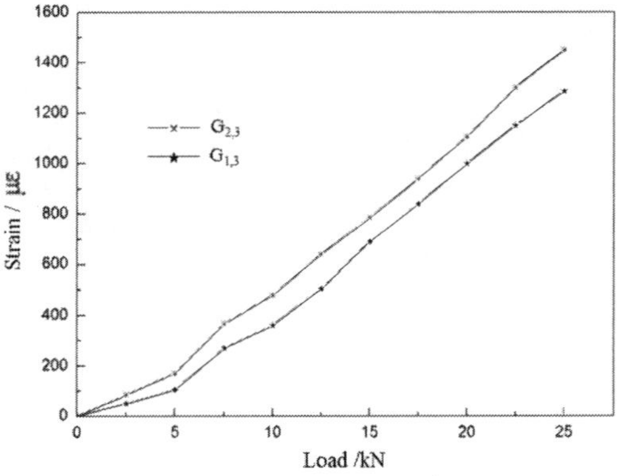

Figure 7. Load—strain curve of $G_{1,3}$ and $G_{2,3}$.

sor is larger than that of the FRP-encapsulated fiber Bragg grating strain sensor. In other words, the strain sensitivity of the former is higher. However, due to the fragile, brittle, and other characteristics of the bare fiber materials, it cannot be applied in a rough engineering environment. Although the sensitivity of the FRP-encapsulated fiber grating strain sensor is slightly lower than that of the bare FBG strain sensor, due to its stable performance, high strength, and other characteristics, it is considered to be more suitable for engineering applications, especially in long-term health monitoring projects, than the bare FBG strain sensor.

5 CONCLUSIONS

With the objective of identifying high-performance sensing elements in a structural health monitoring system, we studied the application of fiber Bragg grating sensors in civil engineering. Based on the experiment of the application of the quasi-distributed FBG strain sensor in the monitoring of the crack in a reinforced concrete structure, the following main conclusions are drawn:

1. Under the action of load, the quasi-distributed FBG strain sensor can perceive and locate the occurrence of cracks in the concrete member through the load-strain curve of the internal grating. If the load-strain curve of the grating has different degrees of mutation in the strain at a certain level of load, it can be concluded that the concrete member has occurrence of cracks. Moreover, the size of the degree of mutation of each strain curve can be analyzed, and the approximate location of cracks can be achieved.
2. The monitoring of the strain in the failure process of reinforced concrete members can be realized by the fiber Bragg grating and electric resistance strain gauge. However, the optical fiber grating has the advantages of convenient embedding, zero adjustment, quasi-distributed measurement, compact signal transmission line, and long service life. Thus, it is more suitable for the needs of long-term, online structural health monitoring when compared with the strain gauge.
3. The FRP-encapsulated optical fiber grating sensor overcomes the shortcomings of the bare fiber grating which is fragile and brittle. Nevertheless, due to the packaging technology, there is a certain loss in the process of strain transfer, resulting in reduced strain sensitivity relative to the bare fiber grating.

ACKNOWLEDGMENT

This research was supported by the Natural Science Foundation of Jiangxi Province (20141BBE50022, 20133BBE50032) and the Science and Technology Planning Project of Jiangxi Provincial Education Department (GJJ14496), China.

REFERENCES

Grattan S.K.T., Basher P.A.M. Fibre bragg grating sensors for reinforcement corrosion monitoring in civil engineering structures. *Journal of Physics*, 2007, 76(1): 0120181-7.

Kamath G.M., Sundaram R., Gupta N. Damage studies in composite structures for structural health monitoring using strain sensors [J]. *Structural Health Monitoring*, 2010, 9(6): 497–512.

Kirikera G.R., Balogun O., Krishnaswamy S. Adaptive fiber Bragg grating sensor network for structural health monitoring; applications to impact monitoring [J]. *Structural Health Monitoring*, 2011, 10(1): 5–16.

Mehta P.K., Burrous R.W. Building durable structures in the 21st Century, *Concrete International*, 2001, 23(3): 57–63.

Suzhen Li, Zhishen Wu. Development of distributed long-gage fiber optic sensing system for structural health monitoring [J]. *Structural Health Monitoring*, 2007, 6(2): 133–143.

Vikram K.K., Atul S. Ganpatye and Konstantin Maslov. Ultrasonic Ply-by-Ply Detection of Matrix Cracks in Laminated Composites. *Journal of Nondestructive Evaluation*, 2006, 25(1): 39–51.

Withers P.J., Bennett, J., Hung Y.C., Preuss M. Crack opening displacements during fatigue crack growth in Ti-SiC fiber metal matrix composites by X-ray tomography. *Materials Science and Technology*, 2006, 22(9): 1052–1058.

Advanced Engineering and Technology III – Xie (Ed.)
 ISBN 978-1-138-03275-0

Drained effects of geotextile filter materials in extremely soft soils

Yuan Zhou, Di Wang, Lingyun Zhang & Yufeng Gao
Key Laboratory of Ministry of Education for Geomechanics and Embankment Engineering, Hohai University, Nanjing, China
Geotechnical Research Institute, Hohai University, Nanjing, China

Zhihao Lu
East China Electric Power Design Institute, China Power Engineering Consulting Group, Shanghai, China

ABSTRACT: An experimental study on drain effects of different geotextile filter materials in extremely soft soils is carried out. The research is performed for a Wenzhou engineering project. Under the condition of vacuum consolidation, movements of soil particles are investigated, particle size distributions are analyzed, and changes of water content in drain process around the filter materials are measured. Drain effects of five different materials are compared. From these results, PVA material is suggested as a good filter material.

1 INTRODUCTION

Geotextile filter materials have been used now in extremely soft soils for drain purposes. Traditionally, sand drains can be employed for soil consolidation. However, in extremely soft soils, sand drains are very difficult to construct (Ye et al. 2010, Gao & Liang 2013). And sometimes sand drains do not work because too many fine particles will flow into sand drains and block them entirely. In these cases, geotextiles are a good choice.

It is important to select an effective material to construct a drain path. Some research works have been already conducted to study drain effect of geotextile materials (Xu 2000, Dong-sheng 2007, Ingold & Miller 1988). But very often blocking problems are also encountered in geotextile filter materials (Shan et al. 2001, Huang & Koerner 2005, Novak et al. 1999, Cazzuffi et al. 1999, Fannin & Pishe 2001, Zhuang et al. 2008). Drain mechanism of filter materials is analyzed by some researchers (Schober & Teindl 1979, Hoare 1982, Lawson 1982, Palmeira & Gardoni 2002, Fannin & Pishe 2001, Zhou Yuan 2010).

In this paper, an experimental setup is designed. Soil samples are taken from Wenzhou engineering project site. Five kinds of geotextile filter materials are employed and they are put into soils as drains. Movements of soil particles and changes of water content are measured. Based on the experimental study, drain effects of different geotextile materials are investigated.

2 GEOTEXTILE FILTER MATERIALS AND VACUUM DRAIN TESTS

2.1 *Geotextile filter materials*

In this experimental research, different geotextile materials are employed. They are PVA material-1 (deerskin towel), PVA material-2 (absorbent cotton), super-fine fiber material, coral fiber material and bamboo fiber material. The basic characteristics of these materials are as follows:

1. PVA (polyvinyl alcohol) Material: This material consists of two types: PVA material-1 (deerskin towel) (Fig. 1), and PVA material-2 (absorbent cotton) (Fig. 2). PVA fiber enjoys some good characteristics such as high strength and good hydrophilic property, so it can be used as filter material. Currently, artificial deerskin towel and absorbent cotton squares are widely used.
2. Super-fine fiber material (Fig. 3): Super-fine fibers have many excellent properties like small diameter, large specific surface area, high aspect ratio, etc. The main features of super-fine fiber material are high water absorption and fast-drying ability, which make it very suitable as a new filter material.

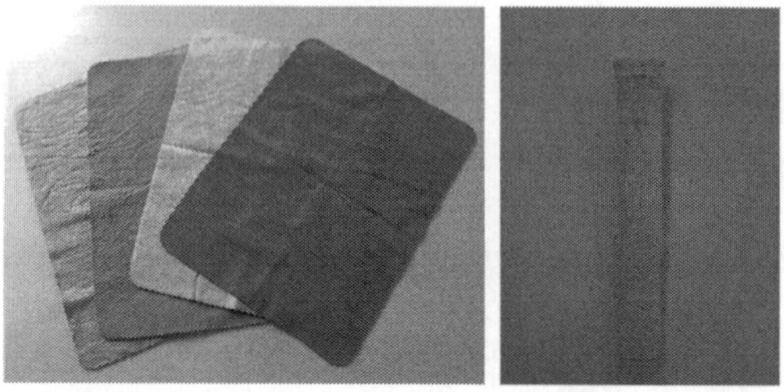

Figure 1. PVA material-1 (deerskin towel).

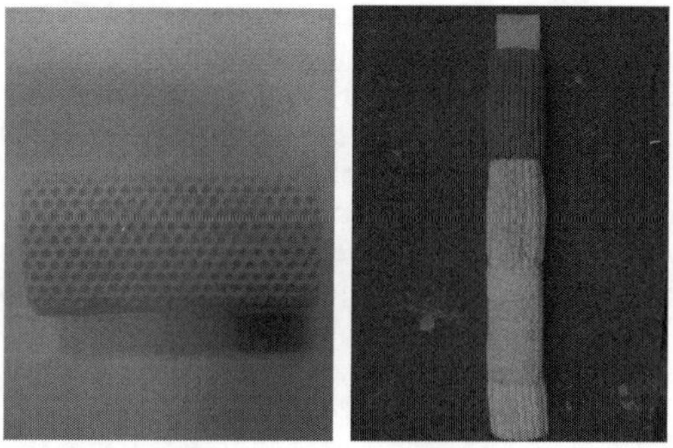

Figure 2. PVA material-2 (absorbent cotton).

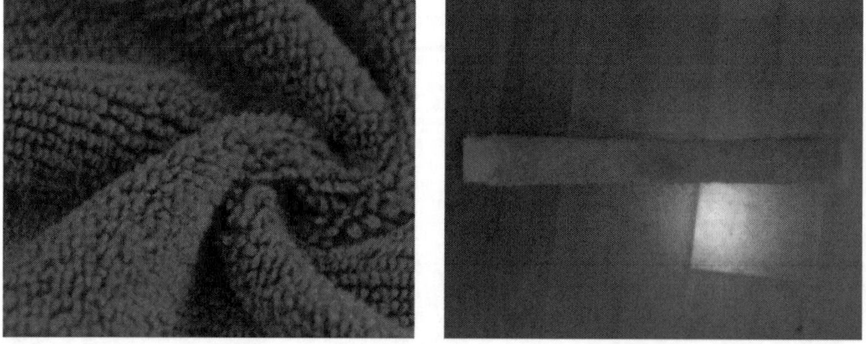

Figure 3. Super-fine fiber material.

3. Coral fiber material (Fig. 4): Coral fiber material is a kind of material that combines coral velvet fabric and super-fine fiber material, which has both a strong absorbent capability of super-fine fiber material and a good softness of coral velvet material.
4. Bamboo fiber material (Fig. 5): Bamboo fiber has a strong capillary effect, which can absorb and evaporate moisture quickly.

2.2 *Experiment design*

This experiment is based on a self-developed setup for extremely soft soils. The experimental apparatus is suitable for drain tests on dredged silt samples under vacuum loads. The instrument unit contains 7 main parts: model boxes, drain system, pressure system, vacuum pumping system, valve control system, tail water collection system and measurement system. The schematic diagram of testing equipment is shown in Figure 6.

The initial water content of soil sample is double as its liquid limit (106%). The vacuum degree is 60 kPa.

Turn on the vacuum pump, and then start the experiment. During the test, record the soil top sedimentation, tail water mass and the vacuum degree. At the end of the experiment, collect samples along specific longitudinal and radial locations around the filter material to analyze particle size distribution and water content. In the meanwhile, conduct particle size analysis in the tail water.

Perform tests on 5 groups of geotextile filter materials. In the meanwhile, conduct a test on 1 group of ordinary non-woven fabric material as a comparison. The stop time of testing is determined by accessing the drainage is very slow to ignore. When the ordinary non-woven

Figure 4. Coral fiber material.

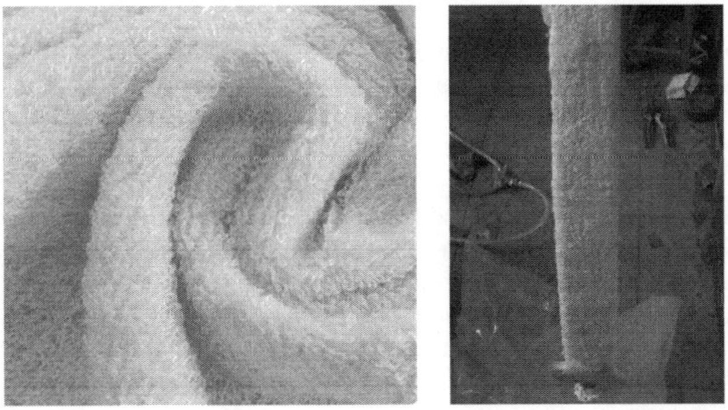

Figure 5. Bamboo fiber material.

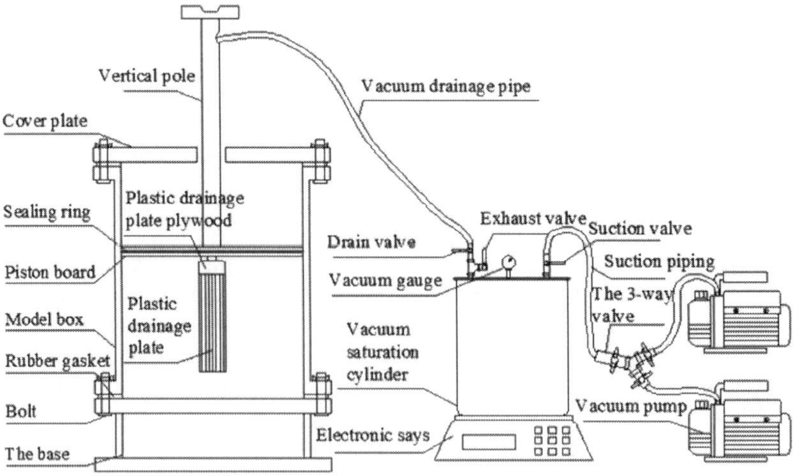

Figure 6. Schematic diagram of testing equipment.

fabric drain board is tested, the stop time for it is 200 hours. Take this time as a reference for other geotextile filter materials.

3 CHANGES OF PARTICLE SIZE DISTRIBUTION AND WATER CONTENT AROUND FILTER MATERIALS

To study changes of soil particle size distribution and water content around filter materials, collect samples at 4 points along the longitudinal direction and 3 points in the radial direction. The positions of sampling are shown in Figure 7. At the same time, conduct particle size analysis of tail water at the end of test.

All the soil samples are from Wenzhou project. Their particle sizes belong to groups less than 0.075 mm. Qian and Wan (Qian & Wan 1983) suggested the sediment particles with less than 0.01 mm sizes would flocculate and coagulate significantly. Therefore, soil particles can be divided into three groups, which are 0.075 mm~0.01 mm, 0.01 mm~0.005 mm, and less than 0.005 mm.

3.1 *Radial movement of soil particles*

3.1.1 *Radial movement of soil particles for PVA material, coral fiber material, and super-fine fiber material*

Take the radius originated from the center of the sample as the reference coordinate. Assume the distance from the center to the edge of the sample as 1. Generally speaking, during the vacuum drain test, soil particles will move with flowing water. Particle size distribution will be changed. For PVA material test, the percentages of various particle sizes are shown in Figure 8.

Figure 8 shows that, in the radial direction, the percentages of various groups do not change basically, and the total fine particle content is essentially the same. It can be considered that the particles do not move significantly in the vacuum drain process for PVA material. The distribution of particle sizes in the tail water is close to the sample.

The radial movement of soil particles for super-fine fiber material is similar to PVA material. The fine particle content almost does not change.

In the test for coral fiber material, the particle percentage changes a little bit. The fine particle content decreases near the center of the sample.

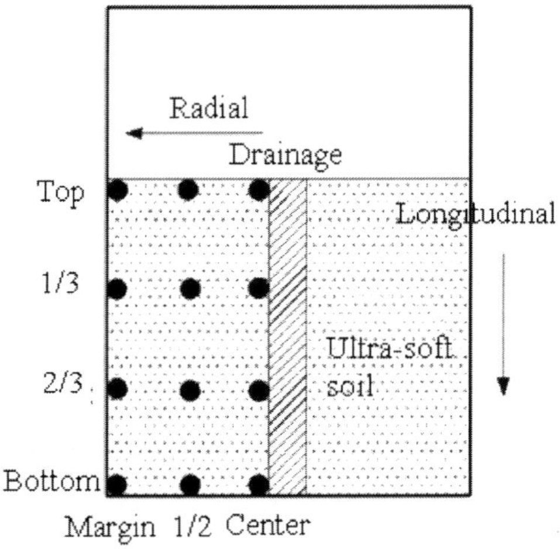

Figure 7. Positions of sampling.

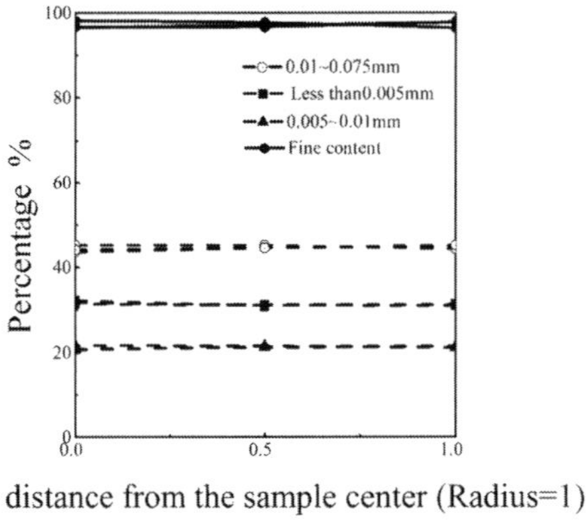

Figure 8. Radial movement of soil particles for PVA material.

3.1.2 *Radial movement of soil particles for non-woven fabric material*

The radial movement of soil particles for non-woven fabric material is shown in Figure 9. Different lines with the same particle size group represent data from different longitudinal height levels. Overall, the fine particle content is reduced near the center of the sample (radial distance = 0). The closer to the center, the more obvious the percentage change of soil particle is. This displays the fine particle flowing due to vacuum effect. The changes of particle groups smaller than 0.01 mm are relatively large. It means that smaller particles are easier to move. The particle size distribution in the tail water shows that the fine particles smaller than 0.01 mm are much more than that of soil samples.

3.1.3 *Radial movement of soil particles for bamboo fiber material*

The radial movement of soil particles for the bamboo fiber material is shown in Figure 10. The particle movement is more complicated in this case. Fine particles move towards 1/2 of the radius. There are a relatively large number of coarse particles in the center (radial distance = 0) and the edge of the sample (radial distance = 1). It can be found from the particle size distribution of tail water that the content of the particles smaller than 0.01 mm is lower than that of the sample. That means more course particles flow out of the sample.

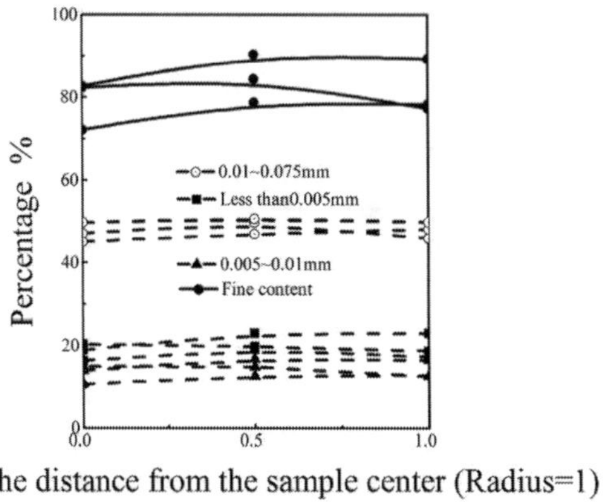

Figure 9. Radial movement of soil particles for non-woven fabric material.

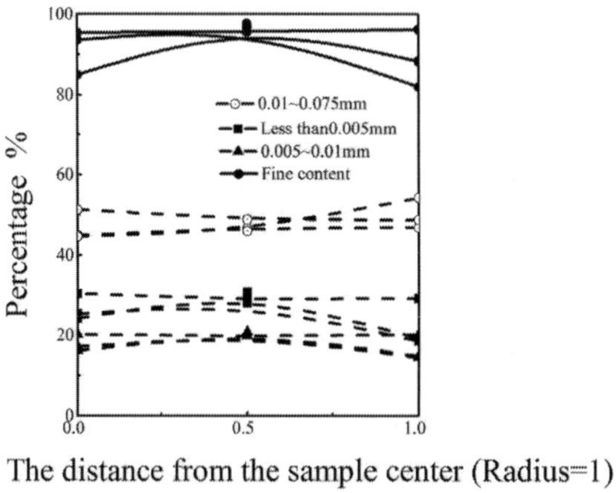

Figure 10. Radial movement of soil particles for bamboo fiber material.

3.2 *Vertical movement of soil particles*

3.2.1 *Vertical movement of soil particles for PVA material, coral fiber material, and super-fine fiber material*

In order to study the vertical movement of soil particles, take the downward distance from the top surface of the sample as the reference coordinate. Assume the distance from the top to the bottom of the sample as 1. The positions of sampling are the coordinates of 0, 1/3, 2/3, and 1. The vertical movement of various particle sizes for PVA material is shown in Figure 11.

As we can see in the figure, soil particles for PVA material hardly move in the longitudinal direction. The total fine particle content is substantially constant. The vertical movement of particles for coral fiber material is similar to PVA material. The vertical movement of particles for super-fine fiber material is also not obvious.

3.2.2 *Vertical movement of soil particles for non-woven fabric material*

The vertical movement of soil particles for non-woven material is shown in Figure 12. The three lines with the same particle size group represent data from the three vertical sections at the radial positions of 0, 1/2, and 1. The fine particle content reaches minimum at 1/3 point of height. The figure indicates that soil particles smaller than 0.01 mm move downward obviously.

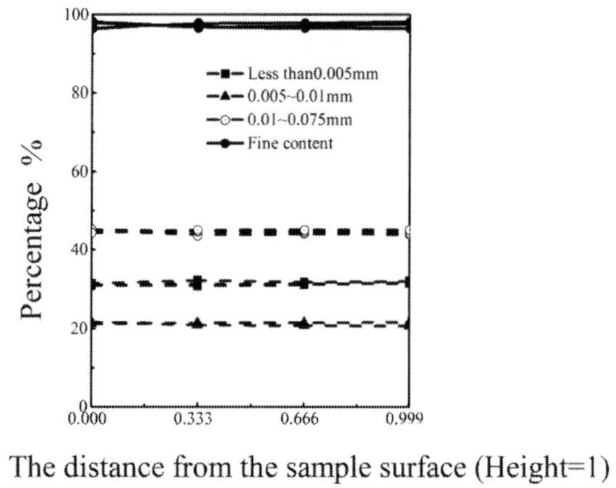

Figure 11. Vertical movement of soil particles for PVA material.

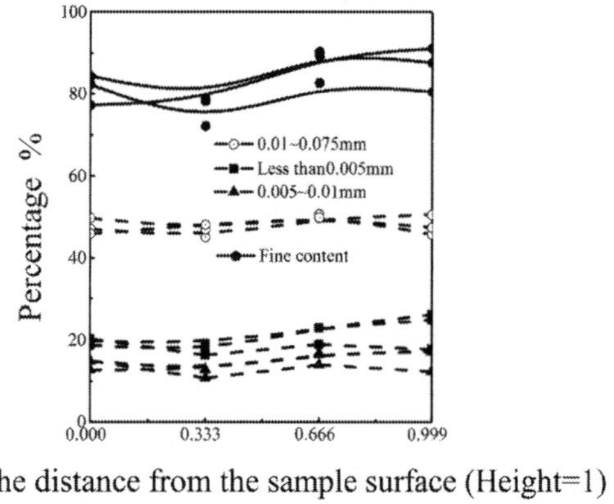

Figure 12. Vertical movement of soil particles for non-woven fabric material.

3.2.3 *Vertical movement of soil particles for bamboo fiber material*

The vertical movement of soil particles for bamboo fiber material is shown in Figure 13. Overall, the soil particles larger than 0.01 mm move downward from top to bottom, while the soil particles smaller than 0.01 mm move upward from bottom to top. The total fine particle content is increased from middle to upper or lower surfaces. So the middle section of the sample is a relatively coarse part.

3.3 *Spatial changes of water content*

To study spatial changes of water content at the end of vacuum drain, take the samples at the points in Figure 7. Conduct water content tests on 5 groups of the samples. They are PVA material-2 (absorbent cotton), coral fiber material, super-fine fiber material, non-woven fabric material, and bamboo fiber material. The initial water content of all the samples is 106%. The spatial changes of water content of the 5 groups are shown in Figure 14. The figure (a)~(e) respectively correspond to the above 5 kinds of materials.

It can be seen from the figure that the overall distributions of water content are lowered near the drain material. This change is more obvious near the bottom. Comparing the five materials, we can see that the water content for PVA material is lowered most and uniformly from left to right in Fig. (a).

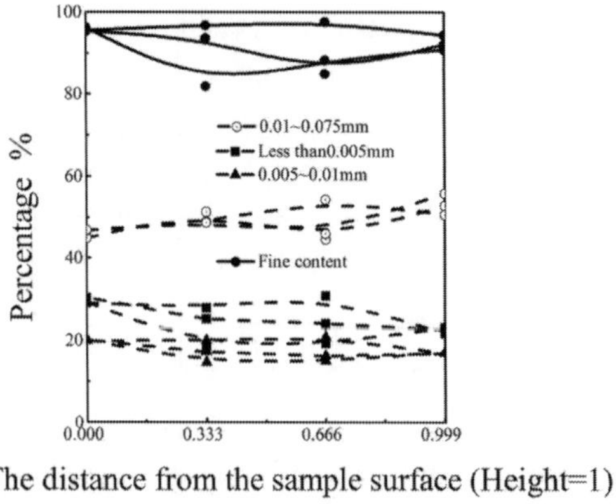

Figure 13. Vertical movement of soil particles for bamboo fiber material.

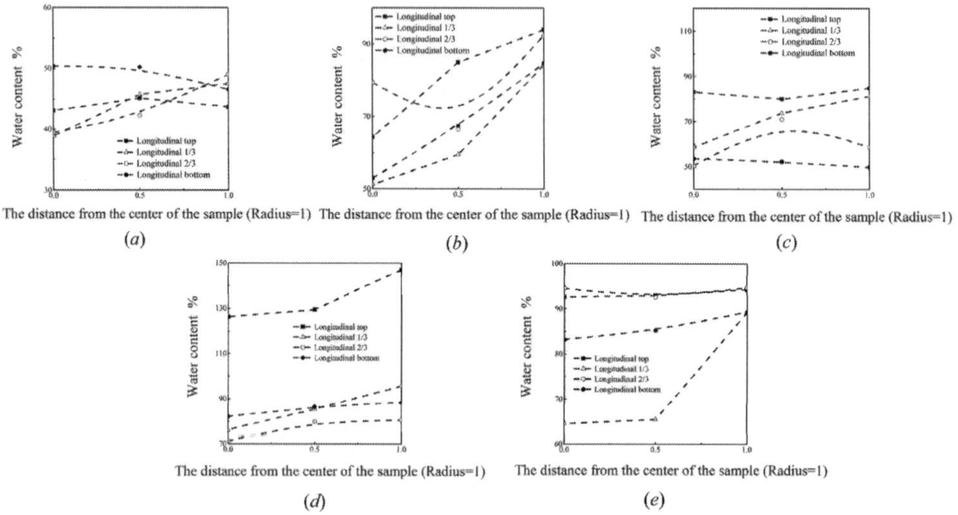

Figure 14. Spatial distributions of water content.

4 CONCLUSIONS

1. In the experimental study, the radial movement of soil particles for various kinds of geotextile filter materials is investigated. For PVA material, soil particles move slightly. Coral fiber material and super-fine fiber material are similar to PVA material. For ordinary non-woven fabric material, soil particles move very obviously to the drain center. For bamboo fiber material, fine particles move toward to 1/2 of the radius.
2. In the meanwhile, the vertical movement of soil particles is measured too. For PVA material, coral fiber material and super-fine fiber material, soil particles basically do not move. For ordinary non-woven fabric material, soil particles obviously move downwards. For bamboo fiber material, course particles concentrate to the middle section.
3. The spatial changes of water content are explored. For non-woven fabric material and bamboo fiber material, the water content is lowered greatly near the drain center. However, for PVA material, the water content is reduced most. And drain process occurs uniformly throughout the soil sample.
4. From all those experimental results, it can be seen that PVA material is the best for drain effect among the five geotextile filter materials. It has already been employed in Wenzhou engineering project and achieved an excellent effect.

ACKNOWLEDGEMENTS

Sponsored by the National Natural Science Foundation of China (Grant No.51109070) First author: Yuan Zhou, Male, Feb.1983, Research Assistant *Corresponding author: E-mail: zdpy_1000@hhu.edu.cn

REFERENCES

Cazzuffi D A, Mazzucato A, Moraci N, et al. A new test apparatus for the study of geotextiles behavior as filters in unsteady flow conditions: relevance and use [J]. Geotextiles and Geomembranes, 1999, 17(5): 313–329.

Dong-sheng D. Comparison of remolded shear strength with intrinsic strength line for dredged deposits [J]. China Ocean Engineering, 2007, 21(2).

Fannin R J, Pishe R. Testing and specifications for geotextile filters in cyclic flow applications [C]// Proceedings of the Geosynthetics Conference, Portland, Oregon, USA. 2001: 423–435.

Gao Zhiyi, Liang Aihua. Definition and Characteristics of Ultra Soft Soil [J]. Petroleum Engineering Construction, 2013, 39(6) (in Chinese).

Hoare D J. A laboratory study into pumping clay through geotextiles under dynamic loading [C]// Proceedings of Second International Conference on Geotextiles, Las Vegas, United States. 1982, 2: 423–428.

Huang W, Koerner R M. An Amendment Strategy for Enhancing the Performance of Geotextile Tubes used in Decontamination of Polluted Sediments and Sludges[C]//Geosynthetics Research and Development in Progress. ASCE, 2005: 1–6.

Ingold T S, Miller K S. Geotextiles Handbook [M]. 1988.

Lawson C R. Filter criteria for geotextiles: relevance and use [J]. Journal of the Geotechnical Engineering Division, 1982, 108(10): 1300–1317.

Novak J T, Agerbæk M L, Sørensen B L, et al. Conditioning, filtering, and expressing waste activated sludge [J]. Journal of Environmental Engineering, 1999, 125(9): 816–824.

Palmeira E M, Gardoni M G. Drainage and filtration properties of non-woven geotextiles under confinement using different experimental techniques [J]. Geotextiles and Geomembranes, 2002, 20(2): 97–115.

Qian Ning, Wan Zhaohui. Advancement of Sediment Transport Mechanics [M]. Beijing: Science Press.1983 (in Chinese).

Schober W, Teindl H. Filter Criteria for Geotextiles [A]. Proceedings of the Seventh European Conference on Soil Mechanics and Foundation Engineering [C]. Brighton: 1979, 261–268.

Shan H Y, Wang W L, Chou T C. Effect of boundary conditions on the hydraulic behavior of geotextile filtration system [J]. Geotextiles and Geomembranes, 2001, 19(8): 509–527.

Xu Youjian. Hydraulic Geosynthetics Applications [M]. The Yellow River Water Conservancy Press, 2000 (in Chinese).

Ye Guoliang, Guo Shujun, Zhu Yaoting. Analysis on Engineering Property of Ultra Soft Soil [J]. China Harbor Engineering, 2010 (5): 1–9 (in Chinese).

Zhou Yuan, Large-scale Model Tests on Dewater of Dredged Clay by use of Ventilating Vacuum Method [D]. Nanjing: Hohai University, 2010 (in Chinese).

Zhuang Yanfeng, Chen Lun, Xu Qi. Research on effect mechanism of cyclic flow on filtration system [J], Rock and Soil Mechanics, 2008, 29(7): 1773–1777 (in Chinese).

Advanced Engineering and Technology III – Xie (Ed.)
© 2017 Taylor & Francis Group, London, ISBN 978-1-138-03275-0

The application of silica fume cement grouting materials in non-uniform settlement of building foundation

LanYun Chen & QuanWei Zhao
Jinhua Polytechnic, Jinhua, China

ABSTRACT: This paper analyzes the chemical and grain composition of silica fume, tests its compressive strength, anti-breaking strength, and impermeability, and studies the solidifying time and the toxicity of the grout. The results indicate that the application by the application of silica fume cement grouting materials is an effective means to address the non-uniform settlement of building foundation. This can be used as a reference for further study of silica fume.

1 INTRODUCTION

Many engineering problems arise from the non-uniform settlement of building foundation, affecting the security and function of the building. Therefore, apart from finding the causes of foundation settlement, scope, active, mechanism, and influence on engineering and evaluation, necessary preventive and control measures should be taken.

Grouting technology is one of the effective means to solve the non-uniform settlement of building foundation. Most of the grouting materials have some disadvantages such as high cost, pollution, and toxicity. Microsilica (silica fume, SF) has found to be one of the best grouting materials (Xian, 2013; Zhang, 1994; Esso, 2002). It is a by-product of ferrosilicon and silicon alloys, collected from the soot dust. It has a particle size of about 1 μm or less, and an average particle size of 0.1 μm, with large specific surface area and activity. The main composition is SiO_2, with a density of 2.2 g/cm^3.

2 BASIC FUNCTION OF SILICA FUME CEMENT

2.1 *Chemical and grain composition (Zhang, 1994)*

The powder particles of silica fume are very fine, with large specific surface area. Slurry requires a large volume of water and has a good water-preserving ability. To reduce the cement ratio of slurry water and improve the fluidity of the slurry, an appropriate amount of water-reducing agents is added. Table 1 and Table 2 list the chemicals and grain composition

Table 1. Chemical composition of silica fume.

Chemical composition	SiO_2	Al_2O_3	Fe_2O_3	CaO	MgO	Loss on ignition
Content/%	92.16	0.44	0.27	0.94	1.37	1.63

Table 2. Grain composition of silica fume.

Particle size/μm	0.2–0.4	0.4–0.6	0.6–0.8	0.8–1.0	1.0–1.2	1.2–1.4	1.2–1.6	1.6–1.8	1.8–2.0
Content/%	18.1	29.9	13.7	9.0	6.7	5.7	3.5	5.7	5.7

of slurry water/cement ratio. From Table 2, it can be seen that the particle size of silica fume is in the range of 0.2 m–2.0 m, with an average size of about 0.6 m.

2.2 *Compressive strength (Zhang, 1994)*

We used 52.5# Portland cement to carry out the formula test in the laboratory. The results indicated that doped silica fume in slurry water could not improve the strength of the grout. On the contrary, the high water/cement ratio decreased. Moreover, a stratified phenomenon appeared. The optimum formula for silica fume slurry was as follows: water/cement ratio of 0.5–0.6, silica fume of 6%–10%, UNF-5 (weight of Portland cement) of 1.2–2.2%, viscosity of cement-based grouts of 1.25×10^{-1} Pa · s, compressive strength of slurry stone body of 55–60 MPa at 28 days of age. The indoor test compressive strength standard values at 1, 3, and 28 days were 45.3, 53.7, and 66.6 MPa, respectively. The test results of the field sampling are summarized in Table 3.

The results of the experiments indicated that pure silica fume cement grout strength was far greater than the same formula of the compressive strength of the stone. The reason is that the grouting pressure increases the density of the slurry.

The experimental results from Norway Elkem Company indicated that the strength of concrete significantly increased after the addition of silica fume (Ran, 2004) as shown in Figure 1. The figure shows the percentage of silica fume (accounted for cement weight). The set of data was obtained for the cement dosage of 390 kg/m^3 and the water/cement ratio of 0.41.

2.3 *Bending strength (P.C, 1991)*

Concerning test methods for strength of T0506-2005 cement mortar, there are three groups and size of the test block was 40 × 40 × 160 mm. After the natural conservation, the flexural strength values at 1, 3, and 28 days were 5.1, 8.2, and 8.3 MPa, respectively.

Table 3. Compressive strength of grout stone (MPa).

Sample number	1-1	1-2	1-3	2-1	2-2	2-3	3-1	3-2	3-3	3-4	3-5	3-6	4-1	4-2
Compressive strength	85.4	88.2	122	79.9	90.4	92.0	46.0	35.0	42.6	34.0	54.5	37.4	21.5	25.5

Notes:
(1) The sample 1-1~1-3 for residual pure silica fume cement grout in boreholes.
(2) The sample 2-1~2-3 for silica fume cement slurry and cushion stone combination (bedding stones are hard granite and the surface is clean).
(3) The sample 3-1~3-6 for silica fume cement slurry and the foundation rock combination (pile of stone contains ooze, low intensity).
(4) The sample 2-1~2-3 for cement slurry and foundation stone (the slurry has low strength and the surface of the stone is not clean).

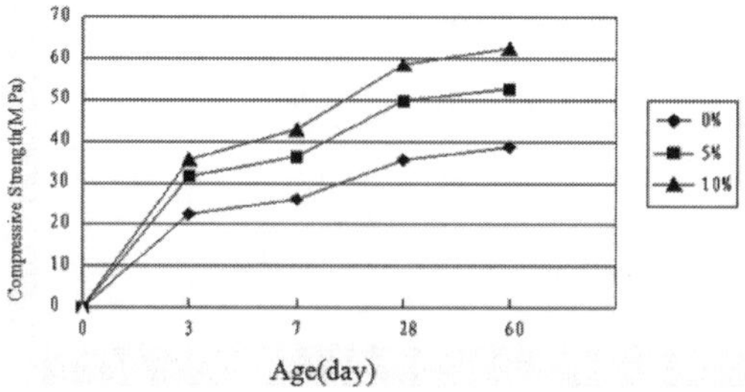

Figure 1. Effect of silica fume on concrete strength.

2.4 *Impermeability (P.C, 1991)*

According to the permeability resistance test of "Standard for test method for long term performance and durability of ordinary concrete" (GB/T 50082-2009), the size of the test block was 175 × 185 × 150 mm. The results of six samples in one group, after 28 days of the natural conservation, indicate that the permeability coefficient of the test piece was less than 0.129×10^{-8} cm/s. The anti-permeability level was greater than S_{12}.

2.5 *Injectivity (P.C, 1991) (mobility)*

According to the "cement mortar sand flow measurement method" (GB/T 2419-2005), when the JZ-2 cement sand flow tester was used, the slurry water/cement ratio was 0.3. In addition, the test results indicated that the fluidity was greater than 300 mm.

2.6 *Slight expansion (P.C, 1991)*

When the amount of micro silicon powder slurry was increased, the inflation rate reached above 0.02%. The shrinkage stress of the grouting concretion body was eliminated to avoid the shrinkage of the slurry concretion body.

2.7 *Gel time (P.C, 1991)*

According to the actual need, the gel time of the slurry consolidation body can be adjusted at any time within 20 to 105 minutes. Moreover, the final setting time was less than 5 hours.

2.8 *Toxicity (P.C, 1991)*

Micro silicon powder cement grouting material is non-toxic, and has no pollution and corrosion resistance, and has a wide range of applications.

3 ENGINEERING APPLICATION

3.1 *Measure of non-uniform settlement of the foundation (Zhang, 1994)*

3.1.1 *Engineering situation*

After the use of a gold ore powder silo for three months, no uniform settlement of the foundation of the ore bin was observed. There were larger cracks found in the wall. On further exploration, it was confirmed that the original earth-rock deposit was mistaken for weathered granite foundation. The higher bearing capacity value was adopted to carry out the design. As the requirement of the upper structure was not met, the bearing capacity of the foundation could not be improved, resulting in non-uniform settlement of the foundation. To meet the needs of production, the earth-rock accumulation body should be strengthened by grouting.

3.1.2 *Reinforcing scheme*

The diameter of the powder silo is 12 m, arranged under seven independent foundations. The basic profile is shown in Figure 2, where the center is arranged on a base, and the rest of the structure is evenly arranged on the circumference of the radius 6 m.

As can be seen from the table, rubble cushion is between the foundation and debris accumulation body. The mortar strength cannot meet the requirements of the bearing capacity of the foundation. According to the characteristics of engineering and conditions, we take the following measures:

1. In the 3 m range, the rubble cushion and cushion debris are piled under the body perfusion of high strength silica fume cement slurry.

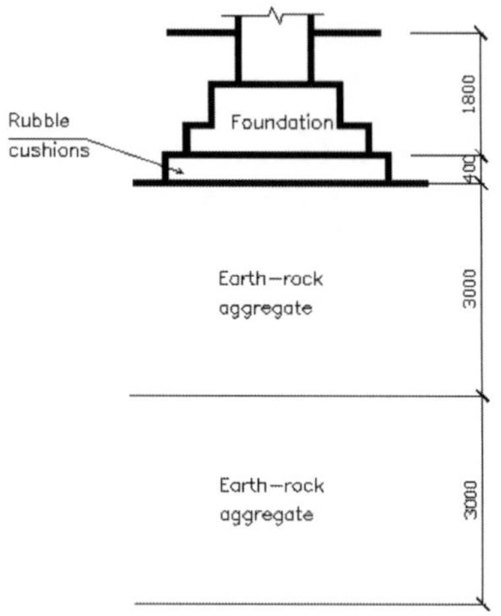

Figure 2. Basic section.

2. Due to the diffusion effect of stress, the stress of the soil rock accumulation body in the 3–6 m range under the cushion layer is less. So it can only be filled with ordinary cement slurry.
3. On the occurrence of 2 column settlement, due to its complex geological condition, the implementation of the above measures is added, under the foundation of the diameter of 150 mm and depth of 8 m root pile, on the basis of 3 m range of epoxy resin grouting.

3.1.3 *Grouting construction*

Determination of slurry formulations: 0.5 of slurry water/cement ratio (proportion of cement) of 0.5, silicon incorporation (proportion of cement) of 8.6%, and the slurry viscosity of about 1.25×10^{-1} Pa · s. The compressive strength at 28 days was about 55~60 MPa.

Construction procedures: first, the root pile is constructed, followed by the inner cushion layer and 3 m debris accumulation body perfusion silica fume cement slurry, and finally the 3–6 m debris accumulation body perfusion, and the common cement slurry.

When the performance of silica fume cement slurry is relatively complex, the cost is relatively high. It is the first application of silicon water slurry in the domestic engineering. We should adopt some pertinent measures in the construction, such as super dense hole, no flush drilling, strictly controlling the grouting pressure and monitoring the change of the ground and wall cracks, accurately calculating the ratio of slurry, addition of warm water (30–40°C) melt UNF-5, micro silicon powder and cement mixing evenly after adding liquid UNF-5.

3.1.4 *Injection volume and analysis*

1. The project within the rubble cushion is injected into micro silicon powder slurry of 19.8 m^3. In fact, it should be considered as a part of the grout diffusion into the earth-rock deposit;
2. Deposit within 3 m. stone and the perfusion of micro silicon powder slurry of 80 m^3. The injection rate is greater than expected, which suggests that there are quite a number of serous spreading to a considerable extent.
3. In the range 3~6 m of stone deposit and additional drilling of 160 m^3, the injected water mud deposit shows that debris is connected with the porosity of the soil at the bottom. So, it is more realistic to consider a grouting at the bottom of the pile of rock body together.
4. Based on the 3 m range of two foundation perfusion of 12.3 m^3 epoxy resin slurry. When micro silicon powder slurry is filled in formation pore, the grout injection rate becomes reasonable.

3.1.5 *Effect test*

Samples were collected from the drill hole after the grouting construction process and sent to the relevant departments for the uniaxial compressive strength test. The results are summarized as follows:

1. Pure stone of micro silicon powder slurry, the strength values of the three samples are 85.4, 88.2, and 122.3 MPa, which are greater than those from the laboratory test results. This can be attributed to grouting pressure slurry solid results.
2. Intensity of micro silicon powder slurry and cushion approximates that of the strength of the cement stone body. The reason is that bedding rock is hard granite, and the surface is clean.
3. Intensity of micro silicon powder water deposit mud and debris is about 50% lower than that of the stone body. The reason is that piles of stone containing silt are not washed before the grouting hole. So, the intensity is low.
4. The strength of the common cement slurry and the foundation pile is further reduced and the fluctuation is larger. The main reason is that the slurry itself is low and the surface of the stone is not clean.

After three months of observing powder silo foundation in the Grouting Reinforcement Project, the maximum settlement of the column value is found to be only 0.36 mm. The data indicate that all the strength values of the sample are much larger than the design requirements. Foundation settlement can be ignored to meet the engineering requirements.

3.2 *Improving the bearing capacity*

3.2.1 *Engineering situation*

In a Bureau of Mines with a fan room base, the cohesive soil water content increases and the bearing capacity decreases because of the wind well drainage infiltration under the fan base. The foundation settlement is about 19 mm. The normal operation of the fan is affected. To ensure that the foundation does not sink, the foundation soil treatment should be reinforced. After the scheme comparison, we decided to adopt a one-way high pressure jet grouting reinforcement.

3.2.2 *Grouting design*

Because the fan room is installed with two desk fans, high spray equipment and drilling becomes hard. It can only work above the roof. Figure 3 shows the arrangement of 20 jet holes. Taking order operations, the sequence is 2, 4, 7 hole, 9, 12, 14, 17, 19 hole. The remainder is a two-order hole. A hole is to take the order of cement water glass grouting material, accelerated slurry

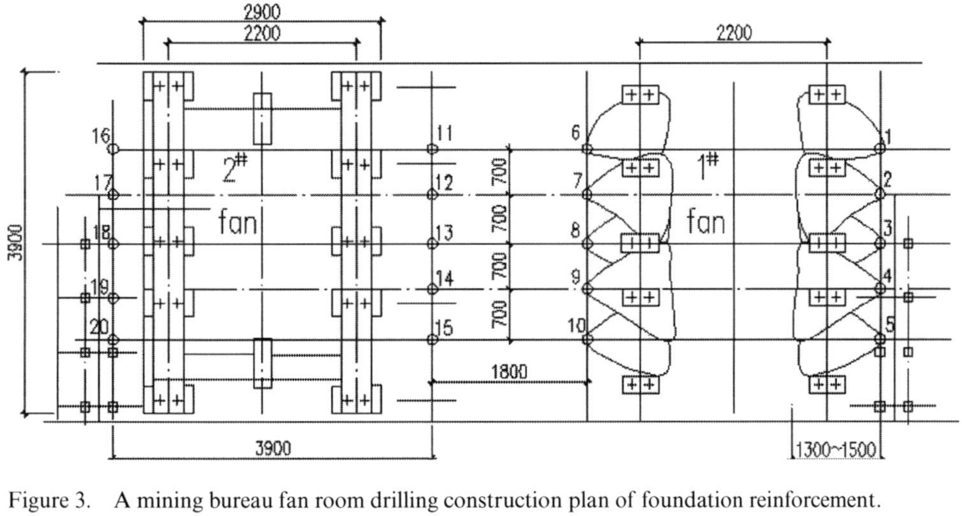

Figure 3. A mining bureau fan room drilling construction plan of foundation reinforcement.

Table 4. Grouting material and dosage.

Name of the material	Specifications, performance	Amount	Manufacturer
Cement	425# (original label) slag cement	143 (t)	Datong Cement Plant
Water-glass	M = 2.33,50'Be, proportion: 1.53	5 (t)	Datong Chemical Factory
Silica fume	SiO_2: 91%, incorporation: 5~8%	3 (t)	Datong XinGuang Ferroalloy Co. Ltd
Calcium lignosulphonate	proportion: 91%, incorporation: 0.2%	300 (kg)	Complete Chemical Fiber Factory
Expansive agent	incorporation: 8~12%	350 (kg)	Beijing Sulfur Glass River Cement Plant

Table 5. Micro silicon powder mixing ratio of cement grout material.

Grouting materials	Water	Cement	Expansive agent	Silica fume	Water-reducing (calcium lignosulphonate)	Water-glass
Mix proportion	100	100	10	5	0.2	2

Table 6. Parameters of grouting technology.

High-pressure water		Compressed air		Cement slurry					
Pressure/ MPa	Flow/ L/min	Pressure/ MPa	Flow/ L/min	Pressure/ MPa	Flow/ L/min	Proportion	Swinging angle/°	Hoisting speed/ cm/min	Rotational velocity/ cm/min
20~24	80~85	0.6~0.7	0.6~0.8	1.0~1.5	90~130	1.3~1.4	45~60	8~15	8~20

gel, prevent jet, and fan-based instantaneous settlement. The slurry water/cement ratio of the second-order hole is 1:1. The admixture is micro silicon powder, calcium lignosulphonate and sodium silicate. Close to the bedrock injection slurry spray (not ascend pendulum) 3–5 min, 2 m within the scope of the rubble, and concrete foundation repeats the injection.

3.2.3 *Grouting materials*

Grouting material and dosage are presented in Table 4, and the ratio of slurry is given in Table 5.

3.2.4 *Parameters of grouting technology*

The one-way swing grouting technology is used in the construction of water vapor, coaxial jet, one-way nozzle, and nozzle swing 8–20 times per minute. The technical parameters are presented in Table 6.

3.2.5 *Effect of grouting*

We have observed the settlement of the foundation in the construction process of pendulum jet grouting, and found that there is no generation of additional settlement. In addition, when checking grouting construction three months after the test, we found that there is no foundation settlement to meet the design requirements.

4 CONCLUSION

Micro silicon powder is a high degree of dispersion. It is also a kind of highly active material. Indoor test and engineering practice show that we can improve the strength of the grouting concretion body when an appropriate amount of micro silicon powder is added to cement slurry. The silicon particle size is very small and the size stability is good, which makes it easy to penetrate into tiny fissure or sand. It can greatly improve the durability of concrete and decrease the permeability of concrete after incorporation of micro silicon powder. By mixing sodium silicate cement grout with an appropriate amount of micro silicon powder, the problem of the instant settlement grouting construction process can be solved. In addition, it has a high rate of stone. So, it can fill the soil hole and improve the bearing capacity of the foundation.

Note: Silica fume is abbreviated as SF, also known as Nanometer silicon, Silicon Powder, Reactive Powder Concrete, etc.

REFERENCES

Aitcin. PC, etc. 1991. The application of silica fume in the grouting liquid, *Modern Grouting Technology*. Beijing: Hydraulic and Electric Power Press.

Esso international trading (Shanghai) co. 2002, 3. The application of Aiken silica fume in concrete. Shanghai.

Ran DF, etc. 2004. The development of geological disaster prevention with high strength grouting materials without pollution. *Exploration Engineering*: 41~42. Beijing.

Xian CJ, etc. 2003. Chemical industry press, *Nano-Materials*: 200~261. Beijing: Chemical industry press.

Zhang ZM, etc. 1994. The application of micro silica fume cement slurry in foundation underpinning. *Practical Rock and Soil Reinforcement Technology*: 208~211. Beijing: Seismological Press.

Advanced Engineering and Technology III – Xie (Ed.)
© 2017 Taylor & Francis Group, London, ISBN 978-1-138-03275-0

Study on the method of crack width monitoring based on circular approaching degree

Zhenping Liu, Genming Du, Jian Liu, Huaijian He, Fei Ai, Kang Bian & Shangge Liu
State Key Laboratory of Geomechanics and Geotechnical Engineering, Institute of Rock and Soil Mechanics, Chinese Academy of Science, Wuhan, Hubei, China

Bo Yang
Construction Headquarters of Xiangsui Expressway, Hubei Provincial Communications Investment Group Co. Ltd., Xiangyang, Hubei, China

ABSTRACT: To avoid carrying heavy instruments, it is efficient to apply the powerful functions of cell phone in engineering practice. The self-designed target patterns on both sides of fracture were shot by cell phone without calibration and the circle and ellipse recognition algorithm was proposed. In this algorithm, circular approaching degree was regarded as the discrimination standard for the circle pattern. Then, the convenient, low-cost fracture width deformation monitoring method was implemented using OpenCV and Python programming. Subsequently, the experimental study was conducted and the results indicated that the method is sensitive to the variation of shooting distance and is insensitive to elliptic recognition and pixel interpolation. Finally, compared with the conventional approach with crack width tester in the engineering example, accuracy of the method proposed in this paper is able to meet the demand of engineering and the approach is convenient for application in engineering practice.

1 INTRODUCTION

In recent years, with the advance of computer science, the shooting function of different kinds of digital cameras is rapidly enhanced. The image processing technology in geotechnical engineering for deformation monitoring is also widely used. Particularly, it is used for the crack deformation monitoring of concrete structure in caverns, which requires high measurement precision and short work hours due to the environmental noise and heavy pollution. Cell phones have become an integral part in daily lives, work and scientific research. Therefore, this paper aims at improving the efficiency of concrete crack width extension monitoring by means of cell phone camera and image processing method.

Sun et al. (2008) proposed an automatic recognition algorithm based on image processing technology to monitor pavement crack diseases. Nie et al. (2012) came up with an innovation on pixel calibration for Nanjing Ming city wall crack deformation monitoring and the value of crack deformation was obtained. Ye et al. (2010) put forward a new method of surface crack width measurement based on image processing technology. Tian et al. (2006) applied photogrammetry and image processing technology to structural deformation in tunnel and underground space and achieved some good results. Zhou et al. (2008) used infrared camera and image processing technique to photograph the constraints of rock tunnel, which successfully overcame the picture noise and the relevant problems caused by the uneven illumination. Liu et al. (2012) considered lining seepage area as the detecting target, and a set of digital image processing algorithm was worked out for removing noise, sharpening, segmentation and correction. Ding et al. (2010) proposed the method of generating a particle flow model by the numerical image combining image processing technology, and a new modeling method for particle flow simulation concerning soil-rock mixture mechanical behavior was provided. Zhu et al. (2011) used image processing technology in numerical simulation of rock failure process analysis, which was a convenient and effective method for the mesoscopic structure of rock damage mechanics behavior research. Allam et al. (1978) applied analytical photogrammetry technology to the open-air mine excavation process for the first time to measure rock fracture spacing and analysis rock stability. Kemeny et al. (1993) photographed the rock fragments after blasting in rock tunnel construction and coded program to analysis image, and successfully identified the size and distribution of rock fragments.

To sum up, image processing technology has been widely used in underground engineering, and lining seepage and numerical simulation and various kinds of image processing algorithms emerged in endlessly. But those methods required high precision digital camera and need to be more professional in calibration, so they have not been widely applied in practical engineering.

Compared with digital cameras, mobile phone is lighter and portable, so it will be useful to apply it to the photogrammetry. However, there remains some obvious shortcomings in its application. Due to the small mobile sensor, high pixel often means complicated interpolation. Moreover, resin lens is different from the optical one in the material aspect, which will eventually lead to differences in the pictures photographed by digital camera and mobile phone. So, to apply the convenient mobile phone camera to crack width monitoring, much efforts need to be made on monitoring means and image processing algorithm.

2 FRACTURE MONITORING METHOD BASED ON MOBILE PHONE PICTURES

2.1 *Image processing*

The abundant image processing function in Matlab was mainly used by researchers, although this method did not facilitate system integration (Tian et al. (2006); Dong, (2006); Xu et al. (2008)). In this article, the OpenCV and the Python language is used for image processing. OpenCV (Open Source Computer Vision Library) is a powerful open-source computer vision library supported by Intel. Python is an interpreted, object-oriented, dynamic semantics, and graceful scripting language (Luo et al. (2008)). It is one of the three most widely used cross-platform languages, namelyTcl and Perl (Zhong et al. (2006)).

2.2 *Monitoring scheme and algorithm*

The target pattern was designed, which was a black box as background with a white circle in the center. According to crack widths and monitoring precision, the target pattern size (box length H and circle radius R) can be adjusted. For monitoring, the target patterns were pasted on both sides of the crack and the picture was taken by mobile phone. Afterwards, image processing techniques were used to identify two target circles on both sides of the crack in mobile photos, which was necessary to get the coordinates of centers of two circles and calculate the length between two target circles. Comparing the length obtained in different time periods, the crack width extension can be monitored. The procedures of this method are shown in Figure 1.

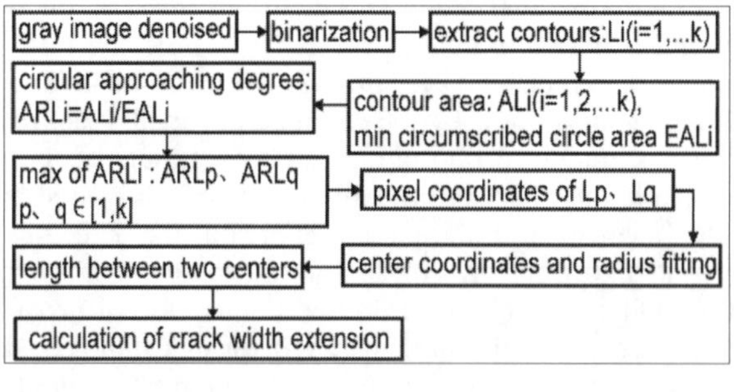

Figure 1. Crack width monitoring flow chart.

To identify two target circles, the least squares fitting method based on circular approaching degree was put forward based on OpenCV.

The circular approaching degree of any extracted contour Li(i = 1,2,...k) is:

$$ARLi = ALi/EALi \tag{1}$$

where ALi is the area of some contour and EALi is the minimum circumscribed circle area of the same contour.

ARLp and ARLq are the most highest circular approaching degree values among all the extracted contours. According to ARLp and ARLq, the sets of image pixel coordinates of contours of Lp and Lq can be obtained.

The equation of least squares analysis for circles is:

$$x^2 - 2Ax + A^2 + y^2 - 2By + B^2 - R^2 = 0 \tag{2}$$

where A is the x-coordinate of circle; B is the y-coordinate of circle; R is the radius of circle.

For calculation convenience, equation (2) can be changed as follows:

$$x^2 + y^2 + ax + by + c = 0 \tag{3}$$

where $a = -2A$; $b = -2B$; $c = A^2 + B^2 - R^2$.

Equation (3) is the objective function of least squares fitting, by which the pixel coordinates of two circle centers and radius length in pixel are available. The average radius of two circles in pixel is:

$$R_{avg} = (R_p + R_q)/2 \tag{4}$$

where R_p is the pixel length of one circle radius and R_q is the pixel length of another circle radius.

The actual length of each pixel is:

$$L_{pra} = R/R_{avg} \tag{5}$$

where R is the actual radius length of the target white circle designed in target pattern.

Finally, the actual distance between the two white target circles is:

$$D = \sqrt{(A_p - A_q)^2 + (B_p - B_q)^2} \times L_{pra} \tag{6}$$

where A_p are pixel x-coordinates of one circle center; A_q are pixel x-coordinates of another center; B_p are pixel y-coordinates of one circle center; B_q are pixel y-coordinates of another center.

3 MONITORING METHOD ACCURACY TEST

In order to examine the accuracy and stability of image processing algorithm, the static target test was conducted considering the influence of shooting distance and pixel interpolation.

3.1 *Test plan*

As shown in Figure 2, the test icon uses the white circle as the goal and the black box as the background. The diameter of the circle is 20 mm and the distance between the centers of two circles is 35.315 mm. In order to study whether the shooting distance between lens and icon would affect the accuracy of the results, five white frames were set by a uniform spacing of 3 mm in the icon. Apple iPhone4S was used; its camera parameters are 8 million pixels, 4 mm focal length, and CMOS sensor.

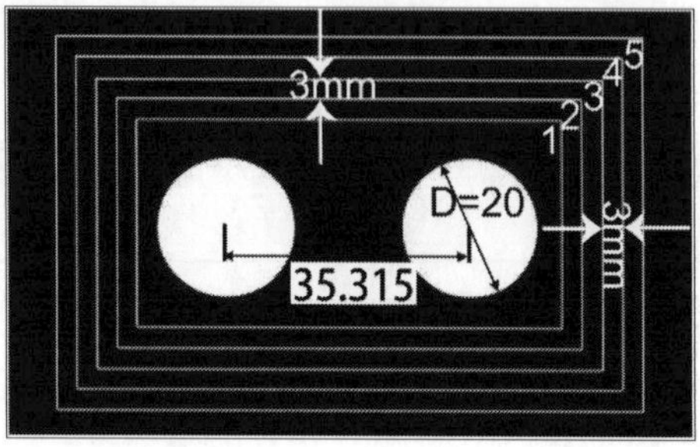

Figure 2. Monitoring accuracy testing icon.

Table 1. The testing results of monitoring accuracy.

		Circle fitting				Ellipse fitting			
		Without interpolation		Interpolation		Without interpolation		Interpolation	
Group number	Picture number	Average (mm)	Standard deviation	Average (mm)	Standard deviation	Average (mm)	Standard deviation	Average (mm)	Standard deviation
1	12	35.589	0.022	35.576	0.021	35.589	0.023	35.577	0.021
2	13	35.556	0.019	35.549	0.030	35.557	0.019	35.550	0.029
3	13	35.561	0.020	35.558	0.019	35.563	0.018	35.558	0.019
4	7	35.555	0.025	35.552	0.027	35.555	0.024	35.552	0.026
5	10	35.588	0.017	35.579	0.018	35.588	0.019	35.578	0.019
6	11	35.563	0.022	35.554	0.017	35.564	0.021	35.554	0.018

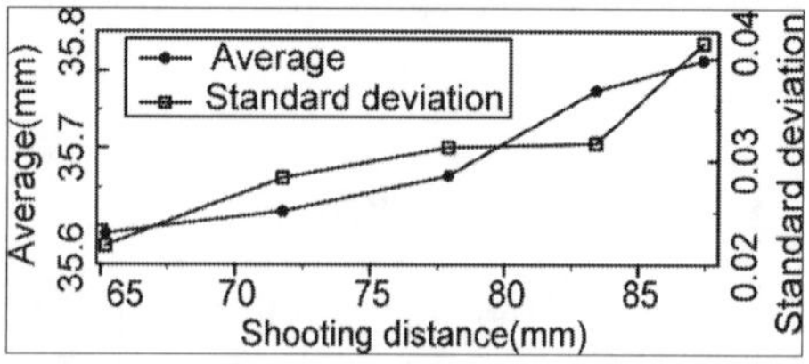

Figure 3. Curves of results with shooting distance.

3.2 *Monitoring accuracy under a certain shooting range*

Under the shooting range constrained by wire frame 1 (Fig. 2), six groups of photographs taken by iPhone 4S in the same brightness were processed by circle fitting, ellipse fitting and interpolation, and the results are listed in Table 1. It showed that in the case of a random shot the standard deviation was almost 0.02 mm, which can meet the requirements of monitoring of crack width extension. Circle and ellipse fitting methods nearly resulted in similar results and interpolation has a little impact on the stability of the crack width monitoring. However, the average distance between two circles after interpolation is closer to the real one, which indicates that the higher pixel is beneficial for the monitoring accuracy.

3.3 *The impact of shooting distance to monitoring accuracy*

Five groups of photographs were obtained by shooting in different wire frame 1, 2, 3, 4, 5 (Fig. 2) during a continuous period of time with 13 to 16 photos in each group. The results are shown in Figure 3. With the enlargement of the shooting distance, the average length between two circles and standard deviation of each group becomes greater and the results are unstable. This suggests that the smaller the shooting range is, the higher the monitoring precision will be.

4 CASE STUDY

4.1 *Project profile*

A short expressway tunnel spans about 400 meters long and its maximum depth is 101.1 m. The tunnel locates in mainly low mountain hills. The second lining cracking caused by bias is serious in the exit of the right tunnel, including transverse and longitudinal cracks (Fig. 4). Therefore, the monitoring of second lining crack width extension is necessary.

4.2 *Monitoring scheme*

For the same crack width extension monitoring in second lining, the data reading method from crack width tester artificially was used to compare with the results obtained by the method proposed in this paper.

According to the crack size and focal length, the diameter of the target circle was set to 20 mm. The target patters were pasted on the flat positions on both sides of the crack, with the direction basically vertical to the trend of second lining fracture. The details are shown in Figure 5. To reduce the error during data collection by mobile phone, each group included more than six photos and the monitoring values in each group were averaged.

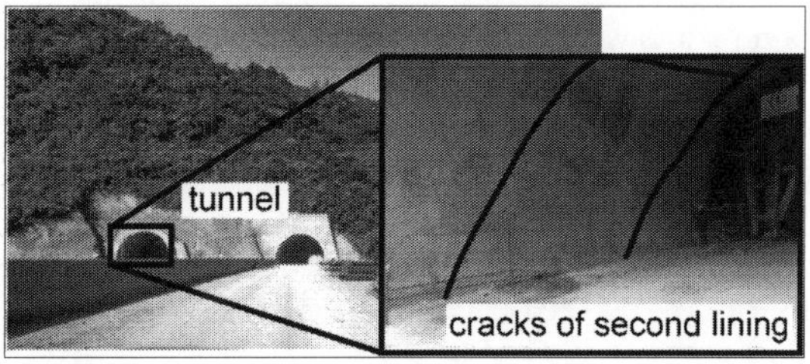

Figure 4. Second lining cracks in right tunnel.

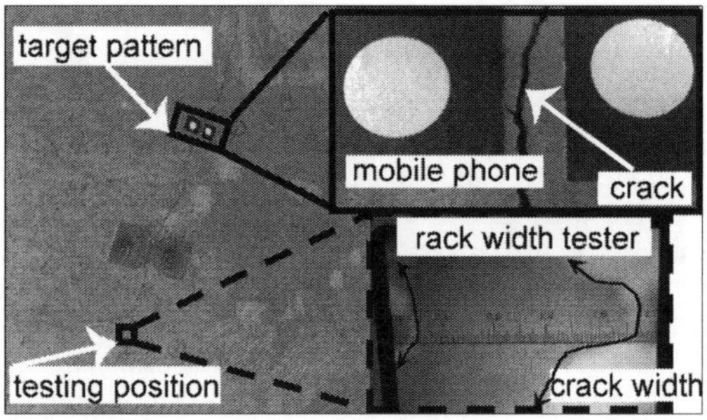

Figure 5. The monitoring methods of crack width.

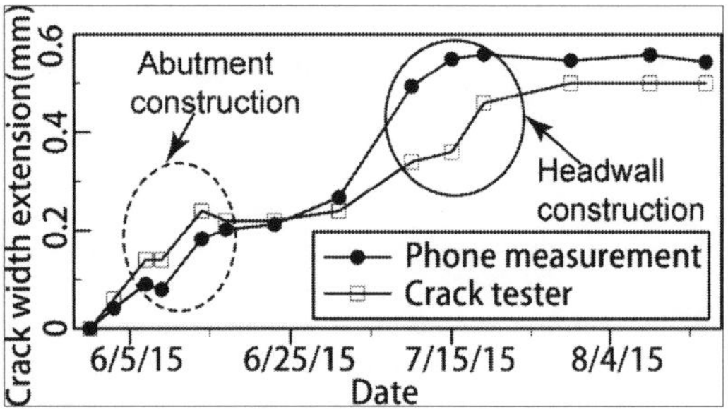

Figure 6. The contrast curves of monitoring results.

4.3 *Monitoring results*

The monitored results are shown in Figure 6. The abutment construction and head wall of foundation excavation has a great influence on crack width extension and the results obtained from the crack width tester the and cell phone method showed good consistency in the two main periods . Eventually, the monitoring data stabilized to a certain value (0.559 mm from the phone and 0.5 mm from the tester). Considering the high coherence in curve change of both methods and the small accumulation of error, it shows that the phone measurement method and the image processing algorithm proposed in this paper are feasible in practical engineering.

5 CONCLUSION

1. A recognition algorithm considering circular close degree was proposed by Python and OpenCV library. This method avoided complex parameter calibrations for pictures.
2. Static target tests indicate that the ellipse fitting and pixel interpolation can slightly improve the monitoring accuracy, while the shooting distance has a greater influence on the results.
3. The results obtained from cell phone camera has good consistency with those obtained from the crack width tester and its precision can satisfy the engineering demand.
4. With improvement of cell phone cameras and image processing techniques, the proposed method will obtain results with better accuracy and be widely applied.

ACKNOWLEDGMENTS

The authors gratefully acknowledge the financial support from the National Science Foundation of China under Grant No. 51204158 and No. 51209198.

REFERENCES

Allam M.M. 1978. The estimation of fractures and slope stability of rock faces using analytical photogrammetry. *Photogrammetria* 34(3):89–99.

Bocheng Sun & Yanjun Q.I.U. 2008. Crack Diseases Recognition Based on Image Processing Algorithm. *Journal of Highway and Transportation Research and Development* 25(2):64–69.

Chunlin Zhou et al. 2008. Application of infrared photography and image processing to tunnel construction with New Austrian Tunneling Method. *Journal of Rock Mechanics and Engineering* 27(a1):3166–3172.

Guiru Ye et al. 2010. Measurement of Surface Crack Width Based on Digital Image Processing. *Journal of Highway and Transportation Research and Development* 27(2):75–79.

Honglin Nie & Wusheng H.U. 2012. Research on deformation monitoring of wall cracks based on digital image processing technology. *Engineering of Surveying and Mapping* 21(4):61–65.

Kemeny J.M. & Devgan A. & Hagaman R M. et al. 1993. Analysis of rock fragmentation using digital image processing. *Journal of Geotechnical and Geoenvironmental Engineering* 119(7):1144–1161.

Shengli Tian et al. 2006. Testing study on digital close-range photogrammetry for measuring deformations of tunnel and underground spaces. *Journal of Rock Mechanics and Engineering* 25(7):1309–1315.

Tongsheng Zhong & Feng WEI. 2006. Second development for fore treatment of ABAQUS using Python language. *Journal of Zhengzhou Univ. (Nat. Sci. Ed.)* 38(1):60–61.

Xiao Luo et al. 2004. Python based mixed-language programming and its implementation. *Computer Application and Software* 21(12):17.

Xiuli Ding et al. 2010. Partical flow modeling mechanical properties of soil and rock mixtures based on digital image. *Journal of Rock Mechanics and Engineering* 29(3):477–485.

Xuezeng Liu et al. 2012. Detection technology of tunnel leakage disaster based on digital image processing. *Journal of Rock Mechanics and Engineering* 31(s2):3779–3747.

Weiguo Dong. 2006. *Matlab 7.X hybrid programming*. Beijing: China Machine Press.

Wenjie Xu et al. 2008. Stability analysis of soil-rock mixtures slope based on digital image technology. *Rock and Soil Mechanics* 29(s1):341–346.

Zeqi Zhu et al. 2011. Numerical simulation of fracture propagation of heterogeneous rock material based on digital image processing. *Journal of Rock Mechanics and Engineering* 32(12):3780–3787.

Advanced Engineering and Technology III – Xie (Ed.)

Mechanical properties of Xigeda soil with different contents of EPS under a consolidated undrained triaxial compression test

Lian Jiang, Xiao Yin & Jianqiu Tian
State Key Laboratory of Hydraulics and Mountain River Engineering, College of Water Resource and Hydropower, Sichuan University, Chengdu, China

Enlong Liu
State Key Laboratory of Hydraulics and Mountain River Engineering, College of Water Resource and Hydropower, Sichuan University, Chengdu, China
State Key Laboratory of Frozen Soil Engineering, Cold and Arid Regions Environmental and Engineering Research Institute, Chinese Academy of Sciences, Lanzhou, China

ABSTRACT: To study the influence of the content of expanded polystyrene on the mechanical properties and strength of Xigeda soil, 16 groups of tests were conducted under a consolidated undrained triaxial compression state. The test results indicate that when the confining pressure is lower, strain softening occurs and the pore water pressure first rises and then decrease. When the confining pressure is higher, strain hardening is observed and the pore water pressure rises and reaches as table state. In this test, with the increase of contents of expanded polystyrene, the deviatoric stress, secant modulus and effective internal friction angle first decrease and then increase, and the effective cohesion decreases linearly.

1 INTRODUCTION

Xigeda soil is widely distributed in Panzhihua region, with smaller preconsolidation pressure and in partly sedimentary diagenesis state. So, its physical and mechanical characteristics is similar to those of soil and rock, having loose structure and low strength. When it encounters water, it may disintegrate and becomes of. Therefore, its engineering properties are poor. Expanded polystyrene (EPS) is a kind of high-performance light subgrade filling, which has many advantages, such as lightweight, small compressibility, good mechanical performance, and simple and convenient construction.

Many scholars have studied the mixed EPS soil. Yajima et al. investigated the deformation and strength properties of foam composite light-weight soil. Kikuchi et al. studied the effect of air foam inclusion on the permeability and absorption properties of lightweight soil. Yoon et al. investigated the effect of initial water content, and silt contents, and cement ratio on lightweight soils. Horpibulsuk et al. studied the compressibility and strength of lightweight clays. Hou et al. investigated the influence of EPS particle size on the shear strength of soil. Dong et al. studied the compression deformation characteristics of light weight sand. In addition, Luo et al. reinforced Xigeda soil with lime; Chen et al. studied the construction technique of Xigeda soil house; and Wang et al. studied the physical and mechanical properties of Xigeda soil. From the existing literatures, the study on the mechanical properties of Xigeda soil with EPS is scare in the literature.

In order to study the influence of different contents of EPS on the mechanical properties and strength of Xigeda soil, 16 groups of tests were conducted under a consolidated undrained triaxial compression state. In addition, the influence of EPS content on the stress strain, pore pressure, and strength of Xigeda soil were analyzed.

Table 1. Physical index of Xigeda soil.

Natural moisture content/%	Natural density/ (g/cm^3)	Specific gravity/ G_s	Plastic limit/ %	Liquid limit/ %	Plasticity index
12.39	1.692	2.672	27	34	7

2 TEST DESCRIPTION

The soil used in the test was taken from a tunnel excavation project in Panzhihua city. The soil was dried, pulverized, and sieved (0.5 mm). The experiment was carried out in accordance with the procedures soil tests. The result indicate that the soil can be identified as a poor grading with fine silty sand. Basic physical indices of the soil are listed in Table 1.

Xigeda soil below 0.5 mm and EPS size under 1~2 mm were used in the test. According to the mass ratio EPS and dry soil, four different levels were used in the experiment: 0.0%, 0.2%, 0.4%, and 0.6%. After the EPS and soil were mixed uniformly, 15% water of the total mass was added and mixed well. The mixing materials were compacted in the four layers with 80 mm in height and 39.1 mm in diameter, with the total dry of 1.4 g/cm^3. Then, the sample was placed in a vacuum saturation container for saturation.

The strain control triaxial compression apparatus TSZ10–1.0 was used in this study. The saturated samples with EPS contents of 0.0%, 0.2%, 0.4%, and 0.6% under consolidated undrained conditions were tested at confining pressures of 50, 100, 200, and 400 kPa. The consolidation time was 16 h and the shear rate applied was 0.08 mm/min.

3 EXPERIMENTAL RESULTS AND ANALYSES

The deviatoric stress–axial strain curves and pore water pressure–axial strain curves of the samples at different confining pressures under consolidated undrained conditions are shown in Figures 1–4. The maximum deviatoric stress and the maximum pore water pressure are presented in Table 2. As shown in Table 2, when the strain softening occurs, the maximum deviatoric stress is taken as the peak. When strain hardening occurs, it is taken as the stress value with an axial strain of 15%. The deviatoric stress and secant modulus at an axial strain of 2% are presented in Table 3.

As shown in Table 2 at the same confining pressure, EPS particles reduces the maximum deviatoric stress. Along with the increase in EPS content, the maximum deviatoric stress decreases first and then increases, and the minimum occurs when the EPS content is 0.4%. The reason for this is that the EPS particle is a lightweight material and the volume replacement effect is very remarkable, which occupies most of mixed soil space, and leads to reducing yield strength with the increasing contents of EPS. For the 0.6% sample, the yield stress increases suddenly, which is caused by the rearrangement of EPS particles in the consolidation process, leading to the increase of EPS particles and the interlocking force among soil particles and, in turn, to the increase in its yield strength. As shown in Table 2, that the pore water pressure of the 0.0% sample is minimum, and the pore water pressure of the samples with different contents of EPS changes very little at the confining pressures of 50 and 100 kPa. However, at the confining pressures of 200 and 400 kPa, the pore water pressure increases with the increase of the EPS content. As shown in Table 3, at the EPS content, the deformation modulus increases with the increasing confining pressure. At the same confining pressure, the deformation modulus decreases with the increasing content of EPS, while for the 0.4% sample, it reduces to the minimum.

As shown in Figure 5, in the $p' - q$ ($p' = (\sigma_1' + \sigma_2' + \sigma_3')/3$, $q = \sigma_1 - \sigma_3$) stress plane, the effective stress path of the four kinds of samples can be drawn. The changes in effective cohesion (c') and effective internal friction angle (φ') for the different contents of EPS are shown in Figures 6 and 7. From these figures, we can see that along with the increasing EPS content,

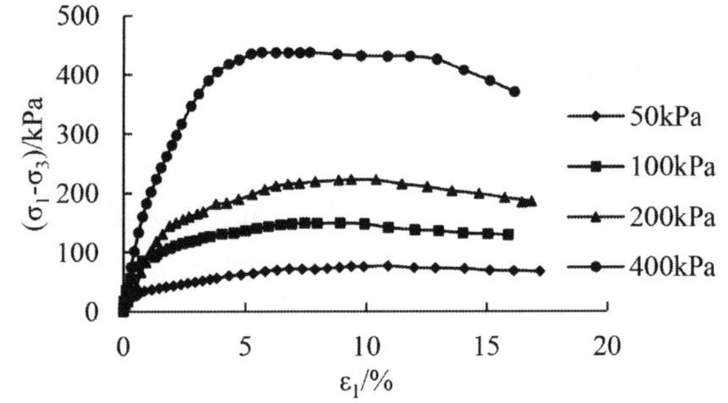

(a) Deviatoric stress and axial strain curves

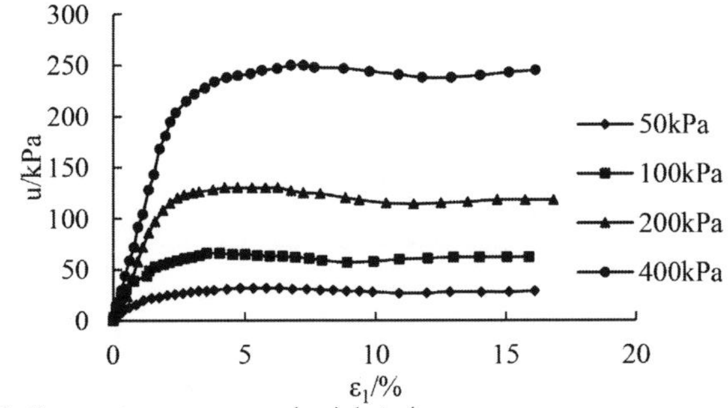

(b) Pore water pressure and axial strain curves

Figure 1. Deviatoric stress, pore pressure and strain characteristics of 0.0% EPS.

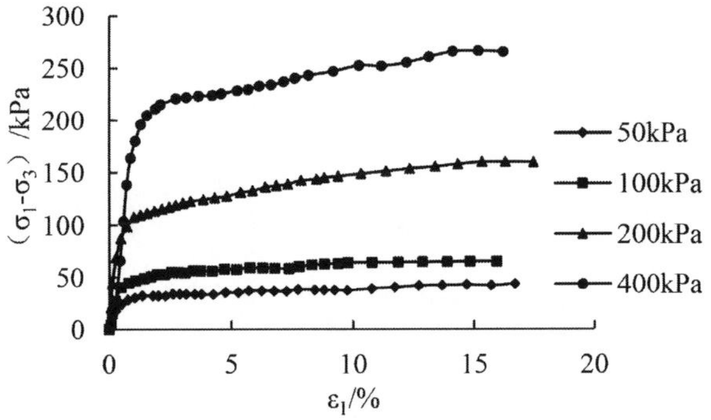

(a) Deviatoric stress and axial strain curves

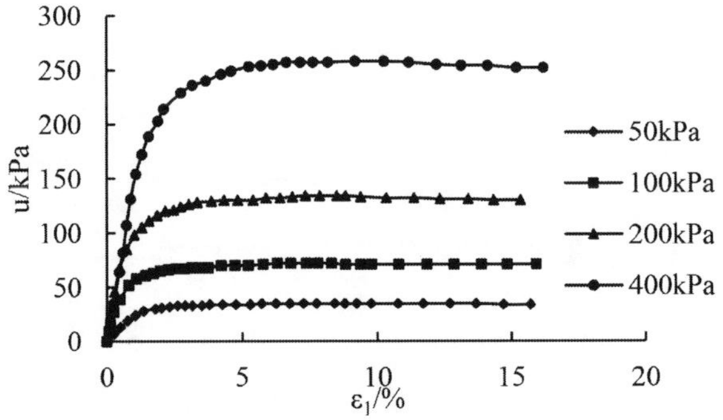

(b) Pore water pressure and axial strain curves

Figure 2. Deviatoric stress, pore pressure and strain characteristics of 0.2% EPS.

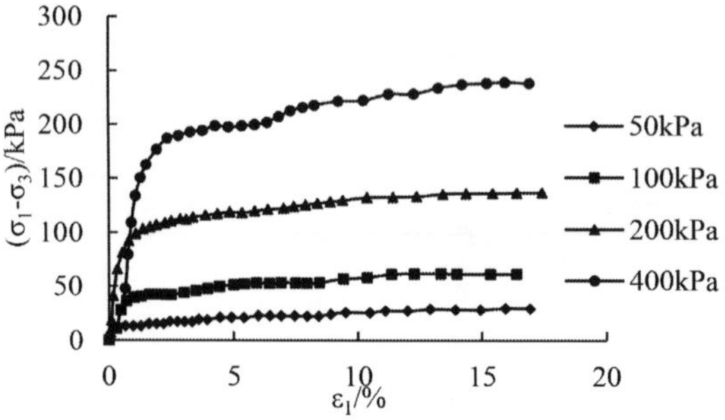

(a) Deviatoric stress and axial strain curves

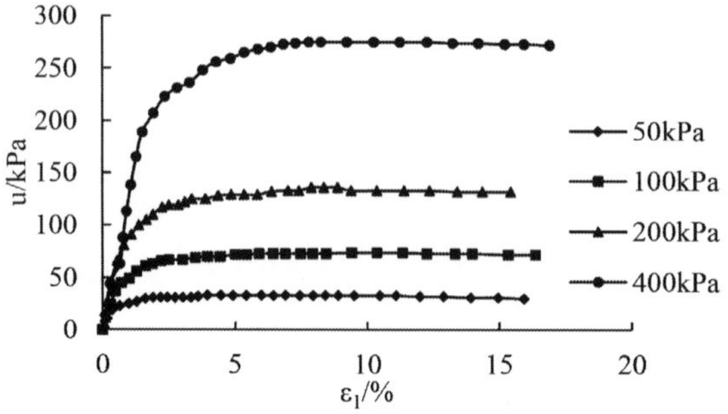

(b) Pore water pressure and axial strain curves

Figure 3. Deviatoric stress, pore pressure and strain characteristics of 0.4% EPS.

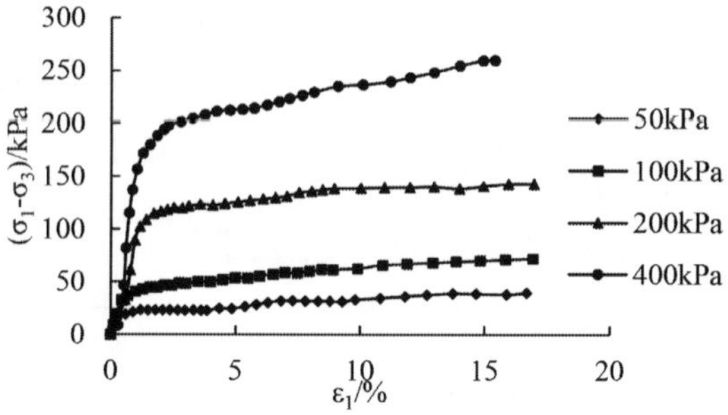

(a) Deviatoric stress and axial strain curves

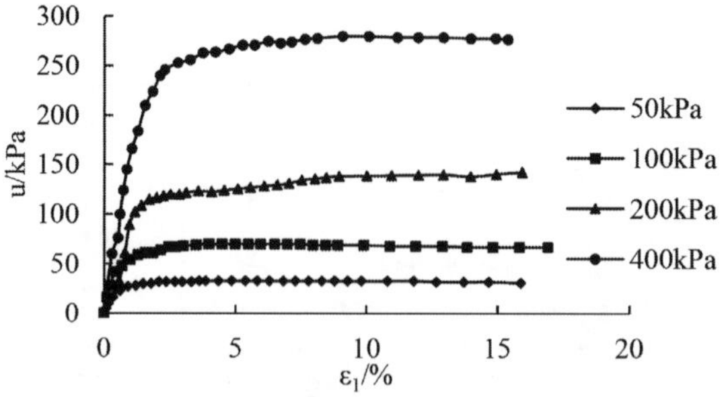

(b) Pore water pressure and axial strain curves

Figure 4. Deviatoric stress, pore pressure and strain characteristics of 0.6% EPS.

Table 2. Maximum deviatoric stress and pore water pressure.

Confining pressure/ kPa	0.0% EPS/kPa		0.2% EPS/kPa		0.4% EPS/kPa		0.6% EPS/kPa	
	$(\sigma_1-\sigma_3)_{max}$	u_{max}	$(\sigma_1-\sigma_3)_{max}$	u_{max}	$(\sigma_1-\sigma_3)_{max}$	u_{max}	$(\sigma_1-\sigma_3)_{max}$	u_{max}
50	76.58	32	43.59	35	30.88	33	39.4	33
100	149.8	66	65.47	72	61.89	74	72.17	70
200	222.51	130	160.35	134	137.02	136	142.8	144
400	437.59	250	266.82	258	239.07	275	260.28	280

Table 3. Secant modulus E (axial strain of 2%).

Confining pressure/ kPa	0.0% EPS		0.2% EPS		0.4% EPS		0.6% EPS	
	$(\sigma_1-\sigma_3)$/ kPa	E/MPa	$(\sigma_1-\sigma_3)$/ kPa	E/MPa	$(\sigma_1-\sigma_3)$/ kPa	E/MPa	$(\sigma_1-\sigma_3)$/ kPa	E/MPa
50	43.22	2.161	32.03	1.602	15.90	0.795	23.15	1.116
100	111.36	5.568	52.87	2.644	45.18	2.259	48.13	2.406
200	139.66	6.9828	115.17	5.759	94.29	4.715	116.67	5.82
400	281.05	14.053	215.34	10.767	182.99	9.15	203.41	10.171

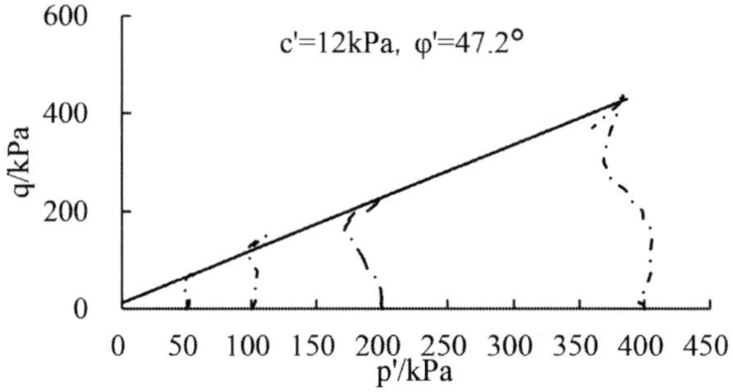

(a) Effective stress path of 0.0%EPS

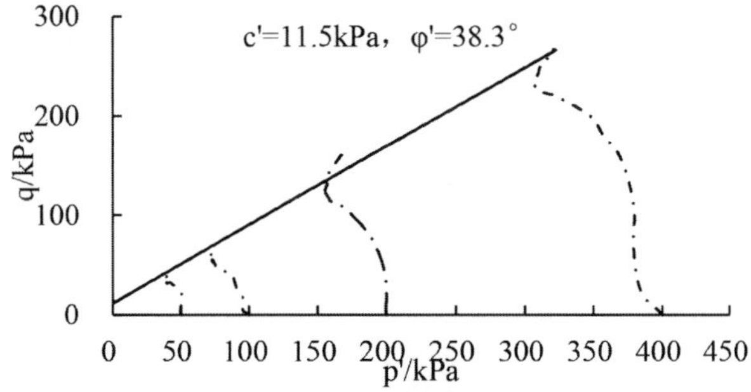

(b) Effective stress path of 0.2%EPS

Figure 5. (*Continued*)

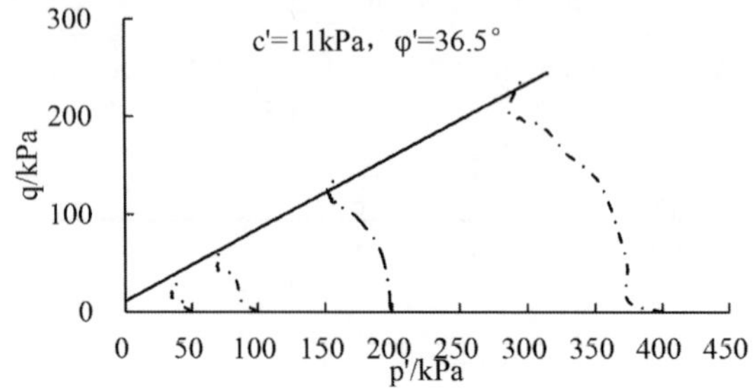

(c) Effective stress path of 0.4%EPS

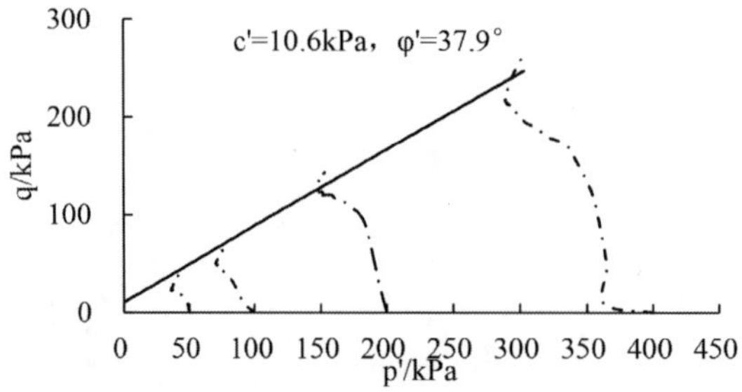

(d) Effective stress path of 0.6%EPS

Figure 5. Effective stress paths.

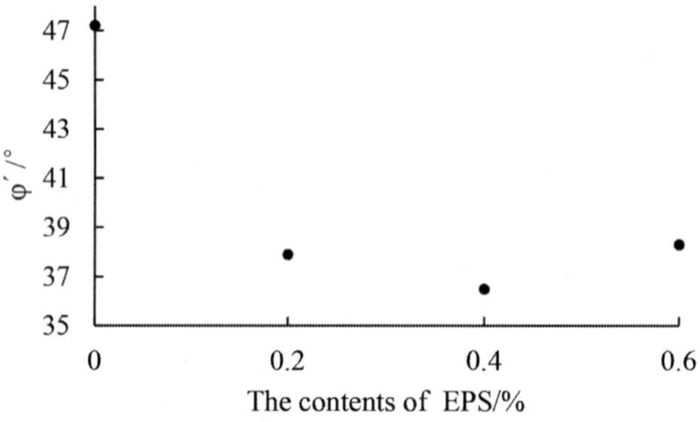

Figure 6. Relationship curves between effective cohesion and EPS content.

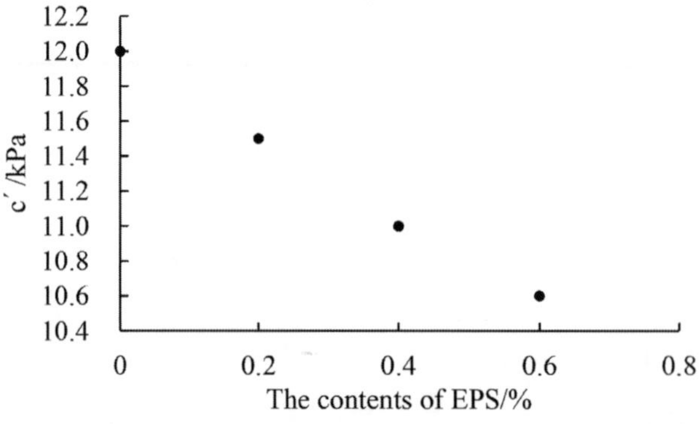

Figure 7. Relationship curves between effective internal friction angle and EPS content.

the effective cohesion (c′) decreases linearly and the effective internal friction angle (φ′) first decreases and then increases slowly.

4 CONCLUSIONS

In this paper, we analyzed the influences of different EPS contents on the mechanical properties and strength properties of Xigeda soil samples under a consolidated undrained triaxial compression state. Based on the analysis, the following conclusions are drawn:

1. In the process of shear, the samples without EPS particles behave as strain softening while the samples with EPS particles behave as strain hardening. The maximum deviatoric stress first decreases and then increases with increasing EPS content.
2. The secant modulus firstly decreases and then increases with increasing EPS content at the 2% axial strain.
3. At the confining pressures of 50 and100 kPa, the content of EPS has little impact on the pore water pressure. In contrast, at confining pressures of 200 and 400 kPa, the maximum pore water pressure increases with the increase of EPS content.
4. Along with increasing EPS content, the effective cohesion (c′) linearly decreases and the effective internal friction angle (φ′) first decreases and then increases slowly.
5. When the EPS content is 0.4%, the maximum principal stress difference, the secant modulus and the effective internal friction angle are lowest. So, there is a sudden change in the mechanical properties of Xigeda soil when the content of EPS is 0.4%.

REFERENCES

The ministry of water resources of the People's Republic of China. GB/T50123–1999 Standard for soil test method. Beijing: Chinese Plans Press, 1999.

Chen, W., Jiao, T., & Li, X.W. 2011. Study on the Construction Technique of Xigeda Soil House. *In Advanced Materials Research* 168: 589–593.

Dong, J., Wang, P., Chai, S., & Zzhu, H. 2013. Compression Deformation Characteristics of Polymer SH Solidified Lightweight Soil. *Journal of Basic Science and Engineering*, 2: 009.

Du, C. & Yang, J. 2001. Expanded polystyrene (EPS) geofoam: an analysis to characteristics and applications. *Journal of Southeast University*, 31(3): 138–142.

Horpibulsuk, S., Suddeepong, A., Chinkulkijniwat, A., & Liu, M.D. 2012. Strength and compressibility of lightweight cemented clays. *Applied Clay Science*, 69: 11–21.

Hou, T.S. & Xu, G.L. 2011. Influence law of EPS size on shear strength of light weight soil. Chinese *Journal of Geotechnical Engineering*, 33(10): 1634–1640.

Kai-tai, L., Yan-lin, Q., Lei, H., En-long, L., & Wei, C. 2011. Mechanical properties of Xigeda soil with different contents of lime. *In Electric Technology and Civil Engineering (ICETCE), 2011 International Conference on:* 2362–2366. IEEE.

Kikuchi, Y., Nagatome, T., Mizutani, T.A., & Yoshino, H. 2011. The effect of air foam inclusion on the permeability and absorption properties of light weight soil. *Soils and foundations*, 51(1): 151–165.

Wang, W., & Fu, X. 2011. Experimental researches on engineering properties of the Xigeda in Southwest China. *In 2011 International Symposium on Water Resource and Environmental Protection.*

Yajima, J., Maruo, S., & Ogawa, S. 1994. Deformation and strength properties of foam composite lightweight soil. *In proceedings japan society of civil engineers: dotoku gakkai,* 197–197.

Yoon, G.L., Kim, B.T., & Jeon, S.S. 2011. Stress-strain behavior of light-weighted soils. *Marine Georesources & Geotechnology*, 29(3): 248–266.

Zhou, H., Cao, P. & Zhang, K. 2014. In-situ direct shear test on Xigeda formation clay stone and siltstone. *Journal of Central South University*, 45(10): 3544–3550.

Advanced Engineering and Technology III – Xie (Ed.)
© 2017 Taylor & Francis Group, London, ISBN 978-1-138-03275-0

Study on mechanical and chemical properties of tailing soils extracted from one dam site in a cold plateau region

Yong Tang, Youneng Liu, Jianqiu Tian & Miao He
State Key Laboratory of Hydraulics and Mountain River Engineering, College of Water Resource and Hydropower, Sichuan University, Chengdu, P.R. China

Enlong Liu
State Key Laboratory of Hydraulics and Mountain River Engineering, College of Water Resource and Hydropower, Sichuan University, Chengdu, P.R. China
State Key Laboratory of Frozen Soil Engineering, Cold and Arid Regions Environmental and Engineering Research Institute, Chinese Academy of Sciences, Lanzhou, China

ABSTRACT: Physical experiments, shear tests, strength tests, salt tests, and chemical experiments were done on tailing soils in this study. The basic physical–mechanical parameters of tailing soils, shear strength under different kinds of water content, and the dynamic strength of the tailing soils under different kinds of deviator stress consolidation were obtained from these tests. Furthermore, the salt content of tailing soils at different temperatures was obtained from the tests. The results indicate that the shear strength of tailing soils decreases with the increase in the water content. The dynamic stress ratios decrease with the increase in cell pressure during the same cycle times. However, the salt content has no significant change at different temperatures and water contents. Through the analysis of the microstructure of tailing soils, we obtain the effects of the microstructure on the physical–mechanical properties of tailing soils.

1 INTRODUCTION

With the development of science and technology, many of the mineral resources have been used. Lots of tailing soils were accumulated by poor efficiency of the mineral resources. As one of the most important potential resources, it is important to do some research on tailing soils (Lei et al, 2008; Meng et al, 2010). The study of tailing soils, as one of the commonly dangerous sources, began in the 1960s: its construction not only is a threat to the health of poor people, but also causes a lot of environmental and ecological problems (Yu et al, 2014).

As one of the special soils, many researchers have performed research on tailing soils. Yin et al. (2007) performed some research on the physical–mechanical properties of tailing soils in the Yang-la Copper Mine. Qiao et al. (2015) researched on the influence of fine particle contents on construction engineering. Zhou et al. (2009) compared the physical–mechanical properties of some tailing soils. Zhang et al. (2003) conducted some conventional triaxial extension experiments on tailing dam sand of a copper mine. By conducting experiments on the physical–mechanical properties of tailing soils, Wang et al. (2005) found that the relationship between dry density of tailing soils and the depth differed greatly from those of common soils. Xiong et al. (2003) researched on the shear strength and the physical state variables of a remold unsaturated clay. Zhang et al. (2014) investigated the physical–mechanical properties in the capillary water zone of a tailing dam. By conducting an experiment on the static–dynamic of a salty tailing dam sand, Liu et al. (2012) found that under static loading conditions, a parabolic curve can well reflect the shear strength characteristics under different kinds of stress. Zhang et al. (2006) conducted the dynamic experiments on one kind of tailing soils, and provided a relatively simple model to describe the pore pressure.

Tan et al. (2014) performed the experiments on one saturated silt, and concluded the relationship between dynamic strength and dynamic pore pressure of the tailing dam silt under different kinds of consolidation ratio. Xu et al. (2014) performed the dynamic triaxial experiments and resonant column tests of one tailing soil under different kinds of highway track loads. By conducting the consolidated drained triaxial experiments and compression tests on one kind of tailing sand, Yang et al. (2014) obtained the change in the relative density and the relationship between the change in compressibility parameters and the vertical pressure and porosity ratio. Yin et al. (2011) researched the seepage on the influence of the microstructure on one kind of tailing soil. Lucas et al. (2016) conducted experiments on the chemical, mineralogical, environmental and physical properties of iron ore tailings. Zhang et al. (2015) analyzed the mechanical features of layered structures in tailings dam from macroscopic and microscopic points of view. According to the physical, chemical, and mineralogical properties of tailings, Kossoff et al. (2014) analyzed the influence of environment, time, and finance on the failure of a tailing dam.

Through the physical–mechanical experiments, we obtained the basic physical–mechanical parameters. Next, we obtained the shear strength at different water contents through consolidated quick shear tests. Dynamic strength was obtained through dynamic triaxial tests. Furthermore, we obtained the characteristics of salt content from the experiments conducted at different water contents and temperatures. Finally, we observed the microstructure, chemical element and mineral content through SEM and the corresponding explanations are provided.

2 PHYSICAL–MECHANICAL PROPERTIES OF THE TAILING SOIL

2.1 *Physical properties*

As one of the fractured waste after the mineral separation, there are lots of soluble salt in tailing soils. So in the specific gravity test, we use a neutral liquid (for example, diesel) to replace the pure water. Besides, we use the vacuum suction method to discharge the air in this soil instead of the soiling method.

From Table 1, we can see that the specific gravity of the tailing soil is larger than the common soils. The main reason for a larger specific gravity is that in tailing soils, there are plenty of minerals, so the specific gravity is mainly controlled by these minerals. In Table 2, we can see that all the soils in the three layers belong to silt. The soil layer with a depth of 10.0–20.0 m has a very distinct characteristic from the other layers. Its dry density is far larger than those of the other layers, while its moisture content and porosity are smaller than those

Table 1. Basic physical parameters of different tailing soil layers.

Depth (m)	Moisture content (%)	Dry density (g/cm^3)	Specific gravity (Gs)	Pore ratio (e)	Saturation state
5.0–10.0 m	34.84	1.492	3.00	1.01	Saturated
10.0–20.0 m	23.90	1.765	3.01	0.71	Saturated
20.0–30.0 m	34.64	1.495	2.97	0.99	Saturated

Table 2. Liquid limit and plasticity index of different tailing soil layers.

Depth (m)	Liquid limits (%)	Plastic limits (%)	Natural moisture content (%)	I_P (%)	I_L
5.0–10.0 m	71	32	35	39	0.07
10.0–20.0 m	79	28	24	51	−0.08
20.0–30.0 m	79	38	35	42	−0.07

Table 3. Cohesion c and internal friction angle φ at different water contents in Group 1.

w (%)	c (kPa)	φ (°)
20.00	27.70	9.00
25.00	22.73	8.20
30.00	19.32	8.14
35.00	13.75	6.54
40.00	7.17	5.96

Table 4. Cohesion c and internal friction angle φ at different water contents in Group 2.

w (%)	c (kPa)	φ (°)
15.00	36.39	8.60
20.00	22.85	7.80
25.00	11.34	7.39
30.00	7.66	6.25

Table 5. Cohesion c and internal friction angle φ at different water contents in Group 3.

w (%)	c (kPa)	φ (°)
10.00	54.48	3.55
15.00	16.22	7.85
20.00	16.24	7.74
25.00	5.41	6.65
30.00	4.79	6.13

Table 6. Cohesion c and internal friction angle φ at different water contents in Group 4.

w (%)	c (kPa)	φ (°)
20.00	19.17	7.57
25.00	17.93	6.76
30.00	9.09	6.99
35.00	8.88	6.76
40.00	1.77	7.34

of the other layers. Because the tailing soil belongs to a porous medium, the coefficient of permeability in horizontal and vertical directions is very different during the formation of sedimentation (2007). According to the physical properties of the soil layer with a depth of 10.0–20.0 m, we can assume that the coefficient of permeability in the vertical direction will be smaller than those of the other two layers.

2.2 *Shear strength properties of tailing soil*

As one of the important parameters to analyze the stability of a tailing dam, we conducted the consolidated quick shear tests of this soil at different depths in order to determine these parameters. We performed the experiments by controlling the dry density and prepared the specimen with different kinds of specified water content. Then, we compacted the soil inside a cutting ring by two times, and finally pushed the specimen into the shear box and fixed it. We then applied different kinds of vertical stress. The stresses applied are 100 kPa, 200 kPa, 300 kPa, and 400 kPa for the four groups of soils at different depths, respectively. Four different water contents were prepared for the samples with the shear rate of 0.8 millimeter per minute. The cohesion c and inter friction angle ϕ are presented in Tables 3 to 6, respectively.

From Table 3 to 6, we can see that the cohesion decreases when the water content increases in all the four groups. According to the analysis of the influence of shear strength parameters on halomorphic soil with chlorine at different water contents. Chen et al. (2006) found that when the soil contains salt, the change in these salts will cause the change in the structure of the soil, and then influence the concentration of the liquid around the soil particles and the thickness of the water membrane. Thus, the strength of the tailing soils will be influenced by both the water content and the salt content. When the salt content is constant, with the increase in water content, the concentration of all ions within a soluble salt will decrease, and then the thickness of the water membrane becomes thicker, which makes the bond effect between particles smaller. This means that the cohesion becomes smaller.

3 ANALYSIS ON DYNAMIC STRENGTH OF THE TAILING SOIL

Static-dynamic GCTS apparatus was used to perform the dynamic triaxial experiments, which were controlled by stress with 1.0 Hz constant amplitude sinusoid. The size of the specimen is 10.00 centimeters high and its diameter is 5.00 centimeters. The specimens were

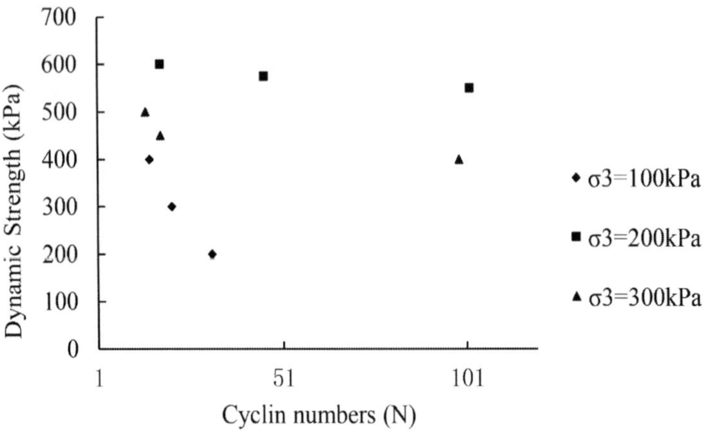

Figure 1. Dynamic strength under cell pressures.

Table 7. Dynamic strength of the tailing soil.

Cycle numbers (N) $K_c = 1.25$		Dynamic shear stress ratio τ_d/σ_{3c}		
		8	12	20
σ_{3c} (kPa)	100	2.338	2110	1.653
	200	1.505	1550	1.488
	300	0.805	0.799	0.786

controlled by dry density at natural moisture content, with the density of the three layers being 1.492 g/cm^3, 1.765 g/cm^3, and 1.495 g/cm^3, respectively. The specimens were compacted in a mold into three same parts, and then we used a vacuum pump to evacuate vacuum. Finally, we put the specimen into a sink for maintenance. Cell pressures were applied under the consolidated undrained condition, and σ_3 = 100 kPa, 200 kPa and 300 kPa were used in this experiment. The coefficient of anisotropic consolidation was Kc = 1.25. The samples were saturated by back pressure, with the coefficient of pore pressure B($B = \Delta u/\Delta\sigma_3$) being larger than 0.97.

The relationship between the strength and cycle numbers for the soils of different layers is shown in Figure 1, and the strengths of this tailing soil are presented in Table 7. As we can see from Table 7, the dynamic strength ratio has the tendency to reduce with the increasing cycle numbers and cell pressure. From these data we can see that the failure pattern of the second layer is mainly tension failure, while that of the first and third layers are dominated by compression. The reason for this phenomenon is that the dry density of the second layer is much larger than the other two layers. So in this experiment the cell pressure is not the main factor while the dry density is the factor to influence the failure form of the specimen. Thus, here we choose two different failure criteria to control the experiment. As for the first and third layers, when the axial dynamic strains reach up to 5%, the specimen fails. But for the second layer, when the axial dynamic strain increases steeply, the specimen fails.

4 ANALYSIS ON THE SALT CONTENT OF THE TAILING SOIL

There are various kinds of minerals in the tailing soil, which contains soluble salts and insoluble salts. As one of the important aspects in tailing soils, its presence has a strong influence on its strength. So, it is necessary to conduct an investigation on it.

First, the experiments on the salt content of this soil at different water contents and normal atmospheric temperatures were performed. The results are shown in Figure 2. As for the fourth group samples, the experiments were conducted at different temperatures and water contents. The results are shown in Figure 3.

From Figure 2, we can see that the salt content first decreases and then increases with the increasing water content for all the three groups. The content of the salt becomes minimum at the natural moisture content for all the three groups. But the content of the salt for these two groups at depths of 10.0 m to 20.0 m is larger than that of the other group. The reason for this is that, under the influence of the discharging method, deposit patterns, the graduation of the tailing soil particles, and other factors, the porosity of the second layer soil is smaller than that of the other groups, which decreases the vertical permeability of the second layer.

The change in the salt content at different temperatures is shown in Figure 3. As we can see, the change in the salt content is not obvious at different temperatures, ranging from 1.5% to 3.4%, and the content is very small. The reason for this phenomenon is that the natural structure of the tailing soil is different, which makes the adherence of salt onto the soil particle different, so the salt content in the leaching solution is different at a certain temperature.

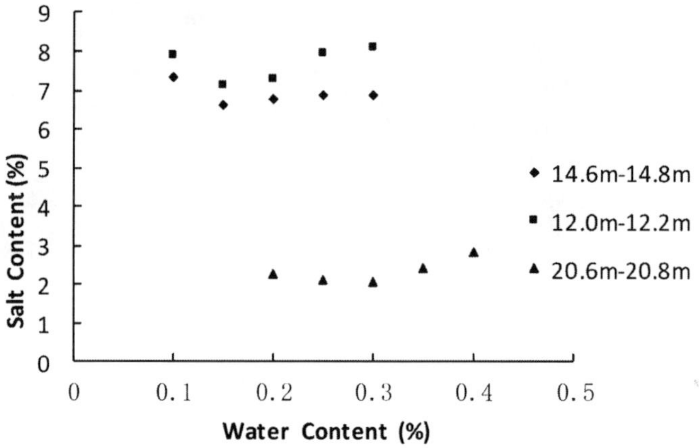

Figure 2. Water content vs salt content for all the three groups.

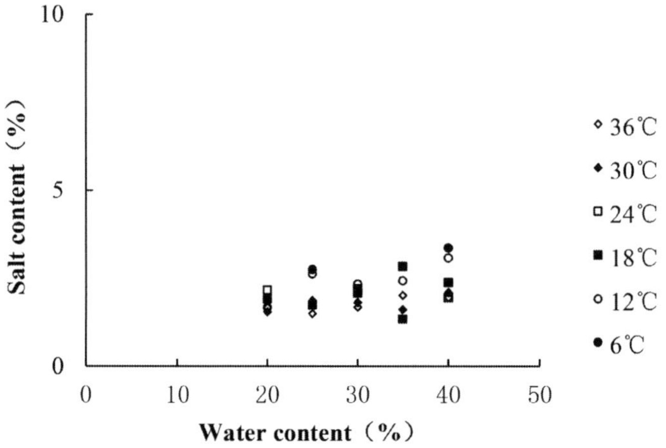

Figure 3. Salt content of layer four at different temperatures and water contents.

5 CHEMICAL ANALYSIS OF TAILING SOILS

Since the tailings material is filtered after the mineral waste, its mineral and chemical elements are very complex. Besides, the mineral content will affect the mechanical properties of the tailing soils. So, it is necessary to conduct a research on the microstructure and chemical component of tailing soils.

We researched the microstructure and chemical elements of the tailing soils by the scanning electron microscope and energy dispersive spectrometer. The microstructures detected by the scanning electron microscope are shown in Figure 4, while the chemical elements analyzed by the energy dispersive spectrometer are presented in Tables 8–10.

From the SEM images, we can see that, the first and the third layer have a very similar microstructure under the same magnification. Both of them have a lot of long strips material with staggered arrangement. Its structure is somewhat loose. But there are long strip materials in the second layer. The microstructure was damaged through the compression, so this could be the reason why the dry density of layer two is larger than those of the other layers.

From Tables 8–10, we see that the main chemical elements for the three layers are almost the same, and the main difference is the contents of Ti and Mn, but the change in the content is very small. Their main elements are O, Si, S, Ca, Fe, and C.

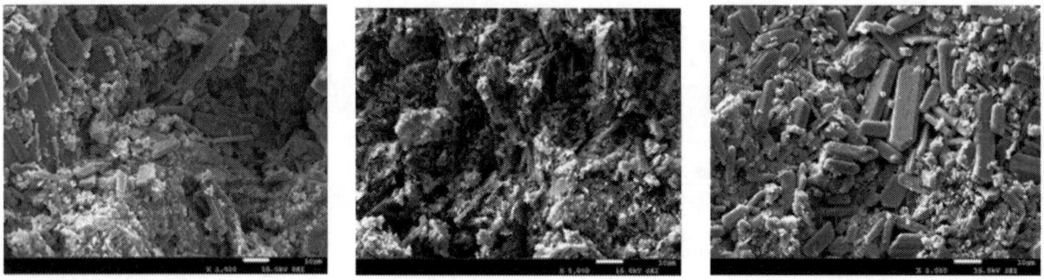

Figure 4. SEM of the three layers (0.0 m–10.0 m, 10.0 m–20.0 m, 20.0 m–30.0 m from left to right).

Table 8. Chemical elements of the first layer (0.0 m–10.0 m).

Element	C	O	Mg	Al	Si	S	K	Ca	Mn	Fe	Zn
Weight percent (%)	5.28	56.28	0.37	1.37	6.03	10.51	0.61	10.46	0.75	5.61	2.74

Table 9. Chemical elements of the second layer (10.0 m–20.0 m).

Element	C	O	Mg	Al	Si	S	K	Ca	Ti	Mn	Fe	Zn
Weight percent (%)	3.71	56.33	0.39	2.54	9.18	8.02	0.92	6.67	0.32	0.37	7.83	3.71

Table 10. Chemical elements of the third layer (20.0 m–30.0 m).

Element	C	O	Mg	Al	Si	S	K	Ca	Ti	Fe	Zn
Weight percent (%)	3.46	57.86	0.4	2.19	9.11	9.16	0.8	8.73	0.24	4.22	3.83

Although the chemical elements in the three layers were almost the same, their minerals elements may have some differences. Thus, we conducted X-ray diffraction experiments on the tailing soils. From the X-ray diffraction experiments, we can see that the main miner of the three layers is the quartz, but the component and the content of other minerals in the three layers are different. The main minerals in layer one are brushite and quartz, the main minerals in layer two are brushite, quartz and glaucocerinite, and the main minerals in layer three are gypsum and quartz. Different layers have different minerals, so the mechanical properties in the three layers are different. The contents of minerals in layer two are more than that of minerals in the other two layers, so the physical–mechanical properties of layer two are different from that of the other two layers.

6 CONCLUSIONS

Through the physical–chemical experiments and chemical experiments of the tailing soils, we summarize the following results:

1. The specific gravity of tailing soils is larger than that of common soils. Besides, the tailing soil has a strong sorting effect, so the coefficient of permeability in the vertical direction is very different.
2. According to the results from the consolidated quick shear test, we can see that the cohesion and internal friction angle have a tendency to decrease with the increasing water content.
3. Soils of different layers have different failure patterns, so we can see that the influence of dry density is more important than the cell pressure on the dynamic strength for the tailing soils. The dynamic strength ratio has the tendency to reduce with the increasing recycle numbers and cell pressure.
4. The salt content in the tailing soils of the three layers first decreases and then increases with the increasing water content. Furthermore, the value of the salt content becomes minimum at the natural moisture content. As for the samples at different temperatures and water contents, there is no obvious change tendency in the salt content.
5. The microstructures in all the three layers are different, but the microstructures of the first and the third layer are similar to each other, and both of them have a long strips material. Although a slightly compacted situation is shown in the second layer, the main chemical elements in all the three layers are almost the same and the content has no major change.
6. All the layers have quartz, but the contents and types of the main minerals are different, and there are more types of minerals in the second layer.

REFERENCES

Chen Weitao, Wang Mingnian, Wang Ying, Li Shu, Wang Yusuo. Influence of Salt Content and Water Content on the Shearing Strength Parameters of Chlorine Saline Soil. *China Railway Science*. 2006. 27(4):1–5.

Guangzhi Yin, Qiangui Zhang, Wensong Wang, Yulong Chen, Weile Geng, Hairu Liu. Experimental study on the mechanism effect of seepage on microstructure of tailings. *Safety Science*. 2011. 50(4):792–796.

Kossoff D., Dubbin W.E., Alfredsson M., Edwards S.J., Macklin M.G., Hudson-Edwards K.A. Mine tailings dams: Characteristics, failure, environmental impacts, and remediation. *Applied Geochemistry*. 2014. 51:229–245.

Lei Li, Zhou Xinglong, Li Jiayu, Cai Xiaowei, Wang Guowen. Status quo and Pondering on Comprehensive Utilization of Mine Tailings Resources in China. *Express Information of Mining Industry*. 2008. 9(9):5–8.

Liu Haiming, Yang Chunhe, Zhang Chao, Mao Haijun. Study on static and dynamic strength characteristics of tailings silty sand and its engineering application. *Safety Science*. 2012.50(4):828–834.

Lucas Augusto de Castro Bastos, Gabriela Cordeiro Silva, Júlia Castro Mendes, Ricardo André Fiorotti Peixoto. Using Iron Ore Tailings from Tailing Dams as Road Material. *Journal of Materials in Civil Engineering*. 2016. 28(5):1–9.

Meng Yuehui, Ni Wen, Zhang Yuyan. Current state of ore tailings reusing and its future development in china. *China Mine Engineering*. 2010. 39(5):4–9.

Qiangui Zhang, Guangzhi Yin, Zuoan Wei, Xiangyu Fan, Wensong Wang, Wen Nie. An experimental study of themechanical features of layered structures in dam tailings from macroscopic and microscopic points of view. *Engineering Geology*. 2015. 195(10):142–154.

Qiao Lan, Qu Chunlai, Cui Ming. Effect of fines content on engineering characteristics of tailings. *Rock and Soil Mechanics*. 2015. 36(4):923–927.

Tan Fan, Rao Xibao, Huang Bin, Wang Zhanbin, Xu Yanyong. Experimental Study of the Dynamic Characteristics of Tailings silts. *China Earthquake Engineering Journal*. 2015.37(3):772–777.

Wang Chonggan, Zhang Jiasheng. Experimental Study on Physical and Mechanical Properties of Tailings. *Mining and Metallurgical Engineering*. 2005. 25(2):19–22.

Xiong Chengren, Liu Baochen, Zhang Jiasheng, Liu Duowen. Relation between Shear Strength Parameters and Physical State Variables of Remolded Unsaturated Cohesive Soil. *China Railway Science*. 2003. 24(3):18–20.

Xu Xutang, Jan Weibin. Analysis of the Stability and Dynamic Properties of Tailings Slopes under High-Speed Train Viberations. *Soil Eng. and Foundation*. 2014. 27(4):106–110.

Yang Kai, Lu Shuran, Zhang Yuanyuan. Experimental Study of Strength Characteristics of Tailing Sand in Tailings Dam. *Metal Mine*. 2014. 2:166–170.

Yin Guangzhi, Yang Zuoya, Wei Zuoan, Tan Qinwen. Physical and Mechanical Properties of YangLacoppe. *Journal of Chongqing university (Natural Science Edition)*. 2007. 30(9):117–122.

Yu Guangming, Song Chuanwang, Pan Yongzhan, Li Liang, Li Ran, Lu Shibao. Review of new progress in taling dam safety in foreign research and current state with development trend in china. *Chinese Journal of Rock Mechanics and Engineering*. 2014. 33(1):3239–3248.

Zhang Chao, Yang Chunhe, Bai Shiwei. Experimental study on dynamic characteristics of tailings material. *Rock and Soil Mechanics*. 2006. 27(1):35–40.

Zhang Chao, Yang Chunhe, Kong Lingwei. Study on mechanical characteristics of tailing dam of a copper mine and stability analysis of tailing dam. *Rock and Soil Mechanics*. 2003. 24(5):858–862.

Zhang Zhijun, Li Yajun, He Guicheng, Zhang Qiucai, Han Yanjie, Chang Jian, Liu Xuanzhao. Study of physico-mechanical properties of dam body materials in capillary water fringe of a certain tailings dam. *Rock and Soil Mechanics*. 2014. 35(6):1561–1568.

Zhou yunqing, Sun Jian, Li Yuhong, Cao Jing. Comparison of physical and mechanical properties of several tailings. *Geotechnical Engineering World*. 2009. 12(4):38–41.

Advanced Engineering and Technology III – Xie (Ed.)
 ISBN 978-1-138-03275-0

Conceptual design and ANSYS analysis on suspended floor damping system under seismic circumstance

Z. Li
Department of Geotechnical Engineering, Tongji University, Shanghai, China

X. Jian
Department of Bridge Engineering, Tongji University, Shanghai, China

L. Xie
Department of Hydraulic Engineering, Tongji University, Shanghai, China

Z. Qin
Department of Civil Engineering, Tongji University, Shanghai, China

ABSTRACT: Tuned Mass Damper (TMD) is a widely-accepted and universally-adopted approach for mega frame structure to reducing vibration effects under earthquakes. However, traditional TMD method requires an additional component with prodigious mass and thus results in excessive investment. This paper introduces the conceptual design of suspended floor damping system which serves as a new vibration mitigating method by setting top floors partly free and enabling them to dissipate seismic energy. Furthermore, the paper illustrates simulated seismic experiments by finite element analysis in ANSYS software. Compared with the blank group, the suspended floor design has its maximum shearing force, peak stress and maximal layer displacement decreased by 16.9%, 20.8% and 35.7%, which demonstrates that the designed suspended floor system achieves damping effect under specific fortification intensity and site condition.

1 INTRODUCTION

Mega frame structure is a relevantly optimal choice for high-rise buildings on the basis that the overall frame is efficacious in transferring and supporting load (Stlouis, N. et al. 1995). Due to the comparatively superior working efficiency, mega frame structure has attained universal application, such as HSBC building in Hong Kong, the federal reserve bank of Minnesota in America, World Trade Center in Taipei and Shanghai Center Tower in Shanghai.

For high-rise buildings, especially mega frame structure, Tuned Mass Damper (TMD) is a widely accepted vibration mitigating measure (Housner, G. et al. 1997, Soliman, I.M. et al. 2015, Tu, H. et al. 2016). This design utilizes an additional component with prodigious mass to adjust the natural vibratory frequency of the building so that the whole building can avert amplification of vibration (Johnson, J.G. et al. 2014, Dinh, V.N. et al. 2015, Greco, R. et al. 2015). For instance, if the natural frequency of the building is coincidently identical with that of the earthquake, the resonance influence will soon multiply vibratory response to many times. Furthermore, the movement of the component facilitates dissipation of seismic energy (Reggio, A. et al. 2015). However, the auxiliary component brings in excessive mass, which connotes heavier load and more investment.

To avert redundant mass in TMD approach, some original parts in buildings, such as water tanks on top floor, are modified to serve as dampers (Veļičko, J. 2015, Wang, J. et al. 2016).

Although some adjustments have proven to be effective, there is hardly any report on the design of suspended floor damping system which partially sets several story slabs on the top of high-rise buildings free so that the suspended floors are able to work as TMD.

This paper introduces the conceptual design of suspended floor damping system for mega frame structure. Moreover, it presents the simulated seismic experimental study by finite element analysis in ANSYS software. By comparing shock absorption performance of the design with the blank group under seismic wave-field, the results of software simulation demonstrate that the designed suspended floor system attains efficacious damping effect under specific fortification intensity and site condition.

2 CONCEPTUAL DESIGN

It is universally acknowledged that story slabs take an overwhelming proportion of a building's gross mass. By suspending several floors at the top of a high-rise building, these stories can function as Tuned Mass Dampers (TMD) owing to their sufficient mass and applicable freedom. To specify, the abundant mass of suspended floors enable them to play an indispensable role in the natural frequency of the whole building. Hence, the response of the building for earthquake is capable of being adjusted in order to escape amplification of vibration. What is more, the motion of suspended floors discharges seismic energy and facilitates its dissipation. Figure 1 presents the train of thought for the design. Whereas, with floors being liberated, their movements tend to be drastic and subversive, thus leading to uncontrolled destruction for the whole building. In this regard, dedicated design should be applied to control the motion of suspended floor while maintaining their original functions as usual.

Figure 2 depicts the overall profile of the design and detailed layouts are illustrated in Figures 3, 4. As Figure 3 portrays, the steel strand ropes hang from the rooftop and are fixed on two crossed horizontal steel tubes. Considering that the mass of three stories is accessible, the two steel tubes virtually can bear the load. Moreover, diagonal bracings are added to ensure the stability of the frame structure under dynamic load.

As Figure 4 presents, the floor has its four corners punched and is hanged by steel strand ropes. To avoid stress concentration on the four holes, steel gaskets are introduced to thicken the openings. Moreover, there are wheels, grooves and small viscous dampers on the four edges of suspended floor, which confine the swaying of suspended floor to governable range and let a portion of seismic energy dissipated by surrounding viscous dampers. In pragmatic application, concrete cover board should be set above the wheels and grooves so that people are able to access the central zone from four margins.

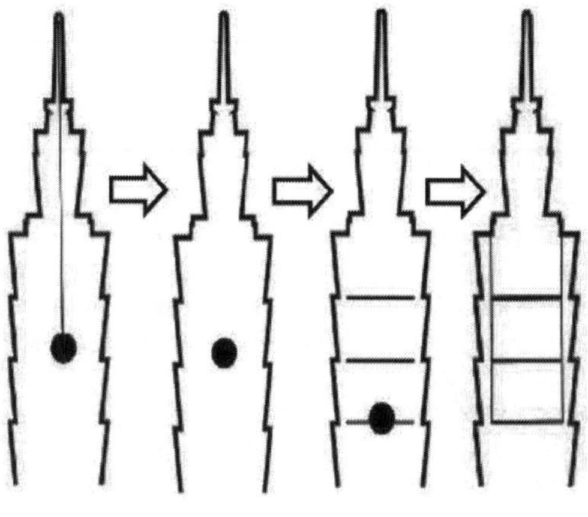

Figure 1. The trace of the design.

Figure 2. Three dimensional view of the design.

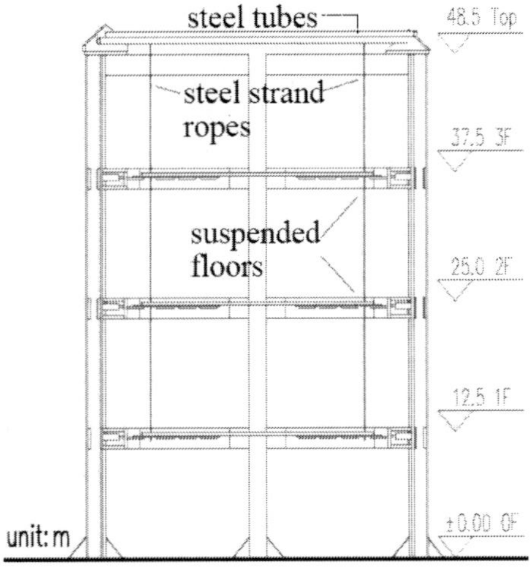

Figure 3. Front view of the design.

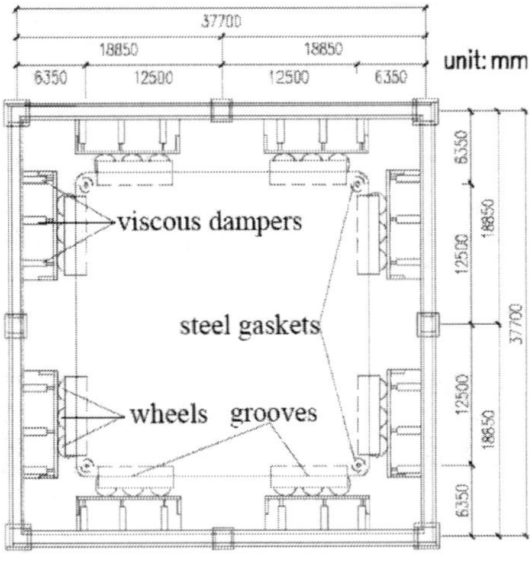

Figure 4. Top view of a suspended floor.

3 ANSYS SIMULATED EXPERIMENTS

3.1 *Modeling assumptions*

To begin with, several hypotheses are determined to simplify calculation.

1. The flexural strength of main pillars is strong enough to embrace the anti-seismic principle of reinforcing columns while impairing beams. Besides, the axial deformation of pillars is neglected because for pillars their modulus of compression is far greater than modulus of bending.
2. The steel strand ropes only transfer axial forces and their axial and tangential deformations are too slight to be taken into consideration.
3. The distribution of mass in suspended floors is ideally average.

3.2 *Modeling process*

The simulated suspended model consists of floors, girders, pillars, steel strand ropes, two crossed steel cubes and viscous dampers. Similarly, the simulated blank model includes identical constituents except for lacking ropes and cubes.

In virtual experiment, the two crossed steel cubes were simulated by Beam No. 189 in ANSYS due to its demand for strength and sectional area. The analogues of floors were shell No. 63 in the software while the steel wire ropes and viscous dampers were respectively substituted by Combin No. 14 in two and three dimensional models.

Next, for girders and pillars, they were defined by referring to properties of commonly-used concrete (elasticity modulus = 35 MPa, Poisson's ratio = 0.1667, density = 2500 kg·m^{-3}). However, the size of the suspended floors was relatively smaller than that of the normal floors owing to application of viscous dampers, wheels and grooves. In this regard, to ensure that the suspended floors and normal floors were of equal mass, for suspended floors the density was multiplied with a factor of 1.2.

Furthermore, the generation of mesh followed automatic distribution by the ANSYS software. Then, grids in the joints of steel wire ropes and two crossed steel cubes were specially enriched on account that the stress conditions were relatively complicated there.

Finally, on account that the seismic effect passed on from the bottom, constrains in horizontal directions were introduced to restrict the displacement of foundation.

3.3 *Spectrum analysis*

3.3.1 *Modal extraction*

The natural frequencies of vibration in diversified modals were calculated by Block Lanczos Method (Golub, G.H. et al. 1997). Table 1 exhibits Natural vibration period T in first six modals. The results of simulated test merely took first three values into account because the top three modals exerted dominant influence which achieved more than 90% of the overall impact.

3.3.2 *Modal analysis*

The analytic target was a typically major earthquake in China. According to relevant specification (DGJ08-9-2013), the seismic influence coefficient α_j of each modal j was figured out.

Table 1. The first twelve modals.

Modal	T_1 (s)	Modal	T_2 (s)	Modal	T_3 (s)
1	0.5482	2	0.5473	3	0.5464
4	0.4019	5	0.3217	6	0.2346

Table 2. Fundamental parameters for specific site condition.

Earthquake intensity I	Characteristic periods of response spectra T_g (s)		Site parameter b	Maximal seismic influence coefficient α_{max}	
9	0.35		0.9	1.4	
Natural vibration period of each modal Ti (s)			Seismic influence coefficient of each modal α_i		
T_1(s)	T_2(s)	T_3(s)	α_1	α_2	α_3
0.5482	0.5473	0.5464	0.9349	0.9362	0.9376

$$\alpha_j = \alpha_{\max} \times \left(\frac{T_g}{T_j} \right)^b \tag{1}$$

where α_{max} indicates maximal seismic influence coefficient, b is a site parameter, T_g and T_j are characteristic periods of response spectra and natural vibration period of modal j respectively. The mentioned parameters were calculated and gathered in Table 2.

For every modal j, the mode of vibration was documented in array $\varnothing_j$.

$$\varnothing_{ji} \in \{\varnothing_{j1}, \varnothing_{j2}, \ldots, \varnothing_{jn}\} = \varnothing_j \tag{2}$$

where n refers to total number of stories in software simulation ($n = 4$).

The participation coefficient for every modal j was computed as expression (3).

$$\gamma_j = \frac{\sum (G_i \varnothing_{ji})}{\sum (G_i \varnothing_{ji}^2)} \tag{3}$$

where G_i means gravity of story i.

The generated force by earthquake on every floor was calculated in formula (4).

$$F_{ij} = \alpha_j \gamma_j \varnothing_{ji} G_i \tag{4}$$

4 RESULTS AND DISCUSSION

After importing a period of the studied seismic spectrum, comparisons between maximum shearing force, peak stress and maximal layer displacement were conducted. The comparisons are respectively exhibited in Figures 5–7. Noting that the top story of suspended floor model contains too many components, the peak stress of the top floor (Fig. 6a) and the maximal layer displacement of the highest story (Fig. 7a) were hidden for aesthetic consideration but the computed results are presented and discussed.

As shown in Figure 5a, b, the maximum shearing force appears at the bottom of the models and the values are 2.07 kN and 2.49 kN respectively. Therefore, the suspended design dwindles summit shear force by 16.9%. Moreover, as Figure 6a, b demonstrate, the suspended floor model deduces peak stress from 48 kN·m^{-2} to 38k N·m^{-2} with droop rate equaling 20.8%. To specify, the peak stress occurs at the edge of floors, which connotes that the connective region of floors and main pillars share more load. Last but as significant, Figure 7a, b demonstrate that the maximal layer displacement attains a reduction of 35.7%, from 28 mm to 18 mm. Besides, the maximum shift appears at the top story due to the fact that the first modal takes a decisive percentage.

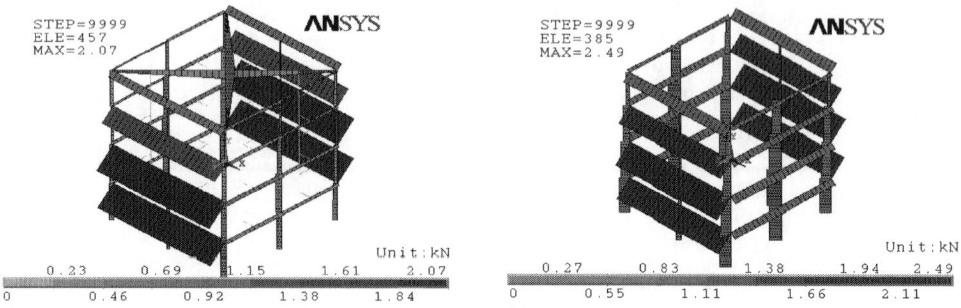

a) b)

Figure 5. Shearing force of a) suspended floor model, b) blank model.

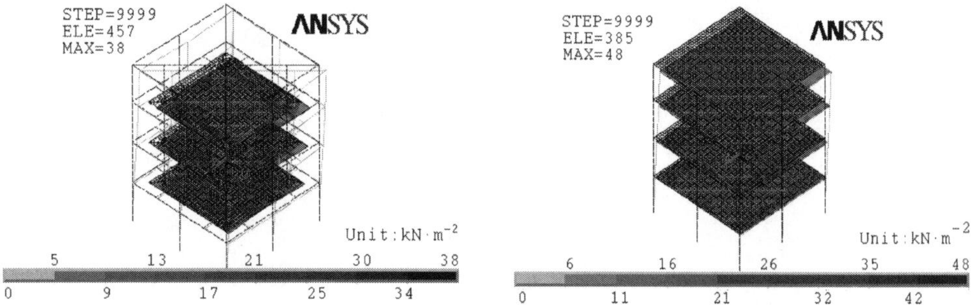

a) b)

Figure 6. Peak stress of a) suspended floor model, b) blank model.

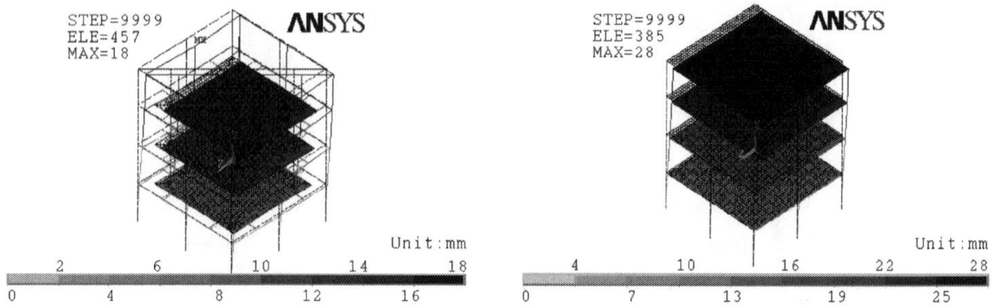

a) b)

Figure 7. Maximal layer displacement of a) suspended floor model, b) blank model.

5 CONCLUSION

This paper introduces the conceptual design of a new system, suspended floor damping system. Furthermore, the paper illustrates the simulated experiments by finite element analysis in ANSYS software, including model generating, spectrum analysis and result discussion. The conclusions come down as followings.

1. Traditional Tuned Mass Damper (TMD) has deficiency in redundant mass and excessive investment. The suspended floor design serves as an economic approach by liberating several top layers to certain degree and enabling them to function as dampers.
2. Compared with the blank group, the suspended floor design has its maximum shearing force, peak stress and maximal story displacement decreased by 16.9%, 20.8% and 35.7% under certain fortification intensity.

REFERENCES

DGJ08-9-2013. *Code for seismic design of buildings in Shanghai*. Tongji University.

Dinh, V.N. & Basu, B. 2015. Passive control of floating offshore wind turbine nacelle and spar vibrations by multiple tuned mass dampers. *Structural Control & Health Monitoring*. 22(1): 152–176.

Golub, G.H. & Underwood, R. 1977. The Block Lanczos Method for Computing Eigenvalues. Mathematical Software: 361–377.

Greco, R., Lucchini, A. & Marano, G.C. 2015. Robust design of tuned mass dampers installed on multi-degree-of-freedom structures subjected to seismic action. *Engineering Optimization*. 47(8): 1009–1030.

Housner, G., Bergman, L., Caughey, T., Chassiakos, A., Claus, R., Masri, S., Skelton, R., Soong, T., Spencer, B. & Yao, J. 1997. Structural Control: Past, Present, and Future. *Journal of Engineering Mechanics*. 123(9): 897–971.

Johnson, J.G., Reaveley, L.D. & Chris, P. 2014. A rooftop tuned mass damper frame. *Earthquake Engineering & Structural Dynamics*. 32: 965–984.

Reggio, A. & Angelis, M.D. 2015. Optimal energy-based seismic design of non-conventional Tuned Mass Damper (TMD) implemented via inter-story isolation. *Earthquake Engineering & Structural Dynamics*. 44(10): 1623–1642.

Soliman, I.M., Tait, M.J. & Damatty, A.A.E. 2015. Development and Validation of Finite Element Structure-Tuned Liquid Damper System Models. *Journal of Dynamic Systems Measurement & Control*: 137(11):1–13.

Stlouis, N., Dalton, M.J., Marchenko, S.V., Moffat, A.F.J. & Willis, A.J. 1995. The IUE mega campaign—wind structure and variability of HD-50896 (WN5). *Astrophysical Journal*. 452(1): L57.

Tu, H., Ducharme, K.T., Kim, Y. & Okumus, P. 2016. Structural impact mitigation of bridge piers using tuned mass damper. *Engineering Structures*. 112: 287–294.

Veļičko, J. & Gaile, L. 2015. Overview of tuned liquid dampers and possible ways of oscillation damping properties improvement. *Earthquake Engineering & Structural Dynamics*. 1(7): 967–976.

Wang, J., Wierschem, N., Spencer, B.F. & Lu, X. 2016. Numerical and experimental study of the performance of a single-sided vibro-impact track nonlinear energy sink. *Earthquake Engineering & Structural Dynamics*. 45(4): 635–652.

Advanced Engineering and Technology III – Xie (Ed.)
 ISBN 978-1-138-03275-0

Field test on the influence of the cemented soil around the pile on the lateral bearing capacity of pile foundation

Jia-jin Zhou, Xiao-nan Gong & Kui-hua Wang
Research Center of Coastal and Urban Geotechnical Engineering, Zhejiang University, Hangzhou, China

Ri-hong Zhang & Tian-long Yan
ZCONE High-tech Pile Industry Holdings Co. Ltd., Ningbo, China

ABSTRACT: To investigate the influence of the cemented soil around the pile on the lateral bearing capacity of pile foundation, a group of fields of the bored piles and the pre-bored grouting planted piles were conducted. The test results showed that: when the applied lateral load is relatively small, the lateral bearing capacity of the bored pile is better than the lateral bearing capacity of the pre-bored grouting planted pile, and the cemented soil around the pile does not have much influence on the lateral bearing capacity; the function of the cemented soil is gradually mobilized with the applied lateral load increasing, and the cemented soil can help increase the lateral critical bearing capacity and restrict the lateral displacement as well, moreover, the cemented soil can also prevent the pile foundation from abrupt failure; the properties (strength and elastic modulus) of the cemented soil is between the concrete pile and the surrounding soil, and it can virtually improve the lateral bearing capacity of pile foundation.

1 INTRODUCTION

Nowadays, with the development of offshore engineering and urban infrastructure construction, the pile foundation is widely used in port engineering, marine platform, high-speed railway and large bridge projects. The pile foundation in the above projects, which is different from the pile foundation in normal buildings, has to resist the lateral load. Therefore, it is of great significance to investigate the lateral bearing capacity of the pile foundation. The mainly four methods to analyze the behavior of the pile foundation under lateral loads are (Gong, *et al.*, 2015): (1) the ultimate subgrade reaction method, this method assumes that the soil around the pile is under the ultimate equilibrium state, and the lateral bearing capacity of the pile foundation can be calculated based on the applied load and the ultimate equilibrium condition of the surrounding soil. This method does not allow for the deflection of the pile shaft, and it is suitable for rigid pile with small embedment. (2) the elastic subgrade reaction method, this method assumes that the soil mass is an elastic material and the lateral bearing capacity of the pile foundation can be calculated by the beam bending theory; moreover, the elastic subgrade reaction method can be divided into *k* method, *m* method, *c* method, etc, based on the subgrade reaction coefficient. The *m* method, which assumes that the subgrade coefficient increases linearly with the soil depth, can represent the behavior of the pile under lateral load when the pile head lateral displacement is small. This method is widely used in actual project analysis because of its relatively simple calculation process, and the technical code in China (MOHURD, 2008) also recommends this method for the calculation of the pile under lateral loads. (3) the p-y curve method, this method is considered to be the most effective way for the analysis of nonlinear lateral behavior of the pile foundation as the non-elastic properties of the soil can be simulated. This method is effective in dealing with the pile foundation under cyclic lateral load and with relatively large lateral displacement. (4) the elastic theory method, this method assumes that the pile is embedded in the isotropy semi-infinite elastic body, and the soil parameters (elastic modulus and Poisson's ratio) can be set as constants or variables which changes with the depth.

The load transfer process of the pile foundation under lateral loads is a complex pile-soil interaction process, and the theoretical methods mentioned above cannot truly reflect the load-displacement response of the pile foundation. Hence, many scholars investigated the lateral bearing capacity of pile foundation through field tests and model tests. Wang *et al.* (2007) investigated the behavior of large diameter rock-socketed piles in harbor engineering under lateral loads through field tests and numerical analysis. Wang *et al.* (2010) investigated the behavior of the large diameter belled concrete piles with manual digging under lateral loads based on field tests and numerical analysis. Wang *et al.* (2007) conducted a group of large-scale model tests to investigate the lateral bearing capacity of large diameter rock-socketed cast-in-place piles as well as the behavior of rock. Liu *et al.* (2009) carried out a series of full-scale model tests to investigate the lateral bearing capacity of cast-in-place concrete thin-wall pipe piles (PCC piles) installed in double-layered soils. Ren *et al.* (2014) investigated the horizontal bearing capacity of the jet grouting soil-cement-pile strengthened pile through a group of model tests, and he pointed out that the soil properties within a certain depth below the soil surface would have a direct impact on the horizontal bearing capacity. Huang *et al.* (2013) investigated the effect of soil-cement pile on the lateral bearing capacity of the bored pile based on the field test results, and the test results showed that setting soil-cement pile around the bored pile can not only control the horizontal displacement, but also improve the lateral bearing capacity. Faro *et al.* (2015) conducted a series of laterally loaded pile tests in natural ground and in cemented treated soil, and the experimental results showed a significant increase in lateral load resistance of short piles when the soil is treated with cement.

The pre-bored grouting planted pile (PGP pile) is a new kind of composite pile foundation which containing a precast pile and the surrounding cemented soil, which is somewhat similar to the composition of the jet grouting soil-cement-pile strengthened pile. The author's team has conducted a detailed research on the vertical bearing capacity of the PGP pile through a series of field tests and model tests (Zhou, *et al.*, 2013, 2015a, 2015b), and the compressive and uplift bearing capacity of the PGP pile is considered to be better than the bearing capacity of the bored pile. Nevertheless, few researches were conducted to investigate the lateral bearing capacity of the PGP pile, and as mentioned above, setting cemented soil or soil-cement pile around the pile is effective in promoting the lateral bearing capacity (Ren *et al.*, 2014; Huang *et al.*, 2013; Faro *et al.*, 2015). Therefore, the lateral bearing capacity of the PGP pile is also probably better than the lateral bearing capacity of the bored pile because of the existence of the surrounding cemented soil. In this study, a group of field tests of the PGP piles and bored piles were carried out to investigate the effect of the cement soil around the pile on the lateral bearing capacity of the pile foundation.

2 TEST INTRODUCTION

2.1 *Test site introduction*

This field test was conducted in a power plant project in Wenzhou. According to the geological investigation data, the test site contains a deep soft soil layer which was of poor engineering properties. The specific soil profiles and properties were listed in Table 1, where ω is the water content; γ is the gravity; e is the void ratio; E_{s1-2} is the compression modulus; c and φ are the cohesion and internal friction angle of the soil, respectively.

2.2 *Test pile introduction*

To investigate the lateral bearing capacity of the PGP pile and bored pile, totally six test piles were constructed for the lateral load tests, including four PGP piles and two bored piles. Two of the PGP piles were 700 mm PGP piles, namely the diameter of the cemented soil is 700 mm, and 600 mm pipe pile and 650 (500) mm nodular pile (the diameter of the pile shaft is 500 mm, while the diameter of the nodes along the shaft is 650 mm) were inserted

Table 1. Soil profiles and properties.

Soil no.	Soil layer	ω/(%)	γ/(kN/m³)	e	E_{s1-2}/(Mpa)	C/(kPa)	φ/(°)
0	Plain fill		18.00				
1	Clayey soil	38.0	18.00	1.067	3.5	25.0	12.8
2–1	Mud	59.4	15.85	1.727	0.9	18.9	10.0
2–2	Mud	54.5	16.11	1.589	1.7	21.2	11.0
3–1	Muddy clay	44.4	16.81	1.319	3.5	26.2	12.4
3–2	Silt sand	22.0	19.01	0.685	5.0	14.6	23.8
3–3	Silt clay	24.8	19.11	0.738	6.9	29.5	17.5
4–1	Silt	25.1	19.22	0.722	6.9	15.0	23.8
4–2	Silt clay	26.4	18.87	0.786	6.7	32.6	17.7
4–3	Silt		19.30		6.9	8.0	27.2
5–1	Gravel	27.0	19.01	0.755	11.0		
6–1	Silt clay	31.0	18.79	0.877	7.4	50.8	11.4
6–2	Silt clay	25.7	19.42	0.737	11.7	48.2	14.7
7	Gravel	23.1	19.60	0.650	12.0		
7–1	Silt		19.00		10.0		
8–1	Silt clay	29.2	18.54	0.884	7.0		37.3
8–2	Clay	31.4	18.64	0.903	10.5		49.1
8–3	Silt clay	27.2	19.03	0.793	8.0		45.7
8–4	Fine sand		19.00				

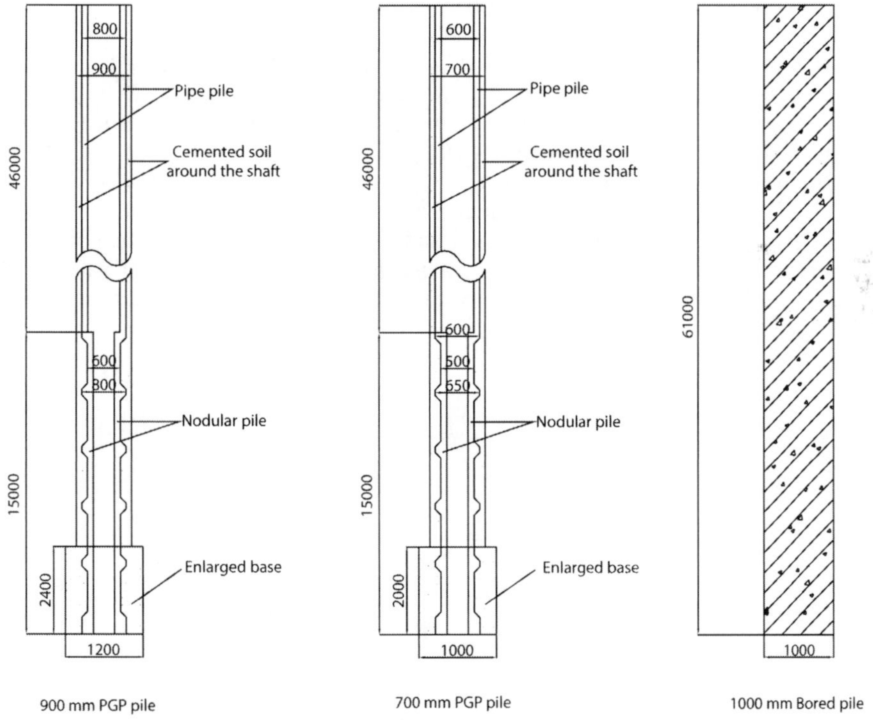

Figure 1. Sketch of test piles (unit: mm).

into the pile hole which was filled with cemented soil; the other two PGP piles were 900 mm PGP piles, that is, the diameter of the cemented soil is 900 mm, and 800 mm pipe pile and 800 (600) mm nodular pile were inserted in to the pile hole. The diameter of the bored pile was 1000 mm. The sketch of the test piles were shown in Fig. 1. It can be seen in Fig. 1

that the length of the test piles were all 61 m, and for the PGP piles, the length of the pipe pile were all 46 m, and the length of the nodular pile were all 15 m, moreover, an enlarged cemented soil base were manufactured at the pile tip. The concrete grade of the precast pile shaft was C80 (the strength of the concrete test block is larger than 80 MPa after curing for 28 days), and the concrete grade of the bored pile shaft is C30.

3 LATERAL LOAD TEST

3.1 *Lateral load test introduction*

The lateral load test was carried out after the cemented soil in the PGP pile being cured for 28 days. The test was according to the Chinese Technical Code for Testing of Building Foundation Piles (MOHURD, 2003), and the monotonic multi-cycle load method was adopted in the test. The field test was loaded by a 600 kN jack, and the displacement was measured by displacement sensors.

3.2 *Test results analysis*

The load-displacement responses of the test piles are shown in Fig. 2. It can be seen in Fig. 2 that the curves of the two test piles with identical diameter are relatively similar to each other, which demonstrates the reliability of the test results. It can also be seen in Fig. 2 that the lateral pile head displacements of the 700 mm PGP piles (PT1 and PT2) increase steadily with the applied load increasing, and the lateral pile head displacement reaches 37.03 mm and 37.69 mm, respectively for PT1 and PT2 when the applied lateral load is 210 kN, which achieves the condition to terminate the loading process given in the technical code (the pile head displacement exceeds 30 mm, MOHURD, 2003), finally, the residual lateral pile head displacements are 12.31 mm and 12.07 mm, respectively for PT1 and PT2 after the unloading process. It can also be seen in Fig. 2 that the lateral pile head displacements of PT3 and PT4 (900 mm PGP pile) reach 42.11 mm and 43.01 mm, respectively when loaded to 350 kN, and the residual displacements are 12.24 mm and 12.91 mm, respectively. For the bored piles (PT5 and PT6), the lateral pile head displacements are also gradually mobilized with the increase of the applied load, moreover, the values of the lateral displacements of the bored piles are smaller than that of the PGP piles. The lateral pile head displacements of PT5 and PT6 reach 45.02 mm and 44.67 mm, respectively when loaded to 350 kN, and the corresponding residual

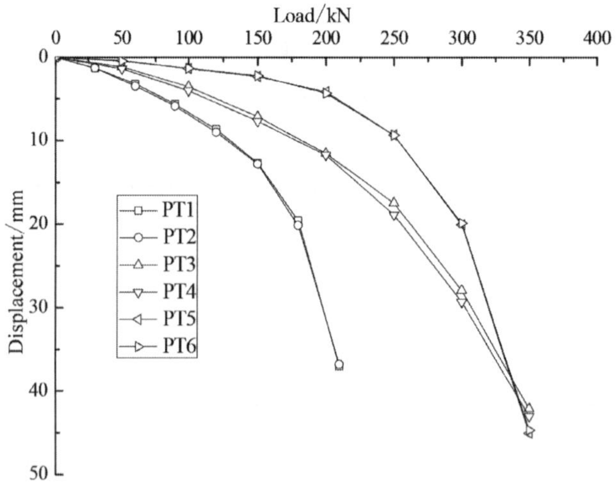

Figure 2. Lateral load-displacement curves of test piles.

displacements are 23.17 mm and 22.89 mm. Hence, the residual displacement of the PGP pile is much smaller than the residual displacement of the bored pile.

The lateral critical load of the pile foundation can be gained based on the lateral load-displacement gradient curve, and the lateral load-displacement gradient curves of the test piles are shown in Fig. 3. It can be seen in Fig. 3 that during the initial loading process, the lateral load-displacement gradient curves of the 700 mm PGP piles is somewhat close to the curves of the 900 mm PGP piles, and the values of the displacement gradients of the bored piles are apparently smaller than that of the PGP piles. The probable reason is that the bending stiffness of the bored pile shaft is larger than the bending stiffness of the PGP pile. The bending stiffness of the bored pile is:

$$E_b I_b = 30 \times 10^9 \times \frac{\pi}{64} \times 1^4 = 1470\ MN \cdot m^2 \tag{1}$$

The bending stiffness of the 900 mm PGP pile is:

$$E_p I_p = 38 \times 10^9 \times \frac{\pi}{64} \times \left(0.8^4 - 0.54^4\right) = 605\ MN \cdot m^2 \tag{2}$$

It can also be seen in Fig. 3 that with the increase of the applied lateral load, the displacement gradient of the bored pile is getting closer to that of the 900 mm PGP pile, moreover, the displacement gradient of the bored pile increases rapidly when the applied load surpassing 300 kN, while that of the 900 mm PGP pile still increases steadily. This is probably because that with the applied load increasing, the soil around the pile reaches the ultimate state, and the behavior of the laterally loaded pile is then controlled mainly by the soil properties. Therefore, the cemented soil around the concrete pile, of which the properties are between the properties of the concrete pile and surrounding soil, acts as a transition layer during the load transfer process. As a result, the cemented soil around the pile can virtually promote the behavior of the laterally loaded piles, and otherwise, prevent the pile from abrupt failure.

The fitting lateral load-displacement gradient curves of the test piles are shown in Fig. 4. According to the technical code in China (MOHURD, 2003), the lateral critical load can be taken as the lateral load of the first turning point in the lateral load-displacement gradient curve.

It can be seen in Fig. 4 that the lateral critical load of the 700 mm PGP pile is 150 kN, and the lateral critical loads of the 900 mm PGP pile and 1000 mm bored pile are both 200 kN. As the bending stiffness and diameter of the 900 mm PGP pile are both smaller than that of the 1000 mm bored pile, it can be considered that the lateral bearing capacity of the PGP pile is better than the lateral bearing capacity of the bored pile, and besides, the cemented soil around the pile can help improve the lateral critical load.

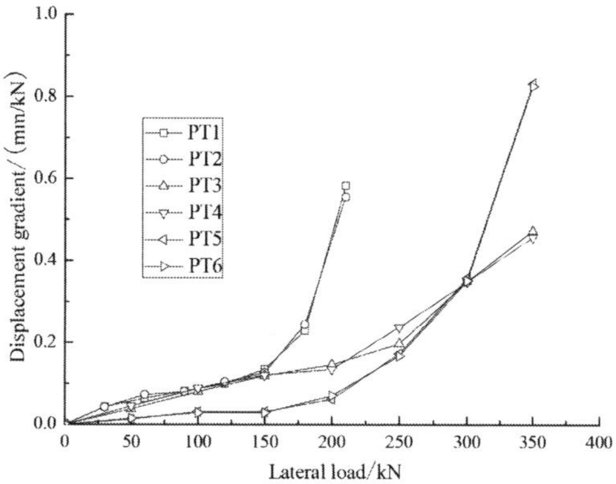

Figure 3. Lateral load-displacement gradient curves of test piles.

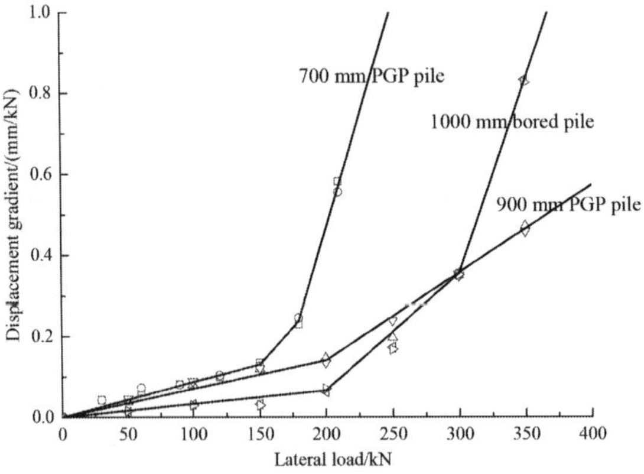

Figure 4. Fitting lateral load-displacement gradient curves of test piles.

4 CONCLUSION

In this study, a group of lateral field tests of PGP piles and bored piles were conducted to investigate the influence of the cemented soil around the pile on the lateral bearing capacity of pile foundation. Based on the results of the field tests presented herein, the following conclusions can be drawn:

1. In the initial loading process, the lateral pile head displacement and displacement gradient of the bored pile are both smaller than that of the PGP pile, and the cemented soil around the pile does not have much influence on the pile behavior.
2. The properties of the cemented soil is between the properties of the concrete and surrounding soil, and the cemented soil around the pile can help promote the lateral bearing capacity of the pile foundation, moreover, abrupt failure can also be prevented because of the existence of the surrounding soil.
3. The soil properties are pretty limited in soft soil areas, and the setting cemented soil around the pile is an effective to promote the lateral bearing capacity of the pile foundation. However, bending moment and soil pressure are not measured in the field tests, and much more tests are needed to give a thorough investigation on the influence of the cemented soil on the pile behavior.

ACKNOWLEDGEMENT

This research is sponsored by the National Natural Science Foundation in China (51578498, 51579217).

REFERENCES

Faro, Consoli, Schnaid, et al. 2015. Field Tests on Laterally Loaded Rigid Piles in Cement Treated Soils[J]. Journal of Geotechnical and Geoenvironmental Engineering, 141(6): 06015003.

Gong, X.N., *et al.* 2015. Pile Foundation Manual. Beijing: China Architecture and Building Press (in Chinese).

Huang, Y.B., Zhao, H.B., Gu, C.C., et al. 2013. Field experimental study of lateral load capacity of filling pile enhanced by soil-cement pile. Rock and Soil Mechanics, 34(4): 1109–1115.

Liu, H.L., Zhang, J.W. & Peng, J. 2009. Full-scale model tests on behavior of cast-in-place concrete pipe piles with large diameter under lateral loads. Chinese Journal of Geotechnical Engineering, 31(2): 161–165 (in Chinese).

Mohurd (Ministry of Housing and Urban-Rural Development of the People's Republic of China), 2008. Technical Code for Building Foundation Pile, JGJ94–2008. China Architecture and Building Press, Beijing (in Chinese).
Mohurd (Ministry of Housing and Urban-Rural Development of the People's Republic of China), 2003. Technical Code for Testing of Building Foundation Piles, JGJ106–2003. China Architecture and Building Press, Beijing (in Chinese).
Ren, L.W., Dun, Z.L., Li, G., et al. 2014. Model tests research on horizontal bearing behavior of JPP pile under different combinations. Rock and Soil Mechanics, 35(S2): 101–106 (in Chinese).
Wang, D.Y., Lan, C., He, G.C., et al. 2007. Researches on lateral support behavior of large diameter rock-socketed cast-in-place piles at river port by laboratory model test. Chinese Journal of Geotechnical Engineering, 29(9): 1307–1313 (in Chinese).
Wang, J.H., Chen, J.J. & Ke, X. 2007. Characteristics of large diameter rock-socketed piles under lateral loads. Chinese Journal of Geotechnical Engineering, 29(8): 1194–1198 (in Chinese).
Wang, J.L., Wang, F.M., Ren, L.W., et al. 2010. Horizontal static load test and numerical simulation of single large diameter under-reamed pile. Chinese Journal of Geotechnical Engineering, 32(9): 1406–1411 (in Chinese).
Zhou, J.J., Gong, X.N., Wang, K.H., et al. 2015a. A model test on the behavior of a static drill rooted nodular pile under compression. Marine Geo resources & Geo technology, 0, 1–9.
Zhou, J.J., Gong, X.N., Wang, K.H., et al. 2015b. Behavior of the static drill rooted nodular piles under tension [J]. Chinese Journal of Geotechnical Engineering, 37(3): 570–576 (in Chinese).
Zhou, J.J., Wang, K.H., Gong, X.N., et al. 2013. Bearing capacity and load transfer mechanism of a static drill rooted nodular pile in soft soil areas. Journal of Zhejiang University-SCIENCE A (Applied Physics & Engineering), 14(10): 705–719.

Advanced Engineering and Technology III – Xie (Ed.)
© 2017 Taylor & Francis Group, London, ISBN 978-1-138-03275-0

A damage model for hard rock under stress-induced failure mode

Zhen Li
Opening Laboratory for Deep Mining Construction, Henan Polytechnic University, Jiaozuo, China
School of Civil Engineering, Henan Polytechnic University, Jiaozuo, China

Qing-hua Zhu
School of Foreign Studies, Henan Polytechnic University, Jiaozuo, China

Bei-lei Tian
Linyi Municipal Bureau of Planning, Linyi, China

Tong-fei Sun & Dong-wei Yang
School of Civil Engineering, Henan Polytechnic University, Jiaozuo, China

ABSTRACT: This paper presents a damage model for hard rock under stress-induced failure. Damage initiation is first proposed based on the study of promoted strain energy and impeded strain energy to damage. Then, simplified damage constitutive law is formulated by assuming that the stress–strain relationship of the intact part conforms Hooke's law. Aimed at expressing the brittle–ductile transition of hard rock under the condition from low confining pressure to high confining pressure, a modified Mazars' damage evolution equation is proposed. The validation results indicate that the proposed model is in accordance with the test data.

1 INTRODUCTION

With the rapid development of underground construction in China, deep-buried rock tunnels encounter a lot of problems. Diederichs M.S. et al. (2004) proposed that in hard rock tunnels at depth, one of the primary design issues is to determine the stress level associated with the onset of wall yield due to boundary compression. Compared with soft rock, hard rock at depth has special mechanical characteristics. For stress-induced failure, flaking, spalling, and possibly bursting of the wall rock can be a costly problem and a major safety concern. Edelbro C. (2009) reported that a large number of fractures may lead to slabbing failure or rock burst.

Stress-induced failure of hard rock can be prevented by using feasible methods such as rapid effective support and prevention of crack propagation under high stress. Untreated hard rock may lead to serious accidents. For example, Martin C.D. & Maybee W.G. (2000) discussed the serious accident that occurred at Coalbrook colliery, South Africa in 1960, in which 900 pillars were destroyed and 437 people were killed due to coal pillar instability under high stress disturbance. Therefore, the study on stress-induced failure of hard rock subjected to the disturbing force should focus on damage initiation threshold and damage process.

The classic work by Griffith A.A. (1924), Horii H. & Nemat-Nasser S. (1986), Kemeny J.M. & Cook N.G.W. (1986), and many other subsequent researchers have used a shearing or sliding crack analogue to simulate the initiation of brittle failure. Two different methods are used for the study of the damage process: micromechanical models and damage mechanics. Shao J.F. et al. (2006) and Jian Y.W. et al. (2006) proposed the implementation of micromechanical models as an alternative approach to overcome the drawbacks encountered with numerical modelling. Mortazavi A. & Molladavoodi H. (2012) proposed the main advantage of continuum damage models, which includes providing macroscopic constitutive equations that can be easily implemented and applied to engineering analyses.

Unlike other areas of engineering, the mechanical behavior of rock structure has the common characteristic of uncertainty. Hajiabdolmajid V. & Kaiser P. (2003) argued that the analysis of mechanical behavior of rock structure relies on empirical judgment in rock engineering due to the lack of understanding. Therefore, considering hard rocks under stress-induced failure mode commonly encountered in deep-buried engineering, this paper presents the research on damage constitutive equations and damage evolution equation in a simplified form. The damage model proposed within the framework of continuum damage is verified at the end by a test.

2 DAMAGE INITIATION CRITERION

Eberhardt et al. (1998) proposed a methodology for determining damage initiation based on acoustic monitoring. As observed, the stationary point of the volumetric strain–axial stress curve nearly equals to the damage threshold. Based on this finding, this paper assumes the damage initiation as the stationary point of the volumetric strain–axial stress curve.

2.1 *The promoted strain energy to damage*

For non-internal frictional material such as metal, the von Mises criterion is given by:

$$(\sigma_1-\sigma_2)^2+(\sigma_2-\sigma_3)^2+(\sigma_1-\sigma_3)^2=C_1 \tag{1}$$

where σ_1, σ_2 and σ_3 are principal stresses and C_1 is the material constant.

Rock is formed by cementation action of particles. The emphasis of failure research should be put on SMP, as shown in Figure 1, which is suggested by Matsuoka H. & Sun D. (1995). From Equation 1, the internal frictional character of rock can be considered as follows:

$$\sum_{\substack{j=1\\3\geq i>j}}^{j=2}\frac{1}{2G}\left(\frac{\sigma_i-\sigma_j}{2\cos\varphi_{ij}}-\frac{\sigma_i+\sigma_j}{2}\tan\varphi_{ij}\right)^2 = w_s \tag{2}$$

where w_s is the promoted strain energy to damage; G is the shear modulus; and φ_{13} is the internal friction angle. From Figure 1, the relation between the angles can be derived as follows:

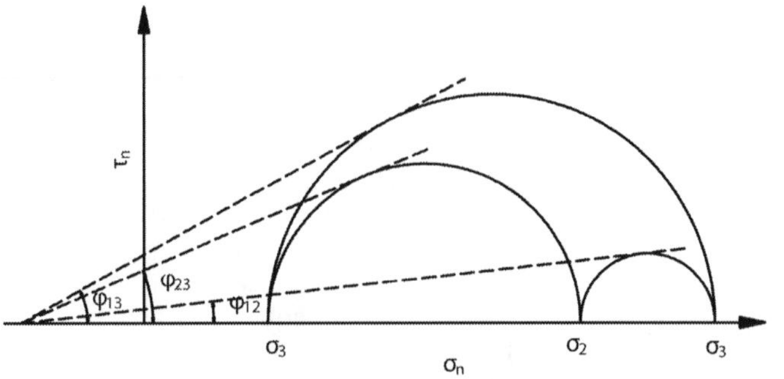

Figure 1. SMP described by the Mohr diagram.

$$\sin\varphi_{12} = \frac{\left(1-\sqrt{3}\tan\theta\right)\sin\varphi_{13}}{2+\sin\varphi_{13}+\sqrt{3}\tan\theta\sin\varphi_{13}} \tag{3}$$

$$\sin\varphi_{23} = \frac{\left(1+\sqrt{3}\tan\theta\right)\sin\varphi_{13}}{2-\sin\varphi_{13}+\sqrt{3}\tan\theta\sin\varphi_{13}} \tag{4}$$

Equation 2 equals to Equation 1 as the internal friction angle φ_{13} is zero.

2.2 *Impeded strain energy to damage*

Within the elastic frame, the relative volume change is called the volumetric strain energy and can be written as:

$$\frac{\left(\sigma_1+\sigma_2+\sigma_3\right)^2}{18K} = w_v \tag{5}$$

where w_v is the impeded strain energy to damageand K is the volumetric modulus.

2.3 *Formulation of damage initiation*

Considering the promoted strain energy and impeded strain energy to damage, in this paper, the damage initiation criterion of rock is simplified to a linear form and can be written as:

$$\sum_{\substack{j=1 \\ 3\geq i>j}}^{j=2} \frac{1}{2G}\left(\frac{\sigma_i-\sigma_j}{2\cos\varphi_{ij}}-\frac{\sigma_i+\sigma_j}{2}\tan\varphi_{ij}\right)^2 - a\frac{\left(\sigma_1+\sigma_2+\sigma_3\right)^2}{18K} - b = 0 \tag{6}$$

where a and b are material constants.

To verify the damage initiation criterion proposed, T_{2b} marble in Jinping II hydro power station was studied. The cylinder samples of size 0.05 m × 0.1 m were fabricated. The conventional compressive test was conducted on an MTS 815.03 mechanical machine as shown in Figure 2, the promoted strain energy and impeded strain energy to damage approximately conform to a linear relationship. Thus, the criterion can be used as the damage initiation criterion.

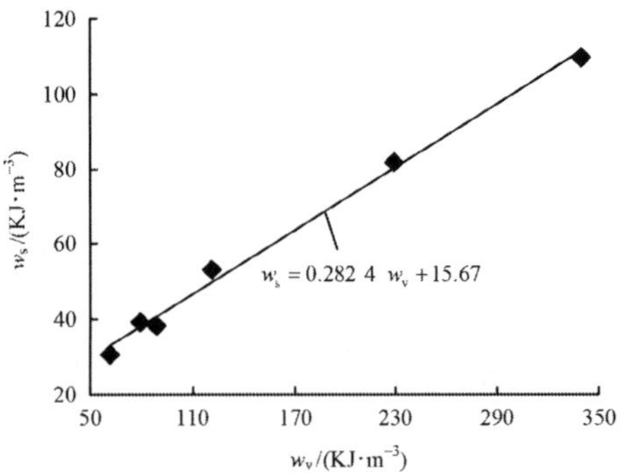

Figure 2. Promoted strain energy and impeded strain energy to damage.

3 DAMAGE MODEL FOR HARD ROCK

A lot of damage models such as elastic and elasto-plastic damage model, meso damage model, and dynamic damage model have been proposed. Damage model can be expressed in an accurate but complex way. For the sake of simplicity, in this paper, the existing damage models are used and developed with the proposed damage initiation criterion. According to Rabotnov Y.N. (1969)'s definition of damage variable D, considering the damaged part not subjected to external loads and the intact part subjected to external loads, we have:

$$\sigma_i = \sigma_i'(1-D) \tag{7}$$

where σ_i is the apparent stress and σ_i' is the stress of the intact part.

3.1 *Damage constitutive equation*

The following presumptions are proposed:

a. The stress–strain relationship of the intact part agrees with Hooke's law. In triaxial stress paths,

$$\sigma_1' = E\varepsilon_1' + 2\nu\sigma_3' \tag{8}$$

where σ_1' and σ_3' are the maximum principal stress and minimum principal stress of the intact part, respectively; ε_1' is the axial strain of the intact part; and E and ν are the elastic modulus and Poisson 's ratio, respectively.

b. The apparent strain in the damage part and the strain in the intact part are in accordance with Lemaitre equivalent strain theory as follows:

$$\varepsilon_1' = \varepsilon_1 \tag{9}$$

where ε_1 is the apparent axial strain.

From Equations 7–9, we get:

$$\sigma_1 = (1-D)E\varepsilon_1 + 2\upsilon\sigma_3 \tag{10}$$

Therefore, the damage constitutive equation for hard rock in the conventional triaxial stress path is given by

$$\sigma_1 - 2\upsilon\sigma_3 = (1-D)E\varepsilon_1 \tag{11}$$

3.2 *Damage evolution equation*

Mazars J. (1986) put forward damage evolution equation of concrete under compression as:

$$D = 1 - \frac{1}{\exp(mY)} \tag{12}$$

where m is the property constant determined by test. Y is the generalized shear strain expressed as:

$$Y = \int\sqrt{\frac{2}{3}\mathrm{d}e:\mathrm{d}e};\mathrm{d}e = \mathrm{d}\varepsilon - \mathrm{tr}(\mathrm{d}\varepsilon/3)\delta \tag{13}$$

where e is the strain deviator tensor and δ is the Kronecker symbol.

To express the brittle–ductile transition, the following damage evolution equation is proposed:

$$D = 1 - \frac{1}{\exp(mY) + nY} \tag{14}$$

where n is the property constant. In contrast to Mazars' damage evolution equation, this article defines the stationary point of volumetric strain–axial strain as the damage threshold. Consistent with this definition, D can be normalized as follows:

$$D_{\mathrm{m}} = \frac{D - D_0}{D_{\mathrm{f}} - D_0} \tag{15}$$

where D_{m} is the apparent damage parameter. For initial damage, $D_{\mathrm{m}} = 0$. For failure, $D_{\mathrm{m}} = 1$. *Here* D_0 is the initial value of D and D_{f} is the residual value of D.

Under the conventional compressive loading path, Equation 13 can be expressed as follows:

$$Y = \int \frac{2}{3}(\mathrm{d}\varepsilon_1 - \mathrm{d}\varepsilon_3) = \frac{2}{3}(1+\nu)\varepsilon_1 \tag{16}$$

Therefore, the damage evolution equation for hard rock is expressed as:

$$D = 1 - \frac{1}{\exp(s\varepsilon_1) + t\varepsilon_1} \tag{17}$$

where s and t are property parameters. From Equations 14–17, we have $s = 2m(1+\nu)/3$ and $t = 2n(1+\nu)/3$. s and t can be determined by test.

4 VALIDATION

4.1 *Model parameters*

The parameters are determined by the triaxial compressive test using two operations. a) The stress by the damage initiation criterion equals to the damage threshold by the damage constitutive equation. b) The root-mean-square error of solved stress relative to the test data is less than the target value. For simplicity, Poisson's ratio is adopted as 0.3. From Equations 18 and 19, the parameters for T_{2b} marble are obtained. Assuming the units of σ_3 as MPa, fitting process is expressed as:

$$E = -32.981\sigma_3^2 + 3215.5\sigma_3 - 2890.7 \tag{18}$$

$$s = -0.1304\sigma_3^2 + 0.8044\sigma_3 + 195.55 \tag{19}$$

$$t = -0.2005\sigma_3^2 + 25.44\sigma_3 - 588.2 \tag{20}$$

4.2 *Numerical validation*

Equation 12 is Mazars' damage evolution equation in the case of compression. The parameters are obtained to contrast with the modified model proposed in this paper. From Equations 11 and 17, the stress–strain curve is simulated by the proposed model. From equations 11 and 12, the stress–strain curve is simulated by the damage model with Mazars' damage evolution equation. Figure 3 compares the curves and the test data. As shown in the figure, the proposed model is in accordance with the test data.

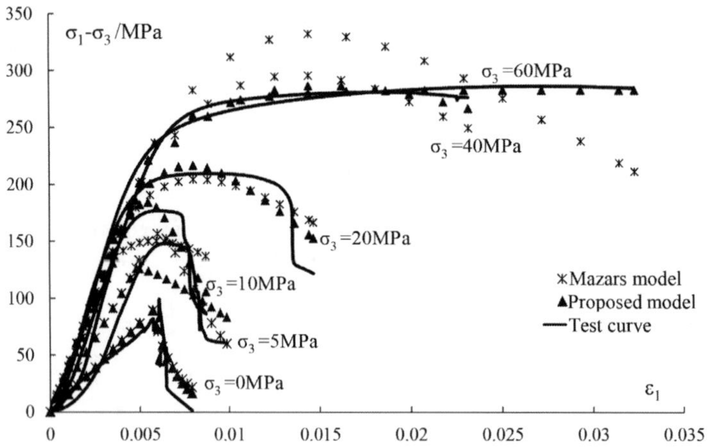

Figure 3. Validation of the damage model proposed in this paper.

5 CONCLUSION AND DISCUSSION

For accurate description of the damage process, the damage initiation criterion is first proposed based on the study of the promoted strain energy and impeded strain energy to damage. Then, a simplified damage constitutive law is formulated. Aimed at expressing the brittle–ductile transition of hard rock under the condition from low confining pressure to high confining pressure, a modified Mazars' damage evolution equation is proposed. As suggested by Trinh N. & Jonsson K. (2013), unlike soft rock the stability monitoring of hard rock tunnel shows that, before destruction, the displacement is minimal or even negligible. Damage becomes a serious concern when the stress exceeds the damage initiation threshold for hard rock tunnel excavation. The damage model is simple and easy to implement. It can express the brittle–ductile transition accurately, which has application for engineering calculation of brittle hard rock.

ACKNOWLEDGEMENTS

This work was supported by the Project of Key Opening Laboratory for Deep Mining Construction at Henan province (2015 KF-06), the Key Scientific Research Project of Institutions of Higher Education at Henan province (16A560004, 16A440008) and Dr. Fund Projects of Henan Polytechnic University (B2016-65).

REFERENCES

Diederichs, M.S., Kaiser, P.K. & Eberhardt, E. 2004. Damage initiation and propagation in hard rock during tunneling and influence of near-face stress rotation. *International Journal of Rock Mechanics & Mining Sciences,* 41(5): 785–812.

Eberhardt, E., Stead, D., Stimpson, B. & Read, R.S. 1998. Identifying crack initiation and propagation thresholds in brittle rock. *Canadian Geotechnical Journal,* 35(2): 222–233.

Edelbro, C. 2009. Numerical modelling of observed fallouts in hard rock masses using an instantaneous cohesion-softening friction-hardening model. *Tunnelling and Underground Space Technology,* 24(4): 398–409.

Griffith, A.A. 1924. Theory of rupture. In: First International Congress of Applied Mechanics, Delft: 55–63.

Hajiabdolmajid, V. & Kaiser, P. 2003. Brittleness of rock and stability assessment in hard rock tunneling. *Tunnelling & Underground Space Technology,* 18(1): 35–48.

Horii, H. & Nemat-Nasser, S. 1986. Brittle failure in compression: splitting, faulting and brittle-ductile transition. Philosophical Transactions of the Royal Society. *A Mathematical Physical & Engineering Sciences,* 319: 337–374.
Jian, Y.W., Jie, L. & Rui, F. 2006. An energy release rate-based plastic-damage model for concrete. *International Journal of Solids & Structures,* 43(3–4): 583–612.
Kemeny, J.M. & Cook, N.G.W. 1986. Crack models for the failure of rocks in compression. In: Desai, Krempl, Kiousis, Kundu editors. Constitutive laws for engineering materials: theory and applications, 2: 879.
Martin, C.D. & Maybee, W.G. 2000. The strength of hard-rock pillars. *International Journal of Rock Mechanics & Mining Sciences,* 37(8): 1239–1246.
Matsuoka, H. & Sun, D. 1995. Extension of spatially mobilized plane (SMP) to frictional and cohesive materials and its application to cemented sands. *Journal of the Japanese Geotechnical Society Soils & Foundation,* 35(4): 63–72.
Mazars, J. 1986. A description of micro- and macro scale damage of concrete structures. *Engineering Fracture Mechanics,* 25(5–6): 729–737.
Mortazavi, A. & Molladavoodi, H. 2012. A numerical investigation of brittle rock damage model in deep underground openings. *Engineering Fracture Mechanics,* 90: 101–120.
Rabotnov, Y.N. 1969. Creep rupture. Proceedings of the XII International Congress on Applied Mechanics. Stanford. Springer Berlin Heidelberg.
Shao, J.F., Jia, Y., Kondo, D. & Chiarelli, A.S. 2006. A coupled elastoplastic damage model for semi-brittle materials and extension to unsaturated conditions. *Mechanics of Materials,* 38(3): 218–232.
Trinh, N. & Jonsson, K. 2013. Design considerations for an underground room in a hard rock subjected to a high horizontal stress field at Rana Gruber, Norway. *Tunnelling & Underground Space Technology,* 38(3): 205–212.

Advanced Engineering and Technology III – Xie (Ed.)

Research on fiber reinforced ultra-lightweight concrete applying Poraver aggregates and PVC fiber

Z. Li
Department of Geotechnical Engineering, Tongji University, Shanghai, China

G. Yang
Department of Civil Engineering, Tongji University, Shanghai, China

L. Xie
Department of Hydraulic Engineering, Tongji University, Shanghai, China

ABSTRACT: Lightweight concrete achieves the merits of light density and low thermal conductivity. However, lightweight concrete suffers from poor strength owing to the inherent porous structure, which confines its further application. In this paper, a kind of recyclable lightweight glass beads named Poraver is adopted as aggregates and its optimal mixture is attained by utilizing Fuller curve and control experiments. Furthermore, Polyvinyl Alcohol (PVA) fiber is employed to enhance concrete strength with comparison tests to probe for its ideal dosage. The Scanning Electron Microscope (SEM) pictures illustrate that the micro structure is compact and integral. With a dry density of merely 650 kg/m^3, the design attains the 28-day compressive strength of 10.3 N/mm^2 and tensile strength of 1.0 N/mm^2, which promises an extensive future application for lightweight concrete owing to its low density, available mechanical properties and environmental-friendly components.

1 INTRODUCTION

Traditionally speaking, concrete can be classified by its density, application, strength and production (ACI 318M-05). In reference to density, the category consists of heavy weight concrete, conventional concrete and lightweight concrete. In recent years, ultra-lightweight concrete with apparent density less than 800 kg/m^3 has received comprehensive researches and heated discussion (Hegyi, P. & Dunai, L. 2016a, b, Yan, J.B. et al. 2016). Many structure designs utilizing ultra-lightweight concrete have come into application such as floating and offshore constructions. However, due to the inherent porous microscopic structure, ultra-lightweight concrete is deficient in strength, which imposes restriction on its universal application (Kidalova, L. et al. 2012, Choi, J. et al. 2014).

It is universally accepted that the performance of lightweight concrete is influenced by distribution of aggregates and dosages of various additives. Marcovic, I. et al. (2006) studied the performance of hybrid-fiber concrete. By trails of metallic and polymeric fiber, the results demonstrated that fiber additives retard the occurrence of cracks and strongly reinforce the strength of concrete. Besides, Wang, H.Y. and Tsai, K.Z. (2006) utilized dredged silt as lightweight aggregates and designed a batch of lightweight concrete with density between 800 kg/m^3 and 1500 kg/m^3 and 28-day compressive strength more than 18 N/mm^2. Moreover, Yu, Q.L. et al. (2015) applied modified Andreasen and Andersen particle packing model for optimal allocation of lightweight aggregates and developed a genre of lightweight concrete with improved mechanical properties and reduced thermal conductivity. Furthermore, Yu, R. et al. (2015) incorporated polypropylene fiber of diversified diameters and lengths into ultra-lightweight concrete. The outcome indicated that the compressive strength strongly

Table 1. Properties of Poraver and PVC fiber.

Poraver				
Diameter (mm)	0.1–0.25	0.25–0.5	0.5–1.0	1.0–2.0
Dry loose bulk density (kg/m³)	400	340	270	230
Apparent density (kg/m³)	850	680	450	410
PVC fiber				
Cut length (mm)	12	Liner density (dtex)	15	
Tenacity (cN/dtex)	12	Elongation (%)	7	

Table 2. Applied components.

Components	Cement	Mineral powder	Aggregates
Materials	Type II Portland cement	Ground Granulated Blast Furnace Slag (GGBFS)	Poraver
Components	Latex	Water reducing agent	Reinforcement agent
Materials	Styrofan ECO 7623	Melflux 4930 F	6 mm polyvinyl alcohol (PVA)

depended on the proportion of fiber. More specifically, long fiber presented an efficacious resistance towards the development of micro fractures while short fiber was efficient in bridging micro-cracks according to Scanning Electron Microscope (SEM) analysis.

Admittedly, numerous efforts have been invested into the development of ultra-lightweight concrete. However, a majority of researches concentrate on its performance of thermal conductivity and fire resistance. The rest of studies, in most cases, probe for either proportion of aggregates or fiber reinforcement effect separately. In respect of mechanical properties, overall researches both on aggregates and fiber are limited.

This paper adopts Poraver as lightweight aggregates and Polyvinyl Alcohol (PVA) fiber as reinforcing agent, respectively. By conceptual design based on Fuller curve, the ideal aggregate allocation is figured out. By conducting a batch of control experiments, the influence of diversified sizes of aggregates and dosages of PVC fiber are discussed both from macroscopic mechanical experiments and Scanning Electron Microscope (SEM) pictures.

2 PREPARATIONS

The baseline cementitious materials combined 77.8% Type II Portland cement with 22.2% Ground Granulated Blast Furnace Slag (GGBFS). Besides, environmental-friendly Poraver (Fig. 1a) were chosen as lightweight aggregates because it derives from industrial waste glass and is capable of recycling. The aggregate sizes consist of 0.1–0.25 mm, 0.25–0.5 mm, 0.5–1 mm, 1–2 mm. To further improve performance of the mixture, Styrofan ECO 7623, a Styrene butadiene latex, was introduced to increase bonding capability and dwindle water seepage phenomenon. In addition, Melflux 4930 F, an effective water reducer, helped to decrease water cement ratio and fortify strength while maintaining desired slump. Moreover, Polyvinyl Alcohol (PVA) fiber (Fig. 1b) as reinforcement agent was employed to enhance concrete strength and crack resistance. All applied components are concluded in Table 1 and the properties of Poraver and PVC fiber are presented in Table 2.

3 AGGREGATE EXPERIMENTS

3.1 *Design procedure*

For the design of ultra-lightweight aggregates, the allocation was supposed to be continuous and compact in order to achieve fine mechanical properties. Fuller W.B. and Thompson

S.E. (1907) proposed an optimal design methodology and have received extensive acceptance according to engineering experience. The Fuller curve is exhibited as following.

$$P = \left(\frac{d}{D}\right)^q \tag{1}$$

where P refers to ratio of the aggregates tinier than size d, d indicates particle size and D means the maximal size of aggregates ($D = 2$ mm). In reference with experience, q equals 0.45.

To validate the effectiveness of Fuller curve, also to probe for how particle size influences concrete mechanical properties, control experiments were conducted with admixing amounts of four aggregate sizes changing respectively. The mixture of aggregate test is presented in Table 3 with the design based on Fuller curve being Mix 9. Also, the performance of nine groups according to Fuller curve is illustrated in Figure 2a, b. From Figure 2a, b, Mix 9 exactly lands on Fuller curve and other eight groups locate at area within 7% deviation of the curve.

3.2 *Density experiments and discussions*

All samples were conserved according to standard method for 28 days (ACI 318M-05). By utilizing projected light system and magnifying lens system, the density test (ASTM C138) was conducted both in oven-dried and ambient conditions with dry and wet density depicted in Figure 3a, b.

Table 3. The mixture of aggregate test.

Component mass		Design (Mix 9)	Mix 1	Mix 2	Mix 3	Mix 4	Mix 5	Mix 6	Mix 7	Mix 8
Cement (g)		200	200	200	200	200	200	200	200	200
Mineral powder (g)		57	57	57	57	57	57	57	57	57
Aggregates	1–2 mm (g)	79	99	59	79	79	79	79	79	79
	0.5–1 mm (g)	31	31	31	51	11	31	31	31	31
	0.25–0.5 mm (g)	40	40	40	40	40	60	20	40	40
	0.1–0.25 mm (g)	55	55	55	55	55	55	55	75	35
Latex ECO (g)		36	36	36	36	36	36	36	36	36
Water reducing agent (g)		1.5	1.5	1.5	1.5	1.5	1.5	1.5	1.5	1.5
Water (g)		130	130	130	130	130	130	130	130	130
PVA (g)		0	0	0	0	0	0	0	0	0

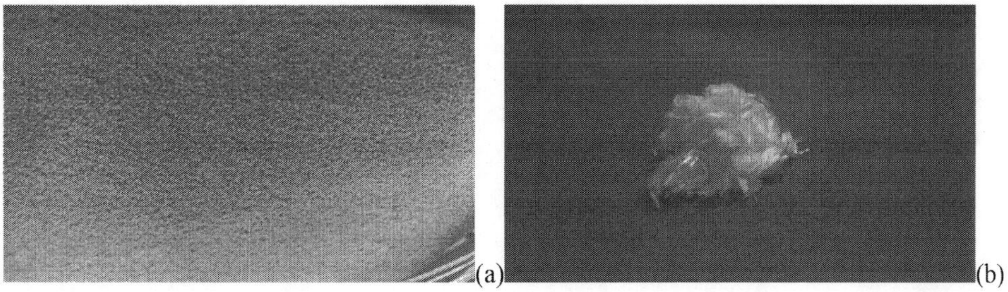

Figure 1. Photo of (a) Poraver aggregates (b) PVC fiber.

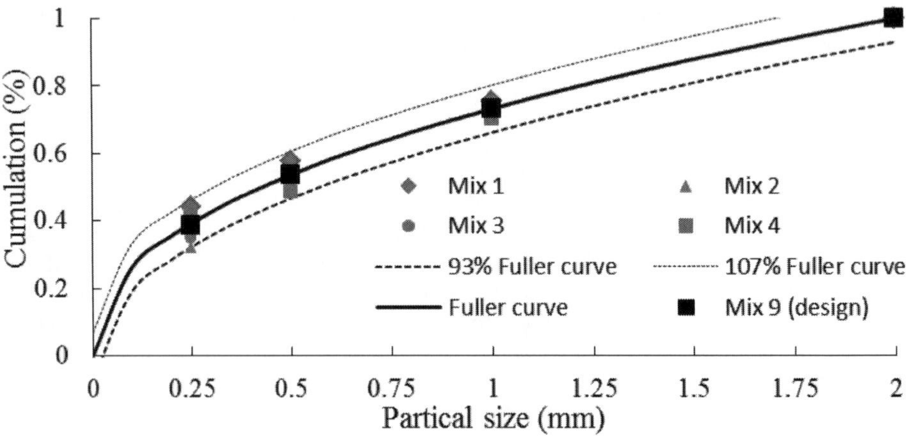

Figure 2a. The performance of mixtures according to Fuller curve (first four groups & design group).

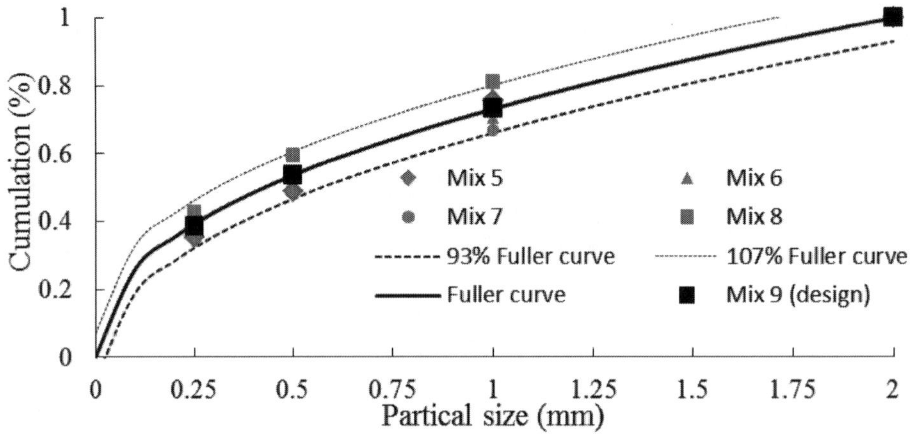

Figure 2b. The performance of mixtures according to Fuller curve (last four groups & design group).

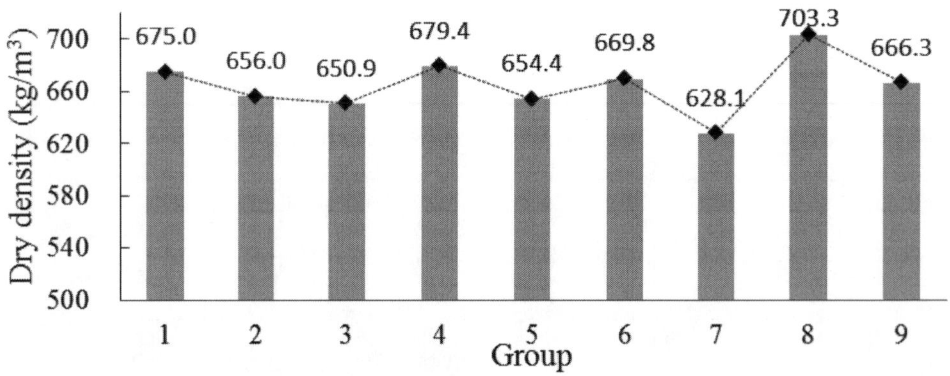

Figure 3a. Dry density of nine groups under oven-dried circumstance.

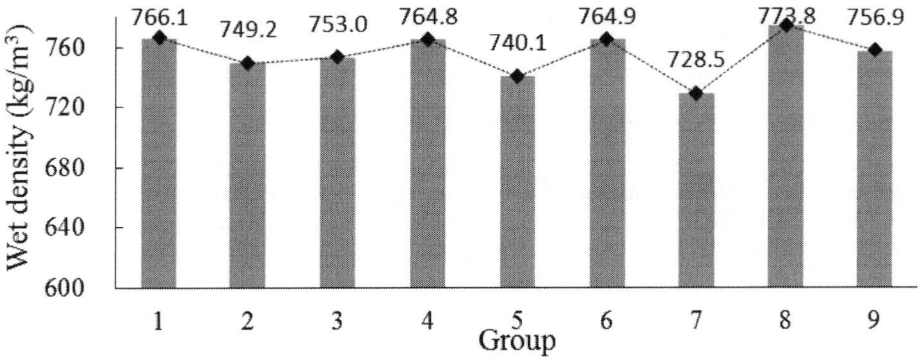

Figure 3b. Wet density of nine groups under ambient circumstance.

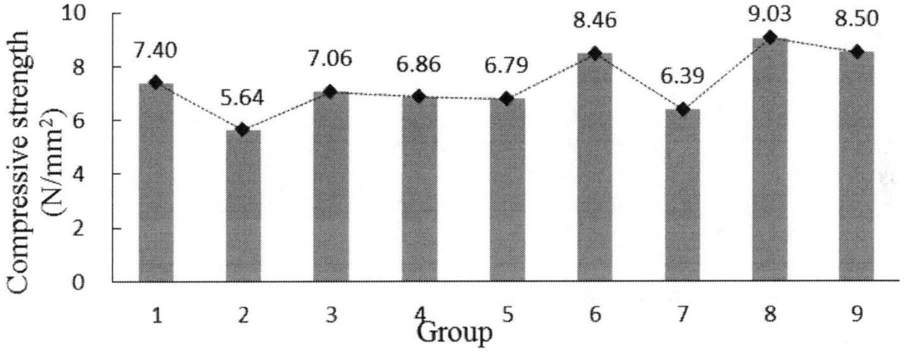

Figure 3c. 28-day compressive strength of nine groups.

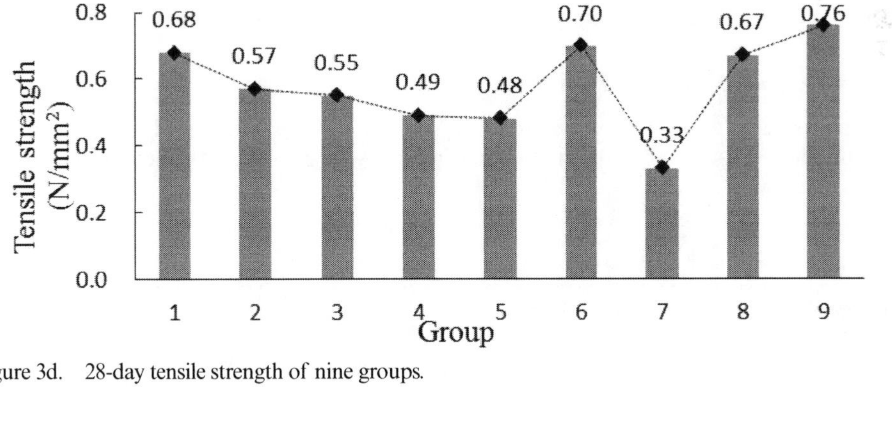

Figure 3d. 28-day tensile strength of nine groups.

As Figure 3a, b illustrate, the difference between Mix 7 and Mix 8 is the maximal, which demotes that granules of large size play a decisive role in the overall density. It is conceivable because large-sized aggregates are particularly porous and uncompact. Besides, for groups with smaller density, the variance between wet and dry density is usually greater than that for denser groups. The regulation is also reasonable due to the fact that denser concrete contains less voids and water among interspace. What is more, although small and medium-sized particles impose limited influence on concrete density, it mainly affects the workability of concrete based on experiments.

3.3 *Mechanical experiments and discussions*

As for compressive (ASTM C39) and tensile (ASTM C496) strength experiments, the 28-day compressive strength and tensile strength of nine specimens are portrayed in Figure 3c, d.

As figure 3c illustrate, in most cases, groups with greater density attain stronger compressive strength owing to their compact structures. Whereas, the growth rate of density and compressive strength is not identical because the continuity of aggregates serves as another factor in determining concrete strength. To specify, groups with poor allocation of granules tend to split and crush, even though the density of which is large enough.

Similarly, figure 3d demonstrates the positive correlation between strength and density. However, the degree of aggregate continuity is more significant for tensile strength than for compressive bearing capacity because there are more gaps among poor distributed aggregates and cracks are easier to develop under tensile conditions.

4 FIBER EXPERIMENTS

4.1 *Design procedure*

Because the design based on Fuller curve (Mix 9) achieved relatively strong compressive strength, fine tensile bearing capacity and low density, the aggregate distribution of Mix 9 was utilized in fiber experiments. Verified dosages of Polyvinyl Alcohol (PVA) fiber were incorporated as Table 4 presents.

4.2 *Density experiments and discussions*

Following relevant standard (ASTM C138), the density test was conducted with measured dry and wet density portrayed in Figure 4a, b. Both two curves exhibit parabolic tendency with density dwindling at the beginning and ascending in the end. As a matter of fact, the introducing of PVC fiber carries bubbles and pores at same time, thus resulting in descending of density. However, the density of PVC itself is large enough (Table 1). In this regard, with the amounts of PVC fiber multiplying, the overall density actually rises.

4.3 *Mechanical experiments and discussions*

After compressive (ASTM C39) and tensile (ASTM C496) tests, the results are documented in Figure 4c,d. Distinct from the regulation in aggregate experiments, the compressive and

Table 4. The mixture of fiber test.

Component mass		Mix 9	Mix 10	Mix 11	Mix 12	Mix 13
Cement (g)		200	200	200	200	200
Mineral powder (g)		57	57	57	57	57
Aggregates	1–2 mm (g)	79	79	79	79	79
	0.5–1 mm (g)	31	31	31	31	31
	0.25–0.5 mm (g)	40	40	40	40	40
	0.1–0.25 mm (g)	55	55	55	55	55
Latex ECO (g)		36	36	36	36	36
Water reducing agent (g)		1.5	1.5	1.5	1.5	1.5
Water (g)		130	130	130	130	130
PVA (g)		0	0.5	1	1.5	2

Table 5. Properties of the optimal design.

Component mass (g)	Cement	Mineral powder	Latex ECO	
	200	57	36	
	Aggregates			
	0.1–0.25 mm	0.25–0.5 mm	0.5–1 mm	1–2 mm
	55	40	31	79
	Water	PVA	Water reducing agent	
	130	1	1.5	
Dry density (kg/m^3)	Wet density (kg/m^3)	28-day compressive strength (N/mm^2)	28-day tensile strength (N/mm^2)	
650	739	10.3	1.0	

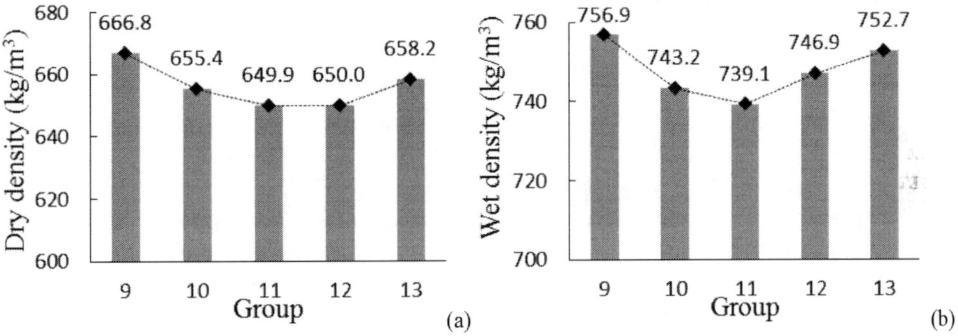

Figure 4. (a) Dry density in oven-dried condition, (b) wet density in ambient condition of five mixtures.

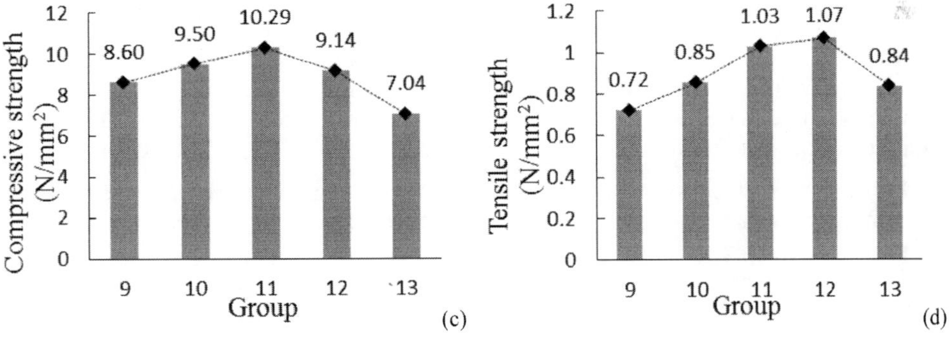

Figure 4. (c) 28-day compressive strength, (d) 28-day tensile strength of five mixtures.

tensile strength actually enlarge with decay of density. In fact, even though PVC fiber gives rise to more voids, it can bridge cracks and retard the development of cracks, which is of superior significance. Besides, from picture 4c,d, PVC fiber imposes a more positive effect on tensile strength than on compressive strength and the phenomena results from the excellent tenacity of the material (Table 1). Eventually, the properties of the optimal design (Mix 11) are concluded in Table 5.

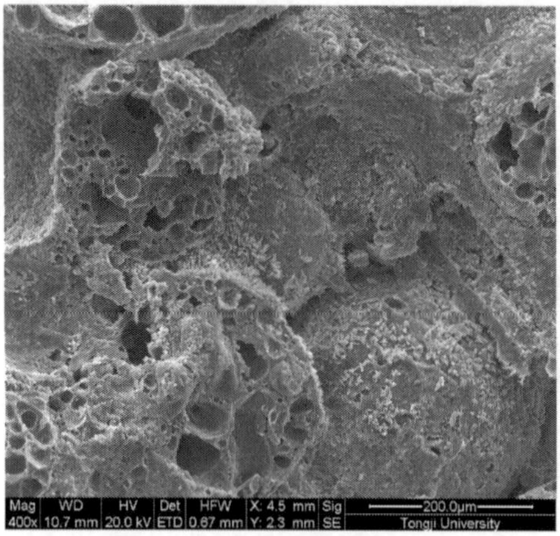

Figure 5a. SEM view of aggregate allocation.

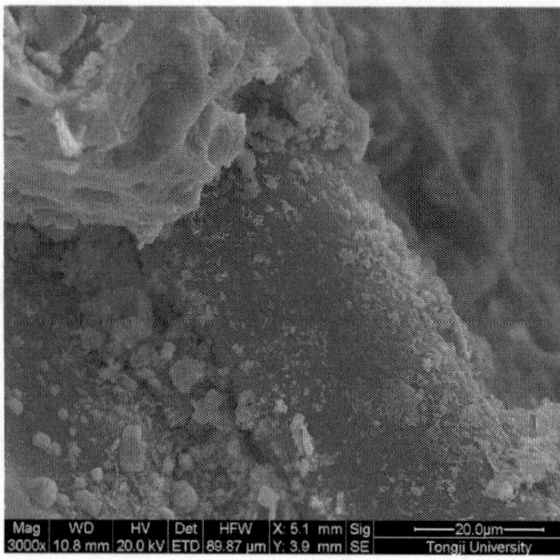

Figure 5b. SEM view of PVC fiber reinforcement.

5 SEM ANALYSIS

By applying Scanning Electron Microscope (SEM), the microscopic view of the optimal design (Table 5) is presented and typical pictures are expressed in Figure 5a,b. As Figure 5a presents, there are tremendous pores within structure, which decrease the density to a large scale. Besides, the combination of various sized aggregates is compact and the coverage is enhanced by cementitious hydration products. In figure 5b, the convex object is PVA fiber and it is embraced by binding materials. The fiber works as adhesive to suppress the occurrence of cracks and fortify the strength of overall structure.

6 CONCLUSION

This paper adopts recyclable Poraver as lightweight aggregates and applies Fuller curve and control experiments to probe for the optimal mixture. More, Polyvinyl Alcohol (PVA) fiber is employed to fortify concrete strength. Besides, the Scanning Electron Microscope (SEM) analysis validates the effectiveness of the design from microscopic perspective. The conclusions are summarized as followings

1. Fuller curve is efficacious in designing aggregate distribution and controlled experiments demonstrates that the design group (Mix 9) attains comparatively favorable mechanical properties and low density.
2. Granules of large size play a decisive role in the overall density. Besides, the introducing of PVC fiber carries bubbles and pores at same time, thus resulting in decay of density.
3. Generally speaking, the density of concrete positively correlates to its strength. However, when introducing PVC fiber, the density dwindles due to bubbles and voids resulted from fiber while the strength augment because PVC fiber is capable of bridging cracks and retarding the development of cracks to a dominating scale.
4. With a wet density of merely 650 kg/m^3, the optimal design attains the 28-day compressive strength of 10.3 N/mm^2 and tensile strength of 1.0 N/mm^2.

REFERENCES

ACI 318M-05. 1995. Building code requirements for reinforced concrete. Detroit, Michigan: American Concrete Institute.

ASTM C39. Standard Test Method for Compressive Strength of Cylindrical Concrete Specimens. ASTM Committee. C39/C39M-15a, West Conshohocken, PA.

ASTM C138. 2010. Standard Test Method for Density (Unit Weight), Yield, and Air Content (Gravimetric) of Concrete. C496/C496M-11. West Conshohocken, PA.

ASTM C496. Standard Test Method for Splitting Tensile Strength of Cylindrical Concrete Specimens. C496/C496M-11, West Conshohocken, PA.

Choi, J. & Zi, G. & Hino, S. & Yamaguchi, K. & Kim, S. 2014. Influence of fiber reinforcement on strength and toughness of all-lightweight concrete. *Constructions & building materials* 63: 132–141.

Fuller, W.B. & Thompson. S.E., 1907, Transactions of the American Society of Civil Engineers. *The Laws of proportioning concrete*: 67–143.

Hegyi, P. & Dunai, L. 2016. Experimental study on ultra-lightweight-concrete encased cold-formed steel structures Part I: Stability behaviour of elements subjected to bending. *Thin-wall structures* 101: 75–84.

Hegyi, P. & Dunai, L. 2016. Experimental investigations on ultra-lightweight-concrete encased cold-formed steel structures Part II: Stability behaviour of elements subjected to compression. *Thin-wall structures* 101: 100–108.

Kidalova, L. & Stevulova, N. & Terpakova, E. & Sicakova, A. 2012. Utilization of alternative materials in lightweight composites. *Cleaning production* 34: 116–119.

Markovic, I. 2006. High-performance Hybrid-fibre Concrete e Development and Utilisation. *Ph.D Thisis*. Delft: Delft University Press.

Wang, H.Y. & Tsai K.C. 2006. Engineering properties of lightweight aggregate concrete made from dredged silt. *Cement & concrete composites* 28(5): 481–485.

Yan, J.B. & Wang, J.Y. & Liew, J.Y.R. & Qian, X.D. & Zhang, W. 2016. Reinforced ultra-lightweight cement composite flat slabs: Experiments and analysis. *Materials & design* 95: 148–158.

Yu, Q.L. & Spiesz, P. & Brouwers, H.J.H. 2015. Ultra-lightweight concrete: Conceptual design and performance evaluation. *Cement & concrete composites* 61: 18–28.

Yu, R. & Onna D.V.V. & Spiesz, P. & Yu, Q.L. & Brouwers, H.J.H. 2015. Development of Ultra-Lightweight Fibre Reinforced Concrete applying expanded waste glass. *Journal of cleaner production* 112: 690–701.

Water science and environmental engineering

Advanced Engineering and Technology III – Xie (Ed.)
© 2017 Taylor & Francis Group, London, ISBN 978-1-138-03275-0

Effects of artificial water supplement on an ancient channel of Yellow River Delta

J. Liu, L. Guo, Q.W. Bu, L. Lin & H.J. Xin
Key Laboratory of Water Resources and Environment, Water Research Institute of Shandong Province, Jinan, Shandong, China

Q.X. Li
East Route Project of South to North Water Diversion, Shandong Limited Liability Company, Jinan, Shandong, China

ABSTRACT: Diaokouhe River is one of the ancient channels of Yellow River Delta. Because of long-term no-discharge, the channel shape and ecosystem degenerate seriously. For saving and protecting the ecology and the environment of the channel, the governmental management department carried out artificial water supplement for the channel in 2010 and 2011. During the supplement period, some important index and data were observed and recorded systematically. The results indicate that the artificial water supplement had positive effects on the local ecology and the environment, such as preventing the saltwater intrusion trend in the degenerative estuarine wetland, controlling the rapid development of soil salinization, optimizing the ecological types and landscape structure, and improving and repairing the wetland biotope and bird habitats. However, some problems also exist, such as difficulty in transferring large amounts of water and huge accumulation of channel sediments.

1 INTRODUCTION

It is well known that water is likely to be the majorenvironmental concern of the 21st century (Diao et al. 2009). The Yellow River, called as Mother Water of China, is the second longest river in China with a length of 5464 km, and it is also noted for its small water discharge and huge sediment load (Yang et al. 2004, Chen et al. 2015, Milliman et al. 1987). Because of the uncoordinated relationship between water and sediment, the river channel continues to be silted, especially in the downstream. Once sedimentation has progressed to a certain level, the estuary is prone to change (Xue 1993, Li et al. 1998).

According to historical records, the Yellow River changed its channel into the sea for more than 26 times and has formed several flow paths in the past 2600 years (Yang et al. 1999, Wang et al. 2006). In January 1964, the Yellow River flowed into the Bohai Sea through the Diaokouhe deltaic channel instead of the Shenxiangou deltaic channel, and the ninth channel in the Yellow River Delta was formed. The Diaokouhe channel with a total length of 55 km locates at the north of the present Yellow River northern levee. In May 1976, the Diaokouhe channel was replaced by the Qingshuigou channel (Chang et al. 2004, Pang et al. 2005). From 1964 to now, the Diaokouhe channel flowed for more than 12 years and stopped running water for more than 30 years. The locations of rivers and deltaic channels in the Yellow River Delta are shown in Figure 1.

During the period of no-discharge, great changes, such as channel cross-section shape and ecological environment, occurred because of the influences of water silt condition, marine dynamic interactions, and human activities in the Diaokouhe channel (Chen et al. 1996).

1. Without the supplement of sediments, the coastline along the Diaokouhe channel began to erode, and the maximum erosion length was more than 10 km.
2. Because of the lack of fresh water, the surrounding wetland began to shrink and its biodiversity developed by water–sediment resources was destroyed.

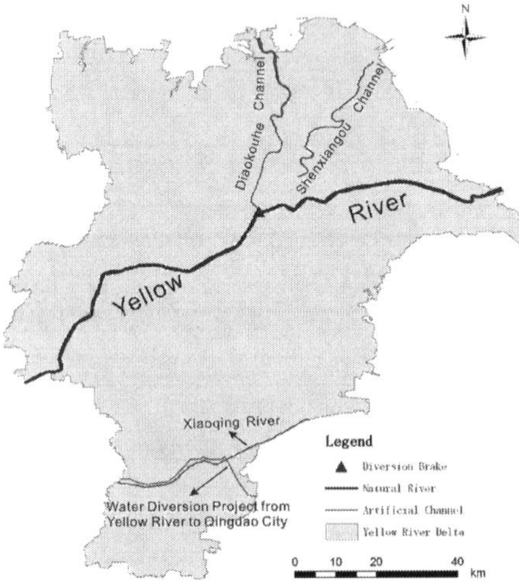

Figure 1. Locations of rivers and deltaic channels in the Yellow River Delta.

3. The channel began to atrophy seriously and the discharge capacity was reduced greatly.
4. Due to the impacts of human activities, wind and rain erosion, and lack of management, the embankment engineering was incomplete along the Diaokouhe channel (Zheng et al. 2012).

According to the Comprehensive Control Planning of Yellow River Estuary, the Diaokouhe channel is considered as an important backup option for the current Qingshuigou channel (Yellow River Water Resources Commission 2009). For improving the regional ecological environment of the Diaokouhe channel and effectively protecting the alternate channel of Yellow River, artificial water supplement experiments were performed in 2010 and 2011, respectively. The results obtained indicate that the water supplement achieved good effects, but some problems also occur. Based on these two artificial experiments, the effects of water supplement on the ancient deltaic channel and ecological environment of Yellow River were analyzed selectively in this paper.

2 COURSE OF ARTIFICIAL WATER SUPPLEMENT

The artificial water supplement experiments were carried out on the Yellow River water and sediment regulation during the period from June to August, and the water was supplied by gravity and pumping ship.

In 2010, the total supplied water was $3.62 \times 10^6\,m^3$. The supplied water stored in the channel, flowed to the nature reserve area, and discharged into the Bohai sea was $25.42 \times 10^6\,m^3$, $8.05 \times 10^6\,m^3$, and $2.73 \times 10^6\,m^3$, which accounted for about 70.2%, 22.2%, and 7.6% of the total supplied water, respectively.

In 2011, the total supplied water was $3.618 \times 10^6\,m^3$, similar to that recorded in 2010. The supplied water stored in the channel, flowed to the nature reserve area and discharged into the sea accounted for about 83.4%, 12.3%, and 4.3% of the total supplied water, respectively.

In these two water supplements, water coverage areas were about 2.333×10^3 hectares in the channel and 1.333×10^3 hectares in the nature reserve.

3 ANALYSIS OF ECO-ENVIRONMENTAL EFFECTS

During the course of the two water supplements, the regional eco-environmental situations were monitored dynamically. The results indicate that supplied water plays positive roles for repairing and improving the local ecosystem and the ecological environment. The positive influences are as follows:

1. The saltwater intrusion trend in the degenerative estuarine wetland was prevented preliminarily. The results observed indicated that the groundwater table rose in the distance range of 1.1 km along the channel, and rose significantly in the range of 0.6 km. The maximum uplift amplitude was about 0.65 m. In the nature reserve, the influence scope of water supplement to groundwater was about 1.5 km. The groundwater table rose significantly in the distance of 0.5 km to the nature reserve area, and the maximum uplift amplitude was 0.45 m.
2. The rapid development of soil salinization was controlled. Soil salinity decreased obviously along the Diaokouhe River. Salinity in the 0–0.3 m soil layer decreased significantly, and the average change in percentage was about 48%. The average decreased percentages were 55% in the 0.1 m layer and 41% in the 0.3 m layer. This laid the foundation for the growth and development of aquatic and wet vegetation in freshwater wetland.
3. The ecological types and landscape structure were optimized in the water supplement area, and the retrogressive succession trend of regional vegetation was controlled preliminarily. The saliferous bare land in the low-lying area was covered by water. The low annual and perennial saline vegetation evolved into a reed marsh ecosystem type, and the tamarix community and hygropium in the high terrain also had the trend of evolving into the reed community. During the course of water supplement and the soil desalting process, reed became the predominant community that adapted to the ecological environment, and the main vegetation habitat of aquatic birds was formed preliminarily.
4. The wetland biotope was improved by the change in bird habitat vegetation, and the habitat of large wader birds was repaired obviously. Aquatic bird populations increased significantly. In 2010 and 2011, the number of waterfowls increased by 11180 and 36798, respectively. Particularly, the effect of rare and endangered bird protection is significant. The number of red-crowned cranes, one of the national key protected birds, increased by 11 and 19, respectively. The number of oriental white storks increased by 18 and 149, and that of black storks increased by 14 and 4, respectively. In 2011, cranes were found with the population size of 14 for the first time, and oriental white storks migrated with a large population size of 152.

4 PROBLEMS DURING THE WATER-SUPPLEMENT

The ecological water supplement proceeded smoothly in 2010 and 2011. According to the measured data during the process, some problems occurred.

1. The water was transferred through the Luojiawuzi diversion brake of Yellow River. Because of high floor elevation, gravity flow could not be achieved when the Yellow River was at a low water level. The time and quantity of water supplement was under restriction. If water was supplied by the pumping station, the operating cost increased. At the same time, obstacles in the channel were not completely cleared. It could only cope with the experimental water supplement, but could not satisfy the long-term need of huge amounts of water transfer.
2. Because of the control of reed and weed, all the sediments were deposited in the channel. The total amount of sediments transferred in the water supplement was more than 5 × 105 tons in 2010 and 2011, and the vast majority of it was deposited 20 km upstream of the channel. The deposition of sediments will cause an adverse effect on the water supplement in the future.
3. Influenced by the high elevation of natural reserve, more fresh water could not be flowed into the core area when the supplied water is small. With the limitation of Yellow River flow and diversion conditions, large capacity of water transfer was unfeasible. So, there was difficulty in transferring water to the core area of the natural reserve.

5 CONCLUSION AND DISCUSSION

According to the research results, water supplement to an ancient estuary channel of Yellow River had promotive effects on protecting and fixing the regional ecological environment. It basically succeeded in: (1) preventing the saltwater intrusion trend in the degenerative estuarine wetland, (2) controlling the rapid development of soil salinization, (3) optimizing the ecological types and landscape structure and controlling retrogressive succession trend of regional vegetation preliminarily, and (4) improving and repairing the wetland biotope and bird habitat obviously. However, it was also prone to some problems, such as restriction of transfer time and quantity for gravity flow, sediment accumulation in the river channel, and difficulty in flowing into the core area of the natural reserve with small water flow rates.

As the succession and development of the ecosystem, it will be a long-term process for the restoration of degraded wetland. The ecological effects of artificial water transfer in the ancient estuary channel of Yellow River Delta will appear because of long-term implementation of water transfer and delay of ecosystem succession. So, the government should establish a long-term working mechanism for artificial water supplement, and carry out scientific investigations, monitoring and research for the transfer process and effect. This will be crucial for understanding the changes in the Diaokouhe deltaic channel, optimizing water allocation and maintaining the virtuous cycle of the ecosystem in the ancient deltaic channel.

ACKNOWLEDGEMENTS

This work was financially supported by the Natural Science Foundation of Shandong Province "Research on variation characteristics and prediction of agricultural drought in Shandong Province under climate change" (ZR2015EQ004) and the National Natural Science Foundation of China "Research on transformation mechanism and prediction of different types of drought under climate change—a case study in Shandong Province" (41501053).

REFERENCES

Chang, J., Liu, G.H., &; Liu, Q.S. 2004. Dynamic monitoring of coastline in the Yellow River Delta by remote sensing. *Geo-information Science* 6(1): 94–98. (in Chinese).

Chen, L.D., & Fu, B.J. 1996. Analysis of Impact of Human Activity on landscape Structure in Yellow River—a Case Study of Dongying Region. *Acta Ecologica Sinica.* 16(4): 337–344.

Chen, Q.F., Ma, J.j., Zhao, C.S., & Li, R.B. 2015. The Spatial and Temporal Variation Characteristics of CH_4 and CO_2 Emission Flux under Different Land Use Types in the Yellow River Delta Wetland. *Journal of Geoscience and Environment Protection* 3: 26–32.

Diao, X.D., Zeng, S.X., & Wu, H.y. 2009. Evaluating Economic Benefits of Water Diversion Project for Environment Improvement: A Case Study. *Journal of Water Resource and Protection* 1: 1–57.

Li, G., Wei, H., Han, Y., & Chen, Y. 1998. Sedimentation in the Yellow River delta, part I: flow and suspended sediment structure in the upper distributary and the estuary. *Marine Geology* 149(1): 93–111.

Milliman, J.D., Yun-Shan, Q., Mei-e, R., & Saito, Y. 1987. Man's influence on the erosion and transport of sediment by Asian rivers: the Yellow River (Huanghe) example. *The Journal of Geology* 95(6): 751–762.

Pang, J.Z., & Jiang, M.X. 2005. On the Evolution of the Yellow River Estuary (Part II), *Transactions of Oceanology and Limnology* (4): 1–13 (in Chinese).

Wang, S.J., Hassan, M.A., & Xie, X. 2006. Relationship between suspended sediment load, channel geometry and land area increment in the Yellow River Delta. *Catena* 65(3): 302–314.

Xue, C.T. 1993. Historical changes in the Yellow River delta, China. *Marine Geology* 113(3–4): 321–329.

Yang, D.W., Li, C., Hu, H.P., Lei, Z.D., Yang, S.X., Kusuda, T., & Musiake, K. 2004. Analysis of water resources variability in the Yellow River of China during the last half century using historical data. *Water Resources Research* 40(6): W065021–W0650212.

Yang, X.J., Damen, M.C., & Van Zuidam, R.A. 1999. Satellite remote sensing and GIS for the analysis of channel migration changes in the active Yellow River Delta, China, *International Journal of Applied Earth Observation and Geoinformation* 1(2): 146–157.

Yellow River Water Resources Commission. 2009. *Comprehensive Control Planning of Yellow River Estuary*. Jinan: Taishan Publishing House (in Chinese).

Zheng, K., Kai, L.R., & Zheng, Y.C. 2012. Statistic Analysis on Eco-environmental Water Requirements in Diaokou Channel of the Yellow River Delta. *Ground Water* 34(5): 92–94.

Advanced Engineering and Technology III – Xie (Ed.)
 ISBN 978-1-138-03275-0

Policies for protecting and exploiting karst springs in Spring City, Jinan, China

J. Liu
Key Laboratory of Water Resources and Environment, Water Research Institute of Shandong Province, Jinan City, Shandong Province, China

G.Q. Sang
School of Resources and Environment, Jinan University, Jinan City, Shandong Province, China

L. Guo, L. Lin & Q.W. Bu
Key Laboratory of Water Resources and Environment, Water Research Institute of Shandong Province, Jinan City, Shandong Province, China

ABSTRACT: Jinan is famous as "Spring City". Because of unconscionable exploitation of groundwater in the last century, a series of problems appear such as groundwater recession, and stopping spewing of springs, causing huge damage to the regional society and economy. As an unavoidable responsibility for water resources protection, the Jinan Municipal Government (JMG) stipulated and adopted a number of comprehensive policies to protect the springs. In this paper, the governmental policy measures are summarized and reviewed systematically including setting up special management department for spring protection, making related planning, strengthening administrative management of water withdrawal, making reasonable water price, perfecting spring protection laws, and strengthening public opinion and propaganda. All these measures have effectively constituted the governmental management system of spring protection in Jinan, and facts proved that the measures have played a positive and effective role in spring protection.

1 INTRODUCTION

As an important part of water resources, karst water means the groundwater that deposits and migrates in the karst medium or karst aquifer. Because of rich quantity and good quality, karst water is treated as the main water resources of many cities in the world (Zou et al. 2004, Jemcov 2007). In the centralized drainage area of karst water, some springs with large flow are almost exposed, which may become famous scenic sights.

Jinan is the provincial capital of Shandong Province, which is one of provinces with strongest economy and largest population in China (Figure 1). Jinan used to be famous for its karst springs in the world, which is well known as the "Spring City" (Qian et al. 2006). However, it is also a typical city with water resource scarcity. The annual average water resource is about 1.748 billion m^3, and the water resource per capita share of 290 cubic meters, less than one-seventh of the national average. With the rapid development of city construction since the beginning of the 1970s, agricultural and industrial water consumption in Jinan increased sharply. According to statistical data, karst groundwater exploitations were 60000 and 112000 m^3/d in the 1950s and 1960s. In the 1970s, the exploitation increased to 273000 m^3/d, and declined to 177000 and 120000 m^3/d in the 1980s and 1990s, respectively (Wang et al. 2008). Because of excess exploitation in the spring area, the groundwater table in Jinan decreased dramatically, and the local groundwater depression zone developed with increasing ranges. Especially, in the downtown area, groundwater excess exploitation had direct effects on the spring area, which made the spring water table to fall below the spring discharge elevation.

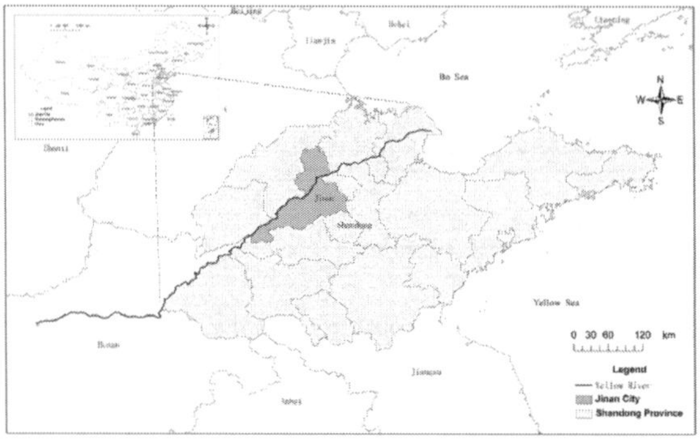

Figure 1. Location of Jinan in Shandong Province and China.

The springs began to stop spewing from spring in 1972, and different degrees of stop-spewing occurred in each successive year. The longest stop-spewing time even reached up to 926 days (from March 2, 1999 to September 17, 2001), causing huge damage to the ecological environment and tourism (Liu & Zhao 1997).

Effective management is an important foundation for guaranteeing the sustainable utilization of water resources, and the government is the policy maker and executor undoubtedly (Jin & Young 2001, Rogers et al. 2002, Loucks et al. 2005, Luzi 2010). For restoring and protecting spring-spewing as soon as possible, the Jinan Municipal Government (JMG) stipulated and adopted a number of comprehensive measures including system, planning, technology, economy, law, and propaganda. Based on these measures, a scientific and long-term mechanism for protecting springs was established. The balance of groundwater exploitation and recharge was achieved gradually. Spouting Spring, one of the most famous springs in Jinan, has kept spewing for more than 12 years until now. In this paper, the management measures for protecting spring stipulated by the JMG were summarized systematically. We hope that these measures can have important values of reference for other nations facing the same problem like Jinan.

2 SETTING UP SPECIAL MANAGEMENT DEPARTMENT

2.1 *Setting up spring management department*

The JMG is solely responsible for spring protection. In the past, the specific work for protecting springs was confined to groundwater monitoring, which were carried out by the Water Conservancy Bureau of Jinan City (WCBJC), an administrative department of the JMG. In order to carry out spring protection comprehensively, three new departments covering Jinan Famous Springs Protection Office (JFSPO), Jinan Urban Water Planning and Saving Office (JUWPSO) and Jinan Southern Mountainous Area Management Office (JSMAMO) were established. The three offices are under the direct leadership of the JMG, and have their own specific tasks. The JFSPO is responsible for organizing and coordinating spring protection. The JUWPSO is responsible for carrying out urban water conservation, and making water use plans and water saving policy. The JSMAMO is responsible for protecting and developing the southern mountain area.

2.2 *Reforming water supply department*

Before reforming the water supply system, urban water supply in Jinan adopted a through-train mode, which means that all the processes including water source, water treating, and water transmission were managed by the former Jinan Water Supply Group (JWSG). In this mode, the JWSG had economic losses for a long time and tended to use groundwater with lower cost. Because the JWSG was in charge of groundwater, the WCBJC was not conducive to groundwater monitoring and control.

In 2005, the JMG reformed the water supply enterprise system, and three new companies including Jinan Qingyuan Water Affairs Group Co. Ltd. (JQWAG), Jinan Hengquan Water Treating Co. Ltd. (JHQT), and Jinan Water Group Co. Ltd. (JWG) were set up separately. The JQWAG is responsible for storing, managing, and supplying urban water sources. The JHQT is responsible for treating urban water. The JWG is responsible for supplying urban tap water.

3 MAKING RELATED PLANNING

3.1 *Strategic planning for famous spring protection*

In 2004, the JFSPO compiled the Strategic Planning for Famous Springs Protection of Jinan City. According to the planning, the springs were divided into absolute protected area and relative protected area. The most famous four spring groups including Spouting Spring, Pearl Spring, Five-dragon Pools, and Black Tiger Spring were involved in the absolute protected area. In this area, some of the behavior is banned explicitly. For example, deep underground engineering below 10 meters is forbidden strictly, and any building preventing spring scenery is forbidden (Jinan Famous Springs Protection Office, unpubl).

3.2 *Discrimination of groundwater functions*

In January 2009, the WCBJC compiled the Discrimination of Groundwater Functions in Jinan City. According to the relatedness of hydrogeological conditions of different areas and spring groups, karst groundwater functions were discriminated into two levels scientifically, which are primary function zone and secondary function district. The primary function zones were divided into three types including reservation zone, protection zone, and development zone, which mainly coordinated the relationship between water utilization for economic and social development and environmental protection (Water Conservancy Bureau of Jinan City, unpubl).

3.3 *Protection and development planning for the southern mountain area*

As the most important recharge area of Jinan springs, the southern mountain area was approved for constructing a special ecological function reserve. In 2010, the protection planning for southern mountain area of Jinan City was approved and implemented. In the planning, key leakage region and source conservation region of groundwater were identified as the construction-forbidden region and the construction-restricted region (Jinan Municipal Government, unpubl).

4 STRENGTHENING ADMINISTRATIVE MANAGEMENT OF WATER WITHDRAWAL

In order to have regulations to comply with the management of water withdrawal licensing, the WCBJC made a series of rules. In daily administration, any new, rebuilt, and extended construction projects must handle water withdrawal licensing formalities, and the WCBJC is in charge of the work. Any water withdrawal licensing that increase groundwater use in the spring area must be restricted strictly. If water withdrawal exceeded the total control index, no application will be approved. At the same time, in the coverage of urban public-supplied water network, any new self-served water well is forbidden to dig, and the existing wells must be designed to seal.

In the approval of water withdrawal licensing, demonstration report of water resource for construction projects is the most important supporting documents. The report must be written by a qualified research institute, and its content should be complete including rationality analysis of water withdrawal, demonstration of reliable water resources, impact analysis of water intaking and recession on the third part, and compensating measures.

If the water withdrawal licensing is approved, project construction should be supervised and inspected by technical staff from the WCBJC. After completing the construction, the project must be checked and accepted by the WCBJC to ensure that water use facilities and process conform to the standard, and no groundwater is over-pumped or stolen.

Besides the strict approval of water withdrawal, the planned water intaking was also carried out strictly in Jinan.

5 APPLYING ECONOMIC MEASURES

To make the best use of the economic leverage function of price, Jinan extensively raised the fees of the water resource due to the practical situation of water resource shortage and previous lower price. In 2001, the water resource fee of self-served wells was 1.2 CNY/m^3, in 2003, the number was raised to 1.8 CNY/m^3, while the public-supplied water resource fee was 0.25 CNY/m^3. In 2005, the JMG increased the fee standard of the water resource again, and split the self-served wells into different forms, namely domestic wells and industrial wells. The water resource fee of domestic wells was adjusted to 2.1 CNY/m^3 and that of the industrial wells to 2.6 CNY/m^3, while the public-supplied water resource fee was raised to 0.4 CNY/m^3.

To protect springs and confining groundwater exploitation, the JMG launched a phased multiple-charge of water resource fee campaign to the public-supplied water network-owned companies which have been unable to shut their self-supplied well simultaneously (excluding domestic water) under the authorization of the Shandong Price Bureau and Financial Department of Shandong Province since January 2009. The specific standards are as follows: taking the self-supplied well groundwater fee of 2.60 CNY/m^3 as the standard, the fee was raised 0.5 times to 3.90 CNY/m^3 from April 1, 2009 and raised 2 times to 5.20 CNY/m^3 from September 1, 2009. Taking advantages of economic leverage, self-supplied well owned units were stimulated to replace groundwater into tap water or surface water. The campaign adjusted the water resource fees, which not only supplied the financial supports for South-North Water Transfer Project (SNWTP) and water resource management, but also strengthened water saving awareness of residents and companies. At the same time, the water use efficiency was improved greatly, and exploitation of groundwater was decreased significantly.

6 PERFECTING LAWS OF SPRING PROTECTION

In July, 1997, the JMG carried out famous spring protection administration measures of Jinan which is the only famous spring protection law throughout the country. It led Jinan famous spring protection to the way of protecting by law. This law included 26 stipulations and the directory of ancient springs, which explained in detail the activities related to destroying the spring field and stipulated the punitive measures to them (Jinan Municipal Government, unpubl).

In September, 29th, 2005, famous spring protection regulations of Jinan were carried out. It was revised and extended on the basic of famous spring protection administration measures of Jinan, which was applied in 1997. One of its most outstanding features is extending its range of application. The protection of spring recharge area and spring source was included in its managing scope (Jinan Municipal Government, unpubl).

In April, 2006, emergency plan was carried out to keep spring spewing of Jinan. The emergency plans were divided into three levels. Taking the Spouting as an example, its flowing level was 26.80 meters. When the groundwater table of Spouting Spring decreased to 28.15 meters, the yellow alert was announced and the emergency plans of the third level were activated. When the groundwater table of Spouting Spring decreased to 28.00 meters, the orange alert was announced and the emergency plans of the second level were activated. When the groundwater table of Spouting Spring decreased to 27.60 meters, the red alert was announced and the emergency plans of the first level were activated (Jinan Municipal Government, unpubl).

7 STRENGTHENING PUBLIC OPINION AND PROPAGANDA

Departments of Jinan such as the WCBJC and the JFSPO took advantage of a variety of channels, for example bulletin boards, brochures, and leaflets, to propagate spring protection knowledge and famous spring protection regulations. They conducted large photography exhibition called "spring water, spring emotion, spring rhyme", recruited retentions, and civilized languages of loving springs. They set up the website and announced the latest dynamic conditions of spring city, spring water, and spring culture throughout the world via the Internet. They convoked the forum of spring and economic development and invited experts to research particularly on protecting and utilizing Jinan springs. Through different kinds of propagating activities, the idea of protecting springs has become well known and deeply impressing, and the consciousness of legal system and responsibilities throughout the society has been strengthened.

8 CONCLUSION

For protecting and exploiting the spring water of Jinan sustainably, the Jinan Municipal Government (JMG) adopted a series of measures including system, planning, technology, economy, law, and propaganda. In this paper, all the measures were summarized fully and reviewed systematically. The measures included setting up special management department for spring protection, making related planning, strengthening administrative management of water withdrawal, making reasonable water price, perfecting spring protection laws, and strengthening public opinion and propaganda. The results and facts indicated that the adopted measures played positive roles in protecting the Jinan.

ACKNOWLEDGMENTS

This work was financially supported by the water resource public-spirited project "Research and Demonstration of Key Technology for Ecological Civilization Construction" (201401003), the Science and Technology Promotion Program "Integration and Demonstration of Key Technology for Ecological Civilization Construction in Jinan" (TG1407), the "948" project "Importation of in-situ online observation technology in shallow groundwater overdraft polluted area" (201319), and the Natural Science Foundation of China "the study for the characteristics of the spatial distribution structure for the fracture network within the rock mass based on the electrical resistivity tomography" (41202174).

REFERENCES

Jemcov, I. 2007. Water supply potential and optimal exploitation capacity of karst aquifer systems. *Environmental geology 51*(5): 767–773.

Jin, L. & Young, W. 2001. Water use in agriculture in China: importance, challenges, and implications for policy. *Water policy 3*(3): 215–228.

Jinan Famous Springs Protection Office. 2004. *Strategic Planning for Famous Springs Protection of Jinan City*. Jinan: unpublished (in Chinese).

Jinan Municipal Government. 1997. *Famous spring protection administration measures of Jinan*. Available at: http://xxgk.jinan.gov.cn/xxgk/jcms_files/jcms1/web1/site/art/2012/10/10/art_8_41679.html (March 16, 2015. in Chinese).

Jinan Municipal Government. 2005. *Famous spring protection regulations of Jinan.* Available at: http://www.jinan.gov.cn/art/2005/10/20/art_679_11877.html (March 17, 2015. in Chinese).
Jinan Municipal Government. 2006. *Emergency plan to keep spring spewing of Jinan.* Available at: http://www.jinan.gov.cn/art/2006/6/12/art_702_12054.html (March 17, 2015. in Chinese).
Jinan Municipal Government. 2010. *Protection and development planning for southern mountain area.* Jinan: unpublished (in Chinese).
Liu, G.A. & Zhao, X.H. 1997. Dynamic characteristics of karst water in Jinan spring field and discussion on some questions related. *Geol Shandong 13*(2): 44–47.
Loucks, D.P., Van, B.E., Stedinger, J.R., Dijkman, J.P. & Villars, M.T. 2005. *Water resources systems planning and management: an introduction to methods, models and applications.* Paris: UNESCO.
Luzi, S. 2010. Driving forces and patterns of water policy making in Egypt. *Water Policy 12*(1): 92–113.
Qian, J., Zhan, H., Wu, Y., Li, F. & Wang, J. 2006. Fractured-karst spring-flow protections: a case study in Jinan, China. *Hydrogeology journal 14*(7): 1192–1205.
Rogers, P., De Silva, R., & Bhatia, R. 2002. Water is an economic good: How to use prices to promote equity, efficiency, and sustainability. *Water policy 4*(1): 1–17.
Wang, M.M., Shu, L.C., Ji, Y.F., Tao, Y.F., Dong, G.M. & Liu, L. 2008. Causes of spring's of flux attenuation and simulation of spring's regime—A case in Jinan karst spring area. *Carsologica Sinica 1*, 004.
Water Conservancy Bureau of Jinan City. 2009. *Discrimination of Groundwater Functions in Jinan City.* Jinan: unpublished (in Chinese).
Zou, S.Z., Liang, B., Zhu, Z.W. & Liang, X.P. 2004. Effect of ecosystem on water resource in karst area—A case study in West Hunan. *Resources and Environment in the Yangtze Basin 13*(6): 599–603.

Advanced Engineering and Technology III – Xie (Ed.)
© 2017 Taylor & Francis Group, London, ISBN 978-1-138-03275-0

Daily Load Model and a novel algorithm for generating flow based on the DLM

Xiangming Tao & Senlin Chen
State Key Laboratory of Water Resources and Hydropower Engineering Science, Wuhan University, Hubei, China

ABSTRACT: With the rapid development of hydropower, the proportion of each hydropower in the power grid has decreased. In addition, the fluctuation in the daily load process has also declined, represented by a much flatter curve in the daily load graph. This paper takes advantage of this change, and proposes the concept of Daily Load Model (DLM), focusing on illustrative characteristics of the daily load process to keep the load process in a stable state. The framework of the model includes the procedure of the model's establishment, the determination of the model's parameters in space-time dimension and the transformation from the DLM to a realistic daily load process. Furthermore, inspired by the model, a novel idea for simplifying generating flow calculation is proposed. The proposed model and simplified algorithm called the period-combined algorithm are successfully applied to a realistic case study and the traditional period-by-period algorithm is used to validate the performance of simplified algorithms based on the DLM. Compared with the classic period-by-period calculating method, the proposed model promotes the efficiency remarkably.

1 INTRODUCTION

In order to meet the needs of the power grid system (especially peak-shaving requirement) and sometimes the great fluctuation of reservoir inflow, the hydropower has to adjust its load of sub-periods frequently. Therefore, the changing rules of the daily load process have significant importance which can be defined as the fundamental scientific subject of short-term optimal operation of hydropower. Some experts have conducted a number of relevant researches previously. Particularly, the Progress Optimality Algorithm (POA) and Genetic Algorithm (GA) are applied to unit commitment problems under a changing daily load process in (Xu, 2002) (He, 2006). A classified load process model adapted to the characteristics of the power grid system and the ability of the hydropower's regulation and generation has been proposed (Cheng, 2011), aimed at the prominent peak-shaving problem. (Tang, 2008) (Ji, 2011) (Huang, 2005) proposed the refined load process model for short-term operation. However, the model could only be used under the circumstances where the scope of peak shaving was already known. The reallocation of daily load was combined with mid-long term optimal operation in (Wu, 2012), which established a model of maximum peak shaving volumes. It can be seen that the changing principles of the daily load process itself are ignored in all of the aforementioned studies.

In this paper, we focus on the characteristics of the daily load process, the principles of the process's change, and its application in hydropower short-term operation. The concept of the Daily Load Model (DLM) is proposed, in which subtle distinctions among the load of successive sub-period are ignored and the characteristics of daily load process are formulated concisely through a series of interactive parameters. The major advantages of the DLM can be discussed as follows: the enormous periods of the realistic daily load process are cut down to only a few combined periods via the generalization of the daily load process structure based on the DLM. (The parameter of DLM could be determined according to the optimal operation, see section 2.) Hence, this model will, to some certain reduce the amount of daily work of generating flow calculation to some extent, bringing benefit to the short-term optimal operation of hydropower.

2 BUILDING DLM

Fig. 1 shows the DLM of a certain hydropower. This model predigests the realistic load process, both timely and spatially. The daily load is constrained to a certain numerical value, particularly, base load P_j, medium load P_y, and peak load P_f. Similarly, a large number of periods are combined and only a few extended periods are left. As shown in Fig. 1, five extended periods are left after combination. Based on the DLM, its characteristics and properties are determined simply by two kinds of parameter: time parameter (t_1, t_2, $t_3 \ldots t_n$) and spatial parameter (P_j, P_y and P_f), representing the newly divided time intervals and the real-time capacity of generation of the objective hydropower, respectively. The whole DLM in Fig. 1 can be numerically formulated as follows:

$$P(t)=\begin{cases} P_f \; t\in\left[t_1,t_2\right]\cup\left[t_3,t_4\right] \\ P_y \; t\in\left[t_2,t_3\right] \\ P_j \; t\in\left[0,t_1\right]\cup\left[t_4,t_5\right] \end{cases} \tag{1}$$

2.1 *Analyzing the parameters of DLM*

2.1.1 *Temporal parameter*

Based on the investigation on the daily load process in the power system, we can summarize a typical day as follows: the time point where the load has changed obviously and the time region of the base load, of medium load and peak load. For example, the time point where the base load turns into the medium load first time in the daily load process often occurs around 7:00. The time intervals of the peak load during workdays are statistically around 8:00–12:00 and 14:00–21:00. All of these time feature points are also the split point of the new combined periods.

2.1.2 *Spatial parameter*

Daily electricity (E_{day}) can be calculated through the realistic daily load process. Once the time parameters are determined using the approach in section 2.1, there can be countless combinations of the three spatial parameters that can match the daily electricity needs. In order to confirm that the combination of spatial parameters costs the least generating flow through optimal operation, this paper introduces a variable called the peak-shaving volume (ΔN_{tf}) into the DLM, which represents the deviation between the peak load and the base load:

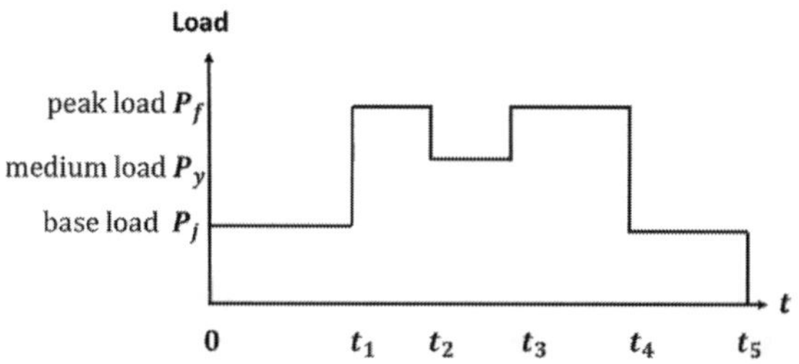

Figure 1. Schematic diagram of the DLM.

$$\begin{cases} P_f = P_j + \Delta N_{tf} \\ P_y = P_j + \lambda * \Delta N_{tf} \end{cases} \tag{2}$$

$$P_f * T_f + P_j * T_j + P_y * \left(T * \Delta t - T_f - T_j\right) = E_{day}$$

where λ is the coefficient.

Based on Eq. (2) and Eq. (3), we can obtain Eq. (4):

$$P_j = \frac{E_{day} - \Delta N_{tf}\left(\lambda * T_y + T_f\right)}{T * \Delta t} \tag{3}$$

where T_f, T_j are the time intervals of the peak load and base load, respectively. Based on the equation used for calculating the base load, the rest of the spatial parameters can be determined correspondingly. Therefore, each peak-shaving volume will correspond to a certain daily load process.

The peak-shaving volume is usually organized in discrete form so that the optimal combination of spatial parameters can be finally found by comparison. If the water level and inflow of the reservoir are already known, the generating flow of every single period under a certain discrete peak-shaving volume value can be calculated by Dynamic Programming (DP). So, the daily generating flow can be obtained through summation. The DP process is as follows:

Objective function: $F = \min \Sigma Q_j(N_j)$

1. Stage variable: the unit number $i = 1,2,3\ldots n$.
2. State variable: the accumulative load $\overline{N_i}$ from the first period to period i.
3. Decision variable: the load of unit i in the current period.
4. State transition equation:

$$\overline{N_t} = \overline{N_{t-1}} + N_i$$

5. Recursive equation:

$$Q_i^*\left(\overline{N_t}\right) = min\left\{Q_i\left(N_i\right) + Q_{i-1}^*\left(\overline{N_{t-1}}\right)\right\}; \quad \overline{N_{t-1}} = \overline{N_t} - N_i$$

where $Q_i^*(\overline{N_t})$ is the optimal generating flow when the total load of the hydropower is $\overline{N_t}$. $Q_i(N_i)$ is the generating flow when the load of unit i is N_i. (N refers to the load of a certain unit, while P refers to the load of the whole hydropower).

The peak-shaving volume value corresponding to the minimum generating flow cost will be chosen as the optimal indirect spatial parameter, leading to the determination of P_j, P_y, and P_f. Obviously, the daily load allocation is also optimized.

3 GENERATING FLOW CALCULATION

Since the periods with the equal load are recognized as one combined new period in the DLM, the generating flow will not be calculated separately in the traditional period-by-period way. The most obvious advance is that the new time interval of calculation based on the DLM will be a combined period. Meanwhile the other procedures will remain the same.

4 CASE STUDY

In this section, numerical results from a realistic case study are presented to illustrate the feasibility of the proposed model and the effectiveness of the algorithm. This case study is based on the Tianshengqiao reservoir whose annual average runoff is 19.3 billion m^3, the total storage of reservoir is 10.256 billion m^3, and regulating storage is 5.77 billion m^3. The hydropower has four generation units with a generation capacity of 300000 kW.

4.1 *Generalization of the daily load process*

Fig. 2 shows six types of DLM obtained by analyzing the reservoir operational data.

4.2 *Determination of parameters*

The inflow, water level, and daily electricity of the calculation are derived from the operational data of Dec 2nd, 2014 and the chosen type of daily load process belongs to (f), as shown in Fig. 2. The process is detailed in Table 1. In addition, the water level is of 753.9-meter height and the inflow of the reservoir was 261 m^3/s on that day.

4.2.1 *Temporal parameter*

Six new combined periods are obtained according to the daily load process presented in Table 1. These time intervals are 0:00–8:00, 8:00–12:00, 12:00–17:00, 17:00–22:00, 22:00–23:00 and 23:00–24:00. Obviously, the total amount of calculating periods is cut down from 24 to 6.

4.2.2 *Spatial parameter*

The daily electricity consumption is estimated as 8740 MWh (E_{day} = 8740 MWh) according to the aforementioned daily load process detailed in Table 1, which will be regarded as the already known conditions along with the water level and inflow of the reservoir. Then, the optimal combination of spatial parameters is determined using the DP method, mentioned in section 2.2, as follows: $P_j = 270\ MW$, $P_y = 362\ MW$ and $P_f = 460\ MW$. Fig. 3 shows the daily load process based on the DLM, presenting a smooth load process graph compared with the realistic load process that shows a zigzag form.

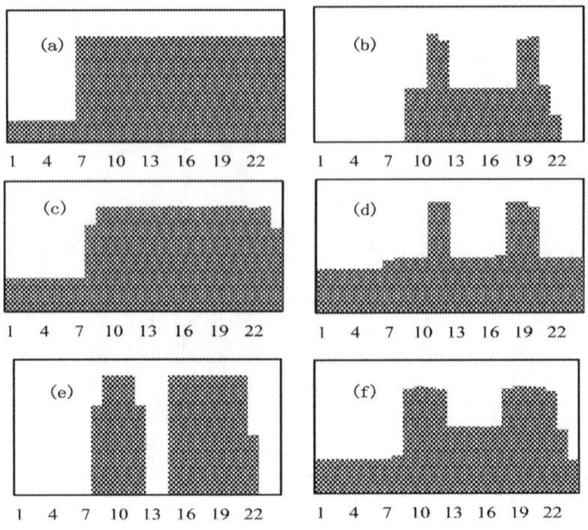

Figure 2. Different types of DLM.

Table 1. Tianshengqiao hydropower daily load process data of Dec 2nd 2014.

1	2	3	4	5	6	7	8	9	10	11	12
183.8	183.6	183.7	183.7	183.6	183.4	183.5	206	493.3	578.9	574.9	491.7
13	14	15	16	17	18	19	20	21	22	23	24
361.2	360.7	361.0	358.9	363.7	494.6	580.2	576.9	575.0	552.2	344.3	178.3

4.3 *Daily generating flow calculation*

Fig. 4 and Fig. 5 shown the comparison between the results of the traditional method and the DLM, indicating that the errors between these two methods can be negligible, which demonstrates the validity of the proposed method.

4.4 *Discussion with respect to the results*

The DLM proposed in this paper optimizes the daily load process by estimating the optimal spatial parameter. The simplified algorithm for daily generating flow based on the DLM makes the period-by-period calculating procedure much easier: the adjacent periods with equal load are put into a combined extended period, which cuts down the calculating work-load remarkably and the error is well controlled.

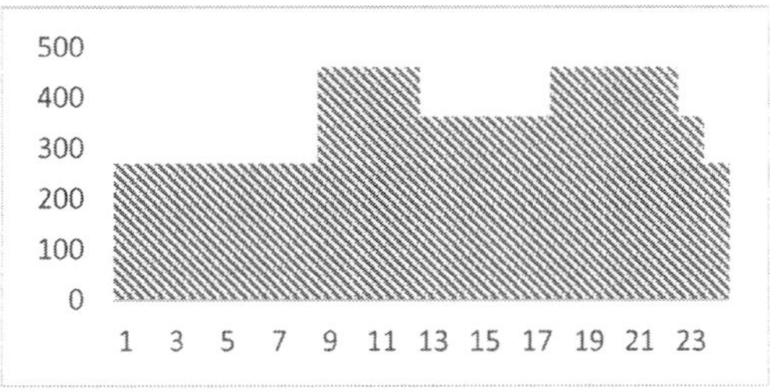

Figure 3. Optimized daily load process based on DLM.

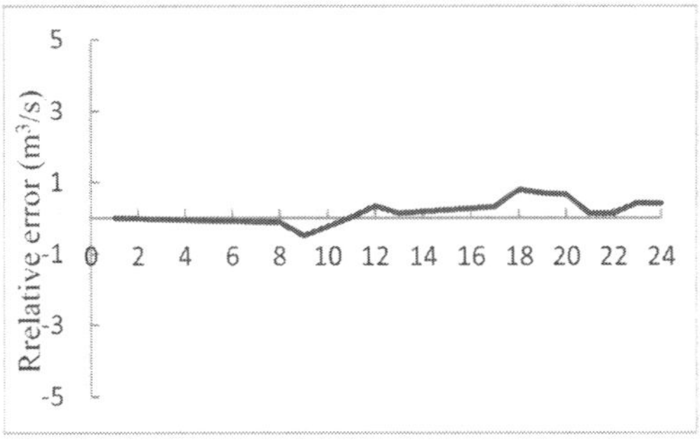

Figure 4. Comparison of daily generating flow.

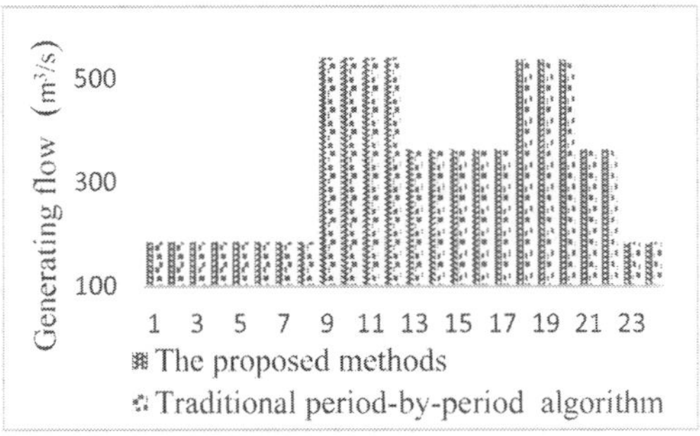

Figure 5. Generating flow of each period.

5 CONCLUSIONS

This paper proposed an innovative concept of Daily Load Model and a novel method for daily generating flow calculation: the period-combined algorithm based on the DLM. The DLM helps the operator to keep the load of hydropower in a stable state without any fluctuations in the daily load process. The proposed algorithm fixes the time-consuming problem of daily generating flow calculation throughout the operational day. The algorithm competes advantageously in terms of computational efficiency and offer benefits to the daily optimal operation of hydropower. The proposed method was successfully applied to a realistic case study. The traditional period-by-period algorithm is used to assess the behavior of the proposed method, demonstrating its accuracy and computational efficiency.

The future direction is the accuracy and flexibility of the proposed model. Moreover the algorithms should have a wide range of applications to cascade reservoirs.

ACKNOWLEDGMENTS

This research was sponsored by the National Natural Science Foundation of China (51479141) for the presentation at the 3rd Annual Congress on Advanced Engineering and Technology (CAET 2016), Hong Kong, China, Oct 22–23, 2016. The authors thank Professor Chen Senlin and Wan Biao for their comments and suggestions, which helped improve the quality of this paper.

REFERENCES

Cheng Chuntian, Li Jianbing, Li Gang, 2011. Study of the method of partition load distribution in hydropower station and its application, *Journal of Hydroelectric Engineering,* 32(2): 38–43.

He Shihua, Zou Jin, 2006. Study on daily load curve optimization of cascaded hydroplants, *Hydroelectric Generation*, 32(7): 67–70.

Huang chun-lei, Zhao yong-long, Guo Xia-ming, 2005. Short-term scheduling for cascade hydropower stations based on typical daily load, *Hydropower Automation and Dam Monitoring,* 29(4): 45–47.

Ji Chang-ming, Yu Shan, Zhang Xiao-xing, 2011. Short-term operation benefit analysis of cascade reservoirs based on power load curve, *Power System Protection and Control,* 39(14): 64–68.

Tang Haihua, Chen Senlin, Zhao Yu, 2008. Model and algorithm of short-term optimal scheduling of three gorges cascaded hydropower stations, *Water Resources and Power*, 26(3): 133–136.

Wu Xin-yu, Cheng Chun-tian, Li Gang, Zhang Shi-qin, 2012. Research on long term typical day peak load regulation energy maximization model for hydropower systems, *Shuili Xuebao,* 43(3): 363–370.

Xu Xin-fa, Li Xiao-wen, Xu-Ming, 2002. A progressive optimization algorithm for optimal dispatch of daily electric load of small cascaded hydropower stations, *International Journal Hydroelectric Energy*, 20 (3): 44–46.

Advanced Engineering and Technology III – Xie (Ed.)
 ISBN 978-1-138-03275-0

The research of Unit Daily Load Process Model of hydropower station

L. Gou & S.L. Chen
State Key Laboratory of Water Resources and Hydropower Engineering Science, Wuhan University, Wuhan, China

ABSTRACT: The concept of the Unit Daily Load Process Model (UDLPM) is proposed in this paper by analyzing the running characteristics of hydroelectric generating sets of a hydropower system. Furthermore, the mode of unit daily operation is described with the temporal characteristic parameters and spatial characteristic parameters. This paper mainly explores the relationship among the characteristic parameters of the model under the relevant constraints of time and space according to the structure of the UDLPM. The UDLPM provides an idea and method to solve effectively the problems of unit daily load changing frequently and unit commitment.

1 INTRODUCTION

The scientific scheduling of a hydropower system is confronted with a lot of basic problems of science and technology, taking into account the multitudinous problems of a hydropower system in China such as huge number of the hydropower stations, hydraulic power and water conservancy closely linked among cascade hydropower stations, and navigation and complicated constraints of the ecological environment comprehensive utilization. The short-term hydro optimal scheduling is a sophisticated, multi-dimensional, multi-objective, non-linear and discrete large-scale, space-time, decision-making optimization problem, and inner-plant economical operation of hydropower station is the foundation of short-term hydro optimal scheduling. For the short-term optimal scheduling and the inner-plant economical operation of hydropower station, the existing research results mainly focus on mathematical models and algorithms. Based on a mathematical model, Chen et al. (1999) proposed a short-term optimal scheduling model of the Mingjiang River basin hydropower station group and solved it by setting up the objective function with minimum water consumption on average; Guo et al. (2011) came up with two real-time load distribution model of cascade hydropower stations. Additionally, many algorithm studies have been performed, such as equal incremental rate, Dynamic Programming (DP), Discrete Differential Dynamic Programming (DDDP), Genetic Algorithm (GA), Artificial Neural Network (ANN), Ant Colony Algorithm (ACO), simulated annealing algorithm, and immune particle swarm algorithm. Ji et al. (2012) proposed a novel algorithm, Virus Particle Swarm Optimization (VPSO), for load allocation optimization. However, there is little research on the unit operation rules of hydropower station.

The hydroelectric generating set generally takes charge of peak load regulation, frequency modulation, and emergency tasks of electric power system because of the flexible reaction between swift start-up and shut-down. This paper mainly studies the temporal and spatial variation characteristics of unit daily load. Furthermore, it presents the Unit Daily Load Process Model (UDLPM) corresponding to the problems of frequent start-up and shut-down, load varying, and throbbing of unit existing in the optimal scheduling of a traditional hydropower system, which incorporates the significance and calibration method of model parameters and the operation characteristics of unit. This model can be used to describe the daily operating mode of unit. Furthermore, the transformation mechanism among characteristic parameters and the spatio-temporal characteristics of the model are analyzed from the perspective of graph theory. This simplified peak-shaving method can avoid the throbbing of unit daily load. Finally, the UDLPM identified in this study provides a novel idea to solve several basic theory and method issues in short-term optimal scheduling of hydropower station.

2 THE CONCEPT OF THE UNIT DAILY LOAD PROCESS MODEL

The nodes of unit start-up and shut-down, load value, and load variation during the scheduling period can be called the unit time characteristic point. In general, there are a series of constraints for a unit, such as the minimum time-consuming of unit start-up and shut-down during the scheduling period, the minimum time-consuming of unit shut-down in the first or last period, as well as the unit vibration area and cavitation area constraints, maximum and minimum load and unit climbing output constraint per unit time, considering the safety operation and life service of the unit.

The concept of the Unit Daily Load Process Model (UDLPM) can be summarized by using the temporal characteristic parameters, which represent the working time area of unit, and by the space characteristic parameters, which represent the unit load in any time period to generally describe the unit daily operating mode. The UDLPM is under the precondition of meeting the unit start-stop time constraints, output, and some related constraints, as shown in Figure 1.

Supposing the unit starts up n times during the total scheduling period T, which, in general, is 24 or 96 in one day, the corresponding unit time Δt is 1 h or 0.25 h. Furthermore, T can be expressed as $T=\sum_{i=1}^{n+1} T_{2i-1}$, where n times of start-up are T_2, T_4,..., T_{2n}, respectively; the outage duration before the first start-up is T_1, which are T_3, T_5,..., T_{2n-3}, T_{2n-1}, respectively, and the outage duration after the last start-up is T_{2n+1}.

The unit start-stop time constraints are as follows:

$$\begin{cases} T_{2i} \geq T_{on} (i=1,2,...,n) \\ T_{2i-1} \geq T_{off} (i=2,3,...,n) \\ T_1 \geq T_{off}^B \\ T_{2n+1} \geq T_{off}^E \end{cases} \tag{1}$$

where T_{on} and T_{off} represent the minimum number of start-up and shut-down time during the scheduling period, respectively; T_{off}^B and T_{off}^E represent the minimum number of shut-down time during the scheduling at the beginning and end, respectively.

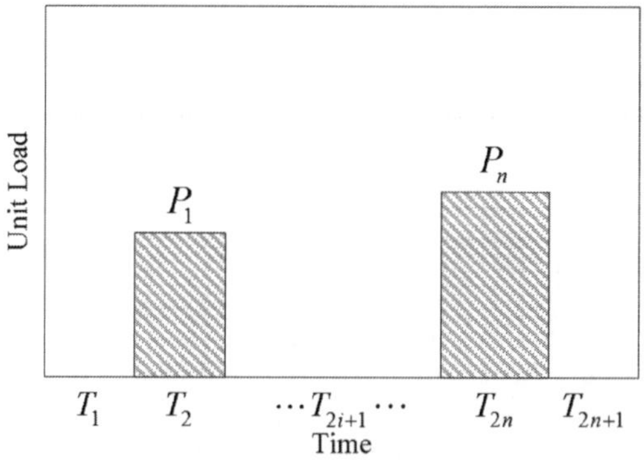

Figure 1. The sketch of the UDLPM.

The temporal characteristic parameters that represent the working time area of unit for i-th start-up are given by:

$$U_i = \left(T_{2i} + \frac{\sum_{j=1}^{i}\left(T_{2(j-1)} + T_{2j-1}\right)}{T} \right) \cdot \Delta t \quad (i = 1,2,\cdots,n) \tag{2}$$

For the traditional method, the short-term hydro optimal dispatching and the inner-plant economical operation need to confirm the load process $P_{j,t}$ of each unit during the scheduling period. Here, j represents the unit number and $t = 1{\sim}T$, because the value of T is very large and there will be "curse of dimensionality" arising from the conditional optimization of the increasing number of unit. However, the UDLPM can simplify optimal variables into three kinds: one is the times of unit start-up n; the other one is the temporal dimension characteristic parameters, namely U_i ($i = 1,2,\ldots,n$), which represent the working time area of unit for i-th start-up; the final one is the spatial characteristic parameters, namely the unit load values P_i ($i = 1,2,\ldots,n$) during each working time area, which comprehensively reflect the basic characteristics, real-time power generation capacity, and the synthetic utilization or idling constraints, sometimes also considering the load difference values and the amplitude difference values of peak-shaving ΔP_{tf} between two adjacent start-up operations.

3 THE RELATED MECHANISM OF TIME DIMENSION CHARACTERISTIC PARAMETERS OF THE UDLPM

3.1 *Single start-up*

There is only a single unit start-up during the total scheduling period T, that is to say, the unit start-up times of the UDLPM $n = 1$, the start-up time is T_2 and the shut-down time of before and after this start-up are T_1 and T_3, respectively. Corresponding to these constraints, $T_1 \geq T_{off}^B$, $T_2 \geq T_{on}$, $T_3 \geq T_{off}^B$, and also $T = \sum_{i=1}^{3} T_i$. Then, we can derive $U_1 = (T_2 + \frac{T_1}{T}).\ \Delta t$.

The upper limit of T_1 is affected by the value of T_2; however, the upper limit of T_1+T_2 is not affected by T_2, so we convert U_1 to the following equation:

$$U_1 = \left(T_2 + \frac{T_1}{T} \right) \cdot \Delta t = \left(\frac{T_1 + T_2}{T} + \frac{(T-1)T_2}{T} \right) \cdot \Delta t \tag{3}$$

Let $V_1 = T_1 + T_2$ and $V_2 = T_2$, then U_1 is derived according to Eqs. (3) as follows:

$$U_1 = \left(\frac{V_1}{T} + \frac{(T-1)}{T} V_2 \right) \cdot \Delta t \tag{4}$$

In accordance with the known conditions, we can derive $T_{off}^B + T_{on} \leq V_1 = T_1 + T_2 = T - T_3 \leq T - T_{off}^E$. So, the value range of V_1 is:

$$T_{off}^B + T_{on} \leq V_1 \leq T - T_{off}^E \tag{5}$$

Similarly, we can derive the value range of V_2 as:

$$T_{on} \leq V_2 \leq T - T_{off}^B - T_{off}^E \tag{6}$$

The value range of U_1 can be obtained according to these above-mentioned formulas:

$$\left(T_{on}+\frac{T_{off}^{B}}{T}\right)\cdot\Delta t\le U_1\le\left(\left(T-T_{off}^{B}-T_{off}^{E}\right)+\frac{T_{off}^{B}}{T}\right)\cdot\Delta t \tag{7}$$

Taking into account $T_1=T_1+T_2-T_2\le T-T_{off}^{E}-T_2$, there are some non-feasible solutions of U_1 in Eqs. (7), namely the feasible solutions distribute within the scope discretely.

3.2 *Two times start-up*

There are two times unit start-up during the total scheduling period T, that is, $n=2$. According to the conclusion made in section 3.1, the value range of U_1 is given by:

$$\left(T_{on}+\frac{T_{off}^{B}}{T}\right)\cdot\Delta t\le U_1\le\left(T-\left(T_{off}^{B}+T_{off}+T_{on}+T_{off}^{E}\right)+\frac{T_{off}^{B}}{T}\right)\cdot\Delta t \tag{8}$$

Similarly, the value range of U_2 is:

$$\begin{aligned}&\left(T_{on}+\frac{T_{on}+T_{off}+T_{off}^{B}}{T}\right)\cdot\Delta t\le U_2\\&\le\left(T-\left(T_{off}^{B}+T_{off}+T_{on}+T_{off}^{E}\right)+\frac{T_{on}+T_{off}+T_{off}^{B}}{T}\right)\cdot\Delta t\end{aligned} \tag{9}$$

4 THE ASSOCIATION AND TRANSFORMATION MECHANISM OF SPATIO-TEMPORAL CHARACTERISTIC PARAMETERS OF THE UDLPM

This section mainly analyzes the relationship between the spatio-temporal characteristic parameters of the UDLPM and the formational mechanism of the unit daily load process. Here, the common two times unit start-up is taken as an example.

Supposing that the unit daily electric energy production is E_{jz}, the amplitude difference value of peak-shaving $\Delta P_{1,2}$ between two adjacent start-up operations is equivalent to P_2-P_1 ($\Delta P_{1,2}$ could be either positive or negative), and then $P_1\cdot\left(\frac{U_1}{\Delta t}\right)\cdot\Delta t+\left(P_1+\Delta P_{1,2}\right)\cdot\left(\frac{U_2}{\Delta t}\right)\cdot\Delta t=E_{jz}$, the value of P_1 is given by:

$$P_1=\frac{\frac{E_{jz}}{\Delta t}-\Delta P_{1,2}\cdot\left[\frac{U_2}{\Delta t}\right]}{\left[\frac{U_1}{\Delta t}\right]+\left[\frac{U_2}{\Delta t}\right]} \tag{10}$$

In general, when the value of $\Delta P_{1,2}$ is determined, different results of unit operation area can be obtained by adjusting the values of U_1 and U_2, and vice versa.

5 CASE STUDY

The Tianshengqiao I Hydropower Station (hereinafter referred to as the "Tian I") is chosen as the case study. It is the first level of Hongshui River cascaded hydropower stations, located in the boundary river, the main stream of Nanpan River of Anlong County in GuiZhou province and Longlin County in GuangXi province. Here, let us suppose tha tonly one unit is in operation in one day, which has two times start-up. We take one hour as the

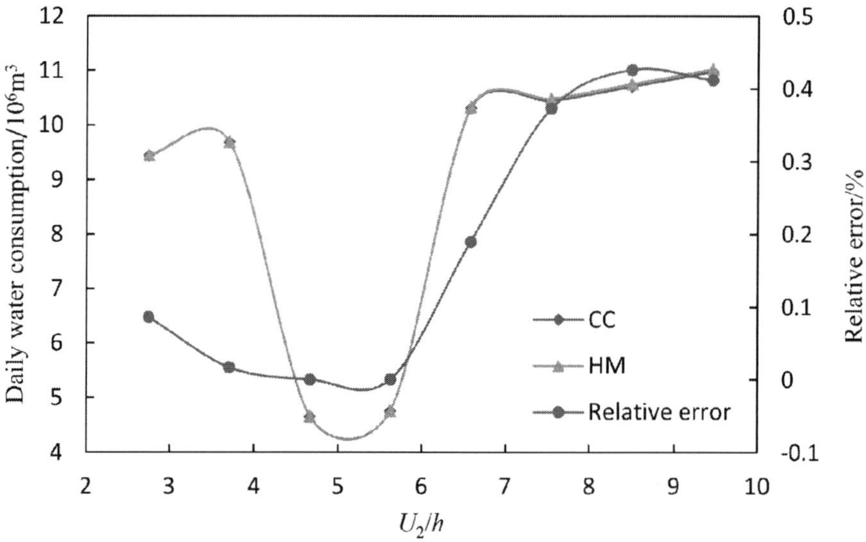

Figure 2. The sketch of relative error statistics of unit daily water consumption. *Here the value of U_1 is predetermined, which is equal to 5.125 h.

calculating time step and select the determining water consumption based on hydropower generation as the operation mode of the reservoir. In addition, the initial reservoir water level $Z_0 = 755$ m, the daily inflow on average $Q_{in} = 500$ m^3/s, and the daily electric energy production $E_{day} = 1200$ MW I h. Based on the UDLPM, the results of daily water consumption using the Conventional Calculation (CC), which successively computes the water consumption of each period and their sum is the unit daily water consumption, and the Homogenization Method (HM), which extends the time step Δ_t of conventional calculation to one time start-up length $\left[\frac{U_i}{\Delta t}\right](i = 1 \sim n)$, are partially shown in Figure 2.

When $U_1 = 5.125$ h, the different values of U_2 represent different unit daily load processes. Compared with CC that iterates T times, the HM that only iterates n times has an advantage of decreasing the number of iterations, so the computation cost is smaller. Meanwhile, it is seen that the relative error of HM is almost in the range of 0.01672% ~ 0.42590%, which indicates that the relative error values are small and satisfy the general accuracy limit. In addition, this simplified method can be adopted in the realistic operation of the hydroelectric power system. So, the optimal operation process of unit could be easily determined according to the minimum daily water consumption with the condition of the same daily hydropower generation.

6 CONCLUSIONS

The emergence of frequent start-up and shut-down, load varying and throbbing of unit has generated unavoidable challenges to the peak load regulation of optimal operation of hydropower station. In this paper, the Unit Daily Load Process Model (UDLPM) is proposed to address these problems related to the conditions of meeting the constraints of unit start-stop time and output. In addition, the association and transformation mechanisms of the spatio-temporal characteristic parameters of the UDLPM are accomplished based on the graph combination theory. In summary, the UDLPM plays a significant role in the development of the short-term hydro optimal scheduling. In addition, it can effectively solve these problems of frequent start-stop, load varying of unit, etc., and has potential theoretical and practical benefits for reservoir operation problems.

ACKNOWLEDGMENTS

This research was supported by the National Natural Science Foundation of China (51479141). The authors thank the associate editor and anonymous reviewers for their comments and suggestions, which helped improved the quality of this paper.

REFERENCES

Chen, S.L. & Wan, J. et al. 1999. Uniform Water Utilization Minimum Model of Hydropower Stations Short-Term Optimization Dispatching in a Basin. *International Journal Hydroelectric Energy* 17(3):9–12.

Fu, X. & Li, A. et al. 2011. Short-term scheduling of cascade reservoirs using an immune algorithm-based particle swarm optimization. *Computers & Mathematics with Applications* 62(6): 2463–2471.

Guo, F.Q. & Guo, S.L. et al. 2011. Real-time load distribution models for the Qingjiang cascade hydropower plants. *Journal of Hydroelectric Engineering* 30(1):5–11.

Heidari, M. & Chow, V.T. et al. 1971. Discrete Differential Dynamic Programing Approach to Water Resources Systems Optimization. *Water Resources Research* 7(2):273–282.

Ji, C.M. & Xie, W. et al. 2012. Optimized load allocation of cascaded hydropower stations based on virus particle swarm optimization algorithm. *Journal of Hydroelectric Engineering* 31(002):38–43.

Li, G. 2007. Short-term energy-saving power generation operation of hydro-thermal power system. *Dalian: Dalian University of Technology*.

Liu, P. & Cai, X.M. et al. 2011. Deriving multiple near-optimal solutions to deterministic reservoir operation problems. *Water Resources Research* 47:1–20.

Teegavarapu, R.S.V. & Simonovic, S.P. 2002. Optimal Operation of Reservoir Systems using Simulated Annealing. *Water Resources Management* 16(5):401–428(28).

Wu, Y. & Ho, C. et al. 2000. A diploid genetic approach to short-term scheduling of hydro-thermal system. *Power Systems IEEE Transactions on* 15(4):1268–1274.

Yi, J. & Labadie, J.W. 2003. Dynamic optimal unit commitment and loading in hydropower systems. *Journal of Water Resources Planning and Management* 129(5): 388–398.

Zhao, J. & Dong, Z.C. et al. 2008. Application of Genetic Algorithm-Ant Colony Algorithm in Optimal Dispatching of Hydropower Station. *Water Resources and Power* 26(5):132–134.

Zhao, X.H. & Huang, Q. et al. 2009. Application of ant colony algorithm for economic operation of hydropower station. *Journal of hydroelectric Engineering* 28(2):139–142.

Zhu, M. & Wang, D.Y. 1999. Daily Optimal Operation of Cascade Hydroelectric Power Stations Based on Artificial Neural Network. *Automation of Electric Power Systems* 23(10):35–40.

Advanced Engineering and Technology III – Xie (Ed.)
 ISBN 978-1-138-03275-0

An experimental study of aerosol generation by showerheads

P.H. Tsui, Y. Zhou, L.T. Wong & K.W. Mui
Department of Building Services Engineering, The Hong Kong Polytechnic University, Hong Kong, China

ABSTRACT: Aerosolization of water from discharging water appliances is a source for Legionellosis. This paper reports the measurement results of aerosol generation from four different showerheads in a mechanically ventilated test chamber. Associations between aerosol generation rate and water supply pressure, flow rate, water velocity, resistance factor K and ratio of total nozzle area to showerhead faceplate area were investigated. The results showed that a higher pressure in water supply mains gave a higher aerosol generation rate, and a water velocity that was very low or very high contributed to the generation of aerosols. Moreover, showerhead designs were found to have influences on aerosol generation. This paper improves our understanding of aerosolization of water from showerheads with running water and provides a source of data for future showerhead designs.

1 INTRODUCTION

Legionella bacteria are transmitted to the human bodies mainly by inhalation of airborne droplets (i.e. aerosols) or particles in fine mist containing them. They can survive and multiply in water in the temperature range of 20°C to 45°C (Prevention of Legionnaires' Disease Committee 2012). As the normal range of operation temperature of water supply systems in buildings is conducive to the growth of Legionella bacteria, aerosolization of water from discharging water appliances such as showerheads with running water is a source for Legionellosis.

In showering, aerosols of suitable sizes will remain in the air for significant lengths of time because of the low settling velocity of small particles and high levels of moisture and humidity in the bathroom (Muhanned 2013). It is believed that particles suspended in the air in the size range from 3 to 100 μm are likely to be inhaled into the human respiratory system (TSI Incorporated 2013). O'Toole et al. (2008) indicated that aerosols of a size less than 10 μm in diameter are potentially associated with inhalation exposure to microorganisms and biological agents which are with their potential deposition in the alveolar region of the lungs.

Yue et al. (2010) reported that at a water temperature of 24–25°C, particle concentration increased with increasing showerhead water flow rate from 5.1 to 9 L/min. It was demonstrated that droplets generated from a high flow showerhead at lower temperature would not have enough time to evaporate. Carson (1996) also showed that particle production rate was linked to the smoothness of the flow stream, i.e. water stream would become more continuous and smoother as the flow rate increased.

In Hong Kong, a voluntary Water Efficiency Labelling Scheme (WELS) on Showers for Bathing has been implemented to encourage the installation of water-efficient showerheads (Hong Kong Water Supply Department 2011). In the scheme, showerheads are rated to 4 grades according to their water flow rates. This study examined the aerosol generation rates of two WELS-rated (i.e. low-flow) showerheads and two conventional ones in a mechanically ventilated test chamber. For each showerhead, characteristics including diameter, sizes and quantities of nozzles, water supply pressure, flow rate, water velocity and resistance factor K were measured against the amount of aerosols produced. The results improve our understanding of aerosolization of water from showerheads with running water and provide a source of data for future showerhead designs.

2 EXPERIMENTAL STUDY

Figure 1 shows a showerhead installed in an experimental glass chamber of size 914 mm(L) × 610 mm(W) × 508 mm(H). An air supply fan with air filter and a suction line at the inlet are also shown. During the experiment, the chamber was mechanically ventilated in order to keep the inlet air moistened at a steady air velocity of 0.25 m/s. The showerhead was fed from a 20 L water tank, which was containing distilled water with 2% salt, and boosted by a pumping system with a working pressure of up to 1.5 bar. Each measurement lasted 3 hours.

Figure 2 exhibits the four showerheads sampled and Table 1 summarizes their characteristics. As the construction of a showerhead may affect water flow rate and pressure loss, it can be characterized by the resistance factor K (Critchley & Phipps 2007) as expressed below, where Q (L/min) is the water flow rate and P (bar) is the pressure loss,

$$K = \frac{Q}{\sqrt{P}} \tag{1}$$

Before the experiment, initial weight of a dry, clean filter paper (designed for particle size ≥0.2 μm) was determined. The filter paper was then installed at the chamber outlet to collect aerosolized salt water from the discharging showerhead. After the experiment, the filter paper sample was placed in an oven at 100°C for 30 minutes to remove redundant moisture. The weight of the salt collected during the measurement was then determined.

Yue et al. (2010) reported that the mass median diameter of an aerosol is 2.5–7.5 μm. In this study, by taking the mean aerosol radius R to be 5 μm, an approximate value of aerosol volume V_0 (m^3) can be determined using the following formula,

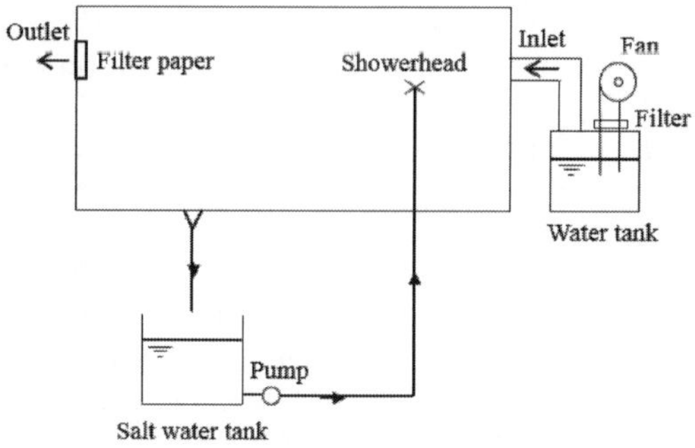

Figure 1. Schematic diagram of experimental setup.

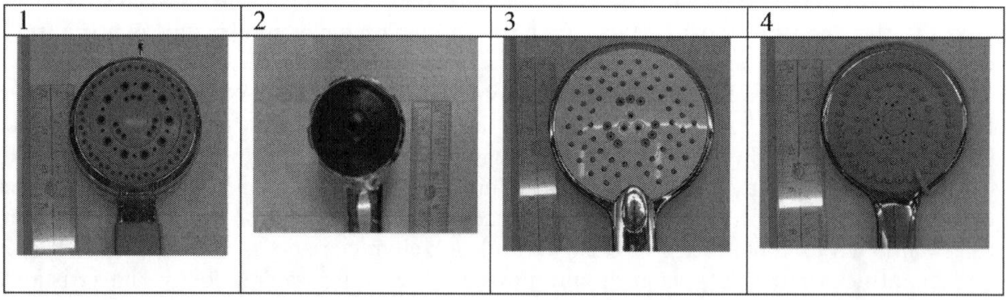

Figure 2. Sample showerheads.

Table 1. Characteristics of showerheads sampled.

	Showerheads			
Specification	1	2	3	4
Type	Conventional	Conventional	WELS*	WELS*
Showerhead diameter (m)	0.065	0.045	0.1	0.075
Number of 1 mm nozzles	48	48	59	53
Number of 2 mm nozzles	19	9	9	15
Number of 5 mm nozzles	10	0	0	0
Resistance factor K	7.82	7	2.54	5.93
Flow rate at 1 bar (L/s)	0.13	0.12	0.042	0.1
Flow rate at 1.5 bar (L/s)	0.16	0.14	0.052	0.12
Average velocity at 1 bar (m/s)	0.44	1.82	0.56	1.13
Average velocity at 1.5 bar (m/s)	0.54	2.12	0.70	1.35
Ratio of total nozzle area to showerhead faceplate area	0.0885	0.0415	0.0095	0.0201

*WELS: Water Efficiency Labelling Scheme rated showerhead.

$$V_0 = \frac{4\pi R^3}{3} \tag{2}$$

By assuming a steady air flow field in the chamber, the aerosol collection rate at the chamber outlet G_n (particles/s) can be approximated by Equation (3), where G_m (g/s) is the salt collection rate at the chamber outlet, C_s is the salt water concentration (= 2% kg/L), M_0(g) and M_e(g) are the mass values of the filter paper at the beginning and at the end of the experiment respectively, and t(s) (= 10800s) is the duration of each measurement.

$$G_n = \frac{G_m}{1\times10^6 C_s V_0}; \quad G_m = \frac{M_e - M_0}{t} \tag{3}$$

Assuming that the aerosol deposition on the chamber wall is steady and linear to the aerosol generation rate, G_n can represent the aerosol generation trend of the discharging showerhead.

3 RESULTS AND DISCUSSION

Results of the weight of salt collected $M_0 - M_n$, salt collection rate G_m and aerosol collection rate G_n are summarized in Table 2. Relationships between aerosol collection rate and showerhead attributes (i.e. water supply pressure, flow rate, water velocity, resistance factor K and ratio of total nozzle area to showerhead faceplate area) are graphed in Figure 3.

Figure 3(a) shows that aerosol generation rate increased with showerhead water pressure in the experiment. With the same water pressure, aerosol generation rates of different showerheads were found to be different. As shown in Figure 3(b), there was no obvious relationship between aerosol generation rate and water flow rate. Figure 3(c) reveals that low water velocities (e.g. <0.56 m/s at 1 bar, <0.7 m/s at 1.5 bar) contributed to the generation of aerosols, and the aerosol generation rate decreased with an increasing low velocity. However, the aerosol generation rate increased with an increasing velocity that was already at a certain level (e.g. 0.56 m/s at 1 bar, >0.7 m/s at 1.5 bar). The relationship between aerosol generation rate and resistance factor K and the one between aerosol generation rate and ratio of total nozzle area to showerhead faceplate area were inconspicuous, as presented in Figures 3(d) and (e) respectively. At a high pressure (e.g. 1.5 bar), the aerosol generation

Table 2. Summary of weights of salt collected and collection rates.

	Showerheads							
	1		2		3		4	
Specification	1 bar	1.5 bar	1 bar	1.5 bar	1 bar	1.5 bar	1 bar	1.5 bar
Weight of salt collected $M_0 - M_e$ (g)	0.0016	0.0022	0.0017	0.0031	0.0008	0.0017	0.0012	0.0019
Salt collection rate G_m (g/s) $\times 10^{-7}$	1.48	2.04	1.57	2.87	0.74	1.57	1.11	1.76
Aerosol collection rate G_n (particles/s)	14154	19462	15039	27424	7077	15039	10616	16808

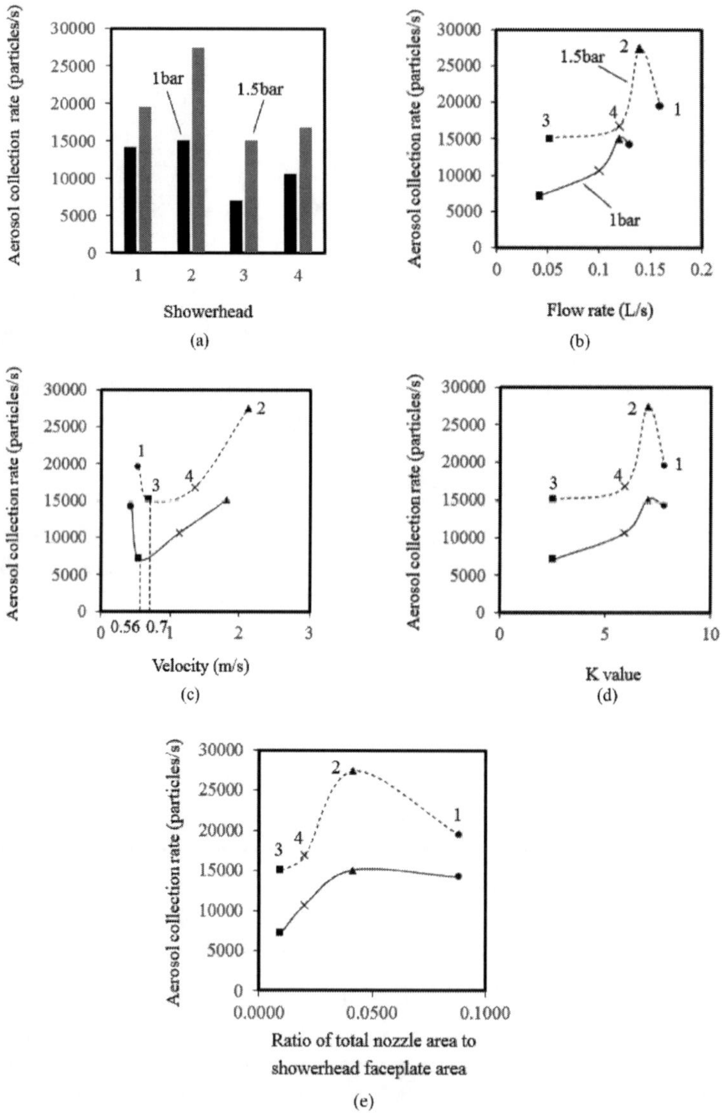

Figure 3. Relationships between aerosol collection rate and showerhead attributes.

rates of Showerheads 1 and 2 were significantly different (29% difference) while their water flow rates and resistance factor K values were similar (12.5% difference and 10.5% difference respectively).

It can be seen in Figure 2 that the nozzles are evenly distributed on the faceplates of all showerheads except for Showerhead 2, where they are mainly distributed on the edge of the faceplate. That may be the reason (besides high water velocity) why the aerosol generation rate of Showerhead 2 was relatively high.

4 CONCLUSION

People get Legionellosis by breathing in the Legionella bacteria, usually carried by a fine mist. Aerosolization of water from discharging water appliances such as showerheads with running water is thus a source for Legionellosis. This study examined the aerosol generation rates of four different showerheads in a mechanically ventilated test chamber. The results showed that a higher water supply pressure gave a higher aerosol generation rate, and a water velocity that was too low or too high contributed to aerosol generation. No obvious relationships were found between aerosol generation rate and flow rate, resistance factor K and ratio of total nozzle area to showerhead faceplate area. Furthermore, showerhead designs were found to have influences on the generation of aerosols. The findings of this study provide a source of data for future showerhead designs.

ACKNOWLEDGMENT

The work described in this paper was partially supported by a grant from the Research Gants Council of the HKSAR, China (PolyU 5272/13E) and by 3 different grants from The Hong Kong Polytechnic University (GYBA6, GYL29, GYM64).

REFERENCES

Carson, W. 1996. Aerosol generation by taps and the resultant exposure of humans to Legionella bacteria. *CIB W062 Symposium.*

Critchley, R. & Phipps, D. 2007. Water and energy efficient shower: project report. *United Utilities: 27.*

Hong Kong Water Supplies Department 2011. Voluntary water efficiency labelling scheme on showers for bathing.

Muhanned, A.R. 2013. Studying the factors affecting the settling velocity of solid particles in non-Newtonian fluids. *Nahrain University, College of Engineering Journal (NUCEJ)* 16(1): 48.

O'toole, J., Leder, K. & Sinclair, M. 2008. A series of exposure experiments—Recycled water and alternative water sources: Part A—Aerosol-sizing and endotoxin experiments. *The Cooperative Research Centre for Water Quality and Treatment* 45: 15.

Prevention of Legionnaires' Disease Committee, Hong Kong 2012. *Prevention of Legionnaires' Disease, Code of Practice.*

TSI Incorporated 2013. *Health-based particle-size selective sampling. Application note ITI-050.*

Yue, Z., Benson, J.M., Irvin, C., Irshad, H. & Yung S.C. 2010. Particle size distribution and inhalation dose of shower water under selected operating conditions. *Inhalation Toxicology* 19(4): 333–342.

Advanced Engineering and Technology III – Xie (Ed.)
© 2017 Taylor & Francis Group, London, ISBN 978-1-138-03275-0

Matter element evaluation on water resources management modernization

Xianfeng Huang, Jingwei Zhong & Guohua Fang
College of Water Resources and Hydropower Engineering, Hohai University, Nanjing, China

ABSTRACT: In order to evaluate modernization level of regional water resources management, the evaluation index system should be established from five aspects, including water resources guarantee, water utilization efficiency, water resources protection, public management ability of water resources and supporting capacity of water resources, then Fuzzy Analytic Hierarchy Process (FAHP) and entropy method are used to determine the index weight. The evaluation model of water resources management modernization is constructed based on the matter element theory, a comprehensive evaluation of water resources management modernization of Gaochun District Nanjing city is studied as an example. The results show that the water resources management in Gaochun District was in the primary modernization stage in 2012, which is consistent with the actual situation and verifies the feasibility and effectiveness of the evaluation model.

1 INTRODUCTION

Water resources management modernization is an important part of water conservancy modernization, is the important support and guarantee of economic and social development modernization. It is a dynamic development process to achieve the maximum efficiency and benefit through the continuous change, innovation, and development (Song 2013). Water resources management modernization must be guided by the modern idea of water conservancy and the theory of sustainable development, transform traditional water conservancy into modern water conservancy, realize harmonious co-existence between human and nature (Zhang et al. 2011).

Water resources management modernization evaluation can analyze modernization level of water resources management and find the gap with the modernization requirements, then promote the construction of regional water resources management modernization. Therefore, water resources management modernization evaluation has important significance. The evaluation modernization model is to integrate the index system through some mathematical and statistical approaches, forming a total index and a number of comprehensive indictors. The total index reflects the level of water resources management modernization, the comprehensive index reflects the realization degree of each part of water resources management.

The methods of modernization evaluation include AHP, matter element analysis method, fuzzy comprehensive evaluation method, AHP and BP neural network. In this paper, matter element analysis method is used to evaluate modernization level of water resources management.

Table 1. Evaluation index system of water resource management modernization.

Target layer A	Criterion layer C	Index layer U
Water resources management modernization	Water resources guarantee (C_1)	Total water control (U_1)
		Guaranteed rate of water supply (U_2)
	Water utilization efficiency (C_2)	Water consumption per ten thousand yuan GDP (U_3)
		Water consumption per ten thousand yuan of industry value-added (U_4)
		Irrigation water utilization coefficient (U_5)
		Reuse rate of reclaimed water (U_6)
	Water resources protection (C_3)	Standard rate of water functional area (U_7)
		Standard rate of centralized drinking water source (U_8)
		Total amount control of pollutant in river (U_9)
		Groundwater management level (U_{10})
	Public management ability of water resources (C_4)	Planning and implementation (U_{11})
		Standard management (U_{12})
		Water resources monitoring ability (U_{13})
		River health assessment (U_{14})
		Management informatization (U_{15})
	Supporting capacity of water resources (C_5)	Organizational institution (U_{16})
		Management facilities (U_{17})
		Fund investment (U_{18})
		Structure of personnel (U_{19})
		Law enforcement ability (U_{20})

2 EVALUATION INDEX SYSTEM OF WATER RESOURCES MANAGEMENT MODERNIZATION

Water resources management modernization evaluation index system is to measure the level of management, also has the function of guiding, monitoring and evaluating. The scientific establishment of the index system should follow the principles of scientificity, feasibility, independence, comparability, guidance and pertinency.

According to the above principles and "Jiangsu province water resources management modernization index system", on the basis of the regional natural and economic conditions, and development target of water resources management modernization construction, the evaluation index system should be established from five aspects, including water resources guarantee, water utilization efficiency, water resources protection, public management ability of water resources and supporting capacity of water resources, a total of 20 indicators, which list in Table 1. Considering regional water resources management features, these indicators can be revised.

3 EVALUATION MODEL OF WATER RESOURCES MANAGEMENT MODERNIZATION

3.1 *Determine the weight of index with combination weight method*

3.1.1 *Fuzzy analytic hierarchy process*

Analytic hierarchy process is a systemic analysis method combined with qualitative and quantitative analysis, it was proposed by the American operational research experts A.L. Saaty in

Table 2. Scale of 0.1–0.9 measure.

Scale	Definition
0.5	a_i is equal important to a_j
0.6	a_i is slightly more important than a_j
0.7	a_i is obviously more important than a_j
0.8	a_i is much more important than a_j
0.9	a_i is extremely more important than a_j

1970s (Xu 1988). But it is difficult to check consistency, and lack of scientificity. In order to improve the traditional AHP, some scholars proposed the fuzzy hierarchy analysis process which can well reflect the consistency and coherence of experts' thought, and simplify the index weight calculation steps (Fan & Jiang 2001, Fan et al. 2001).

In this paper, the FAHP is used to determine the subjective weight of water resources management modernization evaluation indicators, the main steps are as follows:

1. Determine the evaluation target and evaluation index set
2. Establish a fuzzy consistent judgment matrix

The fuzzy consistent judgment matrix which can reflect the relative importance degree of each two elements in $a_1, a_2, a_3, \ldots, a_n$, expressed as:

$$R = \begin{pmatrix} r_{11} & \cdots & r_{1n} \\ \vdots & \ddots & \vdots \\ r_{n1} & \cdots & r_{nn} \end{pmatrix} \tag{1}$$

where, r_{ij} is the membership degree about how much more important a_i is than a_j, using the 0.1~0.9 scale method (Table 2) with the combination of Delphi method to determine it.

The properties of fuzzy judgment consistency matrix R are as follows:

1. $r_{ii} = 0.5, i = 1, 2, 3, \ldots, n$;
2. $r_{ij} + r_{ji} = 1, i, j = 1, 2, 3, \ldots, n$;
3. $r_{ij} = r_{ik} - r_{ij} + 0.5, i, j, k = 1, 2, 3, \ldots, n$;

The necessary and sufficient condition of R being a fuzzy consistent judgment matrix is that there exist n-order non-negative normalized vector $W = (\omega_1, \omega_2, \omega_3 \ldots \omega_n)$ and positive number a make $\forall i, j$ satisfied:

$$r_{ij} = a(\omega_i - \omega_j) + 0.5 \tag{2}$$

In the formula, parameter $a \geq (n-1)/2$ (in this paper, set $a = (n-1)/2$), the value of a reflects the decision makers' personal preference, the smaller the a is, the more attention experts have paid to the difference of weight among elements (Lv 2002).

3. Calculate index weight
   Fix the i in formula (2) and obtain:

$$\omega_i = \frac{1}{a}\left(r_{ik} - \frac{1}{2}\right) + \omega_k \tag{3}$$

Sum of k and obtain:

$$n\omega_i = \frac{1}{a}\sum_{k=1}^{n} r_{ik} - \frac{n}{2a} + \sum_{k=1}^{n} \omega_k \tag{4}$$

From the weight vector normalization condition and obtain:

$$\omega_i = \frac{1}{n} - \frac{1}{2a} + \frac{1}{na}\sum_{k=1}^{n} r_{ik}, \quad i = 1, 2, 3, \ldots, n \tag{5}$$

In the formula, $0 \le \omega_i \le 1$.

3.1.2 *Entropy weight method*

Entropy weight method is used to determine the objective weight, the information entropy is a measure of the disorder degree of the system, and the more orderly a system, the lower the information entropy (Yang et al. 2013). The main calculation steps are as follows:

1. Normalize the judgment matrix *R* established by FAHP, and obtain the standard matrix

$$P = (p_{ij})_{n\times n} \tag{6}$$

In the formula: $p_{ij} = r_{ij} / \sum_{i=1}^{n} r_{ij}, \quad (i, j = 1, 2, \ldots, n)$.

2. Calculate entropy of the *j*-th index

$$e_j = -k\sum_{i=1}^{n} p_{ij} \ln p_{ij}, \quad (i, j = 1, 2, \ldots, n) \tag{7}$$

In the formula: e_j $(0 \le e_j \le 1)$ is the entropy of the *j*-th index; $k = 1/\ln n$, is information entropy coefficient.

3. Calculate entropy weight of each index

$$\omega_j = (1 - e_j)/(m - \sum_{j=1}^{n} e_j), \quad (j = 1, 2, \ldots, n) \tag{8}$$

In the formula: $0 \le \omega_j \le 1, \sum_{j=1}^{n} \omega_j = 1$.

3.1.3 *Combination weight method*

Combine FAHP and entropy weight method to make the determination of the evaluation index weight more scientific and accurate, the two methods complement each other, which can not only fully consider the experts' knowledge and experience, but also reduce the impact of subjective and arbitrary in the process of determining the index weight, so that the evaluation results are more objective and reliable. The calculation formula of the comprehensive weight is as follow:

$$\varphi_i = \omega_i' \omega_i'' / (\sum_{i=1}^{n} \omega_i' \omega_i''), \quad (i = 1, 2, \ldots, n) \tag{9}$$

In the formula: ω_i' is the subjective weight of index; ω_i'' is the objective weight of index.

3.2 *Construct the evaluation model of matter element analysis*

The matter element analysis method is used to deal with the incompatible problem which can't achieve the expected target when using a normal method (Cai 1987). For anything can be described by such three elements-"things *N*, the characteristics of things *C*, the corresponding value of the characteristics *X*", which compose the basic element of ordered triple, and that is matter element. The ordered triple $R = (N, C, X)$ can be used to describe thing as its basic element (Fang & Huang 2011).

1. Matter element matrix
   There are a number of evaluation indicators in the index system, denoted as n characteristics. The matter element matrix is established as follow:

$$R = \begin{pmatrix} R_1 \\ R_2 \\ \vdots \\ R_n \end{pmatrix} = \begin{pmatrix} P_0 & C_1 & X_1 \\ & C_2 & X_2 \\ & \vdots & \vdots \\ & C_n & X_n \end{pmatrix} \tag{10}$$

In the formula: R is n-dimension matter element of water resources management modernization; R_i is the sub matter element of R; P_0 is the unevaluated unit; C_i ($i = 1, 2, \ldots, n$) is the i-th evaluation index; X_i ($i = 1, 2, \ldots, n$) is the corresponding value of the evaluation index C_i.

2. Classical domain and joint domain matter element
   The classical domain matter element can be expressed as:

$$R_j(N_j, C, X_j) = \begin{pmatrix} N_j & C_1 & X_{j1} \\ & C_2 & X_{j2} \\ & \vdots & \vdots \\ & C_n & X_{jn} \end{pmatrix} = \begin{pmatrix} N_j & C_1 & [a_{j1}, b_{j1}] \\ & C_2 & [a_{j2}, b_{j2}] \\ & \vdots & \vdots \\ & C_n & [a_{jn}, b_{jn}] \end{pmatrix} \tag{11}$$

In the formula: N_j ($j = 1, 2, \ldots, m$) is the j-th grade of water resources management modernization evaluation; $X_{ji} = [a_{ji}, b_{ji}]$ is the classical domain, represents the numerical range of each evaluation index corresponding to the j-th grade.

The joint domain matter element of water resources management modernization can be expressed as:

$$R_p(N_p, C, X_p) = \begin{pmatrix} N_p & C_1 & X_{p1} \\ & C_2 & X_{p2} \\ & \vdots & \vdots \\ & C_n & X_{pn} \end{pmatrix} = \begin{pmatrix} N_p & C_1 & [a_{p1}, b_{p1}] \\ & C_2 & [a_{p2}, b_{p2}] \\ & \vdots & \vdots \\ & C_n & [a_{pn}, b_{pn}] \end{pmatrix} \tag{12}$$

In the formula: N_P is the set of evaluation grade; $X_{pi} = [a_{pi}, b_{pi}]$ is the joint domain, represents the sum of the numerical range of the classical domain of each index.

3. The correlation function and the correlation degree
   The correlation function indicates that when the value of matter element is taken as a point on real axis, the conformity degree of matter element with the required range of values. If $|X_0| = |b - a|$ is used to represent the norm of the bounded interval $X_0 = [a, b]$, the distance from x_i to $X_{ji} = [a_{ji}, b_{ji}]$ and x_i to $X_{pi} = [a_{pi}, b_{pi}]$ are respectively:

$$\rho(x_i, X_{ji}) = \left| x_i - \frac{1}{2}(a_{ji} + b_{ji}) \right| - \frac{1}{2}(b_{ji} - a_{ji}) = \begin{cases} a_{ji} - x_i, & x_i \le \dfrac{a_{ji} + b_{ji}}{2} \\ x_i - b_{ji}, & x_i > \dfrac{a_{ji} + b_{ji}}{2} \end{cases} \tag{13}$$

$$\rho(x_i, X_{pi}) = \left| x_i - \frac{1}{2}(a_{pi} + b_{pi}) \right| - \frac{1}{2}(b_{pi} - a_{pi}) = \begin{cases} a_{pi} - x_i, & x_i \le \dfrac{a_{pi} + b_{pi}}{2} \\ x_i - b_{pi}, & x_i > \dfrac{a_{pi} + b_{pi}}{2} \end{cases} \tag{14}$$

Table 3. Rank of water resources management modernization evaluation.

Rank	Initial period	Basic well-off	Overall well-off	Primary modernization	Basic modernization
Range	[0, 40]	[40, 60]	[60, 80]	[80, 90]	[90, 100]

The calculation formula of the correlation function $K_j(x_i)$ is as follows:

$$K_j(x_i) = \begin{cases} \dfrac{-\rho(x_i, X_{ji})}{|X_{ji}|} & x_i \in X_{ji} \\ \dfrac{\rho(x_i, X_{ji})}{\rho(x_i, X_{pi}) - \rho(x_i, X_{ji})} & x_i \notin X_{ji} \end{cases} \tag{15}$$

In the formula: $|X_{ji}| = |a_{ji} - b_{ji}|$.

4. The comprehensive correlation degree

$$K_j(P_0) = \sum_{i=1}^{n} \omega_i K_i(x_i) \tag{16}$$

In the formula: $K_j(P_0)$ is the comprehensive correlation degree of the *j-th* evaluation grade.

5. Grade evaluation
If $K_j = \max\{K_j(P_0)\}$ ($j = 1, 2, \ldots, m$), the regional water resources management modernization level is in grade j. When $0 < K_j(P_0) < 1$, it indicates that the unevaluated unit meets the requirements of the standard object; when $-1 < K_j(P_0) < 0$, it indicates that the evaluation unit can't meet the requirements of the evaluation grade, but it has the condition to transform to which is in accord with the standard, the smaller the value, the easier to transform; when $K_j(P_0) < -1$, it indicates that the evaluation unit can't meet the requirements, and can't transform either.

3.3 *Evaluation criteria*

In order to contact water resources management modernization with the development of economic society and construction of a comprehensive well-off society, according to the realization degree of its modernization, the level of water resources management modernization can be divided into five grades *I, II, III, IV* and *V*, respectively corresponding to the initial period, basic well-off, overall well-off, primary modernization and basic modernization (Ye 2010). According to percentile, the criteria range of water resources management modernization evaluation can reference Table 3.

4 CASE STUDY

4.1 *Water resources and water resources management in Gaochun District of Nanjing City*

Gaochun District is located in the southernmost tip of Nanjing City, land area is 566.5 km^2, water area is 235.3 km^2. The characteristics of the regional water resources are: local water is less than transit water; water in the winter and spring is less than summer and autumn; water in the hilly country in the east is less than polder in the west; quality water is less than polluted water.

In recent years, Gaochun District is generally transforming traditional water conservancy into modern water conservancy, preliminary established modern water management system which can accord with the natural attribute and objective law of water, suit to water resources management, and meet the sustainable utilization requirement.

Based on the water conservancy survey statistics of 2012 and "The implementation plan for water resources management modernization in Gaochun District of Nanjing City", the present value, modernization target value and realization degree of each index in Gaochun District water resources management modernization are shown in Table 4.

4.2 *Calculation of index weight*

According to FAHP, entropy weight method and formula (9), the subject weight ω_i', objective weight ω_i'' and comprehensive weight φ_i can be calculated and then obtain the final weight value of each index corresponds to the target layer:

$$W = (\omega_1, \omega_2, \omega_3 \ldots \omega_{20}) = (0.051, 0.040, 0.022, 0.022, 0.018, 0.019, 0.043, 0.018, 0.019, 0.018, 0.079, 0.102, 0.079, 0.041, 0.060, 0.100, 0.114, 0.052, 0.052, 0.052)$$

4.3 *Determination of the unevaluated matter element, classical domain and joint domain*

Based on the evaluation criteria constructed in section 3.3, combined with Jiangsu Province and the national average level and advanced level, international general indicators and the related references (Wang 2015, Gao et al. 2014), Gaochun District water resources management modernization evaluation grading standards can be determined, so as to establish the unevaluated matter element matrix P_0, classical domain matter element matrix R_I, R_{II}, R_{III}, R_{IV}, R_V and joint domain matter element matrix R_p. Shown in Table 5.

Table 4. Index and index value of water resources management modernization in Gaochun District.

Index		Present value of 2012	Modernization target value	Realization degree of 2012 (%)
C_1	U_1	100%	100%	100.00
	U_2	Agriculture 80%; Industry 95%; Life 95%	Agriculture 85%; Industry 95%; Life 100%	96.06
C_2	U_3	85.4 m^3/million yuan	70 m^3/million yuan	81.97
	U_4	18 m^3	15 m^3	83.33
	U_5	0.62	0.68	91.18
	U_6	12%	30%	40.00
C_3	U_7	75%	85%	88.24
	U_8	100%	100%	100.00
	U_9	4.5%/year	5%/year	90.00
	U_{10}	100%	100%	100.00
C_4	U_{11}	65%	85%	81.25
	U_{12}	85%	100%	85.00
	U_{13}	70%	80%	87.50
	U_{14}	22.22%	100%	22.22
	U_{15}	85%	95%	89.47
C_5	U_{16}	75%	85%	88.24
	U_{17}	77%	90%	85.56
	U_{18}	75%	85%	88.24
	U_{19}	60%	80%	75.00
	U_{20}	83%	95%	87.37

Table 5. Unevaluated matter element matrix, classical domain and joint domain of Gaochun Distract water resources management modernization evaluation.

Index	Unevaluated matter element P_0	Five grade classical domain matter element matrix					Joint domain
		R_I	R_{II}	R_{III}	R_{IV}	R_V	R_P
U_1	100.00	[0, 40]	[40, 60]	[60, 80]	[80, 90]	[90, 100]	[0, 100]
U_2	96.06	[0, 40]	[40, 60]	[60, 80]	[80, 90]	[90, 100]	[0, 100]
U_3	81.97	[0, 60]	[60, 70]	[70, 80]	[80, 90]	[90, 100]	[0, 100]
U_4	83.33	[0, 60]	[60, 70]	[70, 80]	[80, 90]	[90, 100]	[0, 100]
U_5	91.18	[0, 40]	[40, 60]	[60, 80]	[80, 90]	[90, 100]	[0, 100]
U_6	40.00	[0, 35]	[35, 45]	[45, 60]	[60, 75]	[75, 100]	[0, 100]
U_7	88.24	[0, 50]	[50, 60]	[60, 80]	[80, 95]	[95, 100]	[0, 100]
U_8	100.00	[0, 60]	[60, 80]	[80, 90]	[90, 95]	[95, 100]	[0, 100]
U_9	90.00	[0, 40]	[40, 60]	[60, 80]	[80, 90]	[90, 100]	[0, 100]
U_{10}	100.00	[0, 40]	[40, 60]	[60, 80]	[80, 90]	[90, 100]	[0, 100]
U_{11}	100.00	[0, 40]	[40, 60]	[60, 80]	[80, 90]	[90, 100]	[0, 100]
U_{12}	100.00	[0, 40]	[40, 60]	[60, 80]	[80, 90]	[90, 100]	[0, 100]
U_{13}	100.00	[0, 40]	[40, 60]	[60, 80]	[80, 90]	[90, 100]	[0, 100]
U_{14}	100.00	[0, 40]	[40, 60]	[60, 80]	[80, 90]	[90, 100]	[0, 100]
U_{15}	100.00	[0, 40]	[40, 60]	[60, 80]	[80, 90]	[90, 100]	[0, 100]
U_{16}	100.00	[0, 40]	[40, 60]	[60, 80]	[80, 90]	[90, 100]	[0, 100]
U_{17}	100.00	[0, 40]	[40, 60]	[60, 80]	[80, 90]	[90, 100]	[0, 100]
U_{18}	100.00	[0, 40]	[40, 60]	[60, 80]	[80, 90]	[90, 100]	[0, 100]
U_{19}	100.00	[0, 40]	[40, 60]	[60, 80]	[80, 90]	[90, 100]	[0, 100]
U_{20}	100.00	[0, 40]	[40, 60]	[60, 80]	[80, 90]	[90, 100]	[0, 100]

Table 6. Correlation degree of evaluation grades of water resources management modernization evaluation index of Gaochun District.

Grade	R_I	R_{II}	R_{III}	R_{IV}	R_V
U_1	−1.0000	−1.0000	−1.0000	−1.0000	0.0000
U_2	−0.9343	−0.9015	−0.8030	−0.6060	0.3940
U_3	−0.5493	−0.3990	−0.0985	0.1970	−0.3081
U_4	−0.5833	−0.4443	−0.1665	0.3330	−0.2858
U_5	−0.8530	−0.7795	−0.5590	−0.1180	0.1180
U_6	−0.1111	0.5000	−0.1111	−0.3333	−0.4667
U_7	−0.7648	−0.7060	−0.4120	0.4507	−0.3650
U_8	−1.0000	−1.0000	−1.0000	−1.0000	0.0000
U_9	−0.8333	−0.7500	−0.5000	0.0000	0.0000
U_{10}	−1.0000	−1.0000	−1.0000	−1.0000	0.0000
U_{11}	−0.6875	−0.5313	−0.0625	0.1250	−0.3182
U_{12}	−0.7500	−0.6250	−0.2500	0.5000	−0.2500
U_{13}	−0.7917	−0.6875	−0.3750	0.2500	−0.1667
U_{14}	0.4445	−0.4445	−0.6297	−0.7223	−0.7531
U_{15}	−0.8245	−0.7368	−0.4735	0.0530	−0.0479
U_{16}	−0.8040	−0.7060	−0.4120	0.1760	−0.1302
U_{17}	−0.7593	−0.6390	−0.2780	0.4440	−0.2352
U_{18}	−0.8040	−0.7060	−0.4120	0.1760	−0.1302
U_{19}	−0.5833	−0.3750	0.2500	−0.1667	−0.3750
U_{20}	−0.7895	−0.6843	−0.3685	0.2630	−0.1723
Comprehensive correlation degree	−0.7202	−0.6485	−0.3791	0.0479	−0.1924

4.4 *Evaluation results and analysis*

Put the data of unevaluated matter element matrix into the matter element model, and calculate the correlation degree of each evaluation index on each evaluation grade. The results are shown in Table 6.

Shown as in Table 6, $K_j = 0.0479$, so Gaochun District water resources management modernization was in the primary modernization stage, and has the tendency to enter the basic modernization stage. From correlation degree of each index can be seen, most of the indicators have reached the primary modernization level, but there are a few indicators, such as reuse rate of reclaimed water, river health assessment, still far from the target value of modernization, the relevant departments should actively promote the reclaimed water reusing project construction work, improve the utilization efficiency of water, actively establish healthy rivers and lakes assessment system, and then improve the public management ability of water resources.

Overall, the evaluation results adapt to the water conservancy modernization evaluation of Gaochun District, and consistent with the evaluation result in "The implementation plan for water resources management modernization in Gaochun District of Nanjing City", which concluded "Water resources management of Gaochun District in 2012 has not yet achieved modernization", thus the evaluation result is reliable.

5 CONCLUSION

The importance of water resources management modernization is more and more outstanding. Water resources management modernization involves many indictors, and there are few researches on the evaluation methods. In this paper, the matter element analysis method is applied to the water resources management modernization, and take the current year realization degree of each index as its value, which can make the matter element analysis model more general, the calculation more convenient, and the result more clear. The combination of subjective and objective weight is used to obtain the comprehensive weight, which make the determination of weight more scientific and reasonable. The water resource management modernization evaluation model is applied to Gaochun District of Nanjing City, the evaluation result is consistent with the actual situation, indicating that the matter element analysis method has good applicability and can be applied as an evaluation method of water resource management modernization.

In the specific application, the water resources management modernization index system should fully consider the actual situation and local water resources management characteristics, and make appropriate adjustments or supplements. In addition, the scope definition of the classical domain and joint domain of evaluation index need to be further discussed and studied.

ACKNOWLEDGEMENTS

This work was Funded by the Natural Science Foundation of Jiangsu Province (BK20130849), and Funded by the Priority Academic Program Development of Jiangsu Higher Education Institutions (PAPD).

REFERENCES

Cai W, 1987. *Matter element analysis*. Guangzhou: Guangdong Higher Education Press.

Fan ZP, Jiang YP, Xiao SH. 2001. Consistency of fuzzy judgement matrix and its properties. *Control and Decision* 16(1): 69–71.

Fan ZP, Jiang YP. 2001. Review on the research of fuzzy judgment matrix ranking method. *Systems Engineering* 19(5): 12–18.

Fang GH, Huang XF. 2011. *Theories, methodologies and applications for multi-objective decision making*. Beijing: Science Press.
Gao YS, Fang GH, Huang XF, Guan YF. 2014. Water resources management modernization evaluation of Xuzhou City based on fuzzy AHP and variable fuzzy sets theory. *Water Resources and Power* 32(4): 155–158.
Song JJ. 2013. Discussion on the methods to realizing water resources management modernization. *Jiangsu Water Resources* (08): 30–32.
Wang D. 2015. Study on assessment of water ecological civilization construction of Nanchang City. *Nanchang University*.
Xu SB. 1988. *Principle of analytic hierarchy process*. Tianjin: Tianjin University Press.
Yang L, Liu CC, Song Li, Sheng Wu. 2013. Evaluation of coal mine emergency rescue capability based on entropy weight method. *China Soft Science* (11): 185–192.
Ye J. 2010. Evaluation and prospect of the development of modernization of water conservancy in Jiangsu. *Water Resources Development Research* 10(6): 11–14.
Zhang L, Zhang JP, Chen WB. 2011. Discussion on issues of modernization of water resource management in China. *Journal of North China Institute of Water Conservancy and Hydroelectric Power* 32(1): 14–16.

Advanced Engineering and Technology III – Xie (Ed.)
 ISBN 978-1-138-03275-0

Dramatic decrease in suspended sediment concentration carried by hyper-concentrated flood in the Lower Yellow River, China

L. He
Key Laboratory of Water Cycle and Related Land Surface Processes, Chinese Academy of Sciences, Institute of Geographic Sciences and Natural Resources Research, Beijing, China

ABSTRACT: The suspended sediment concentration carried by hyper-concentrated flood in the Lower Yellow River shows a distinct decrease from Xiaolangdi to Lijin. The analysis of typical hyper-concentrated flood that occurred in 1977 shows that the decrease in suspended sediment concentration in the reach of Huayuankou–Gaocun accounts for more than 70% of the total decrease observed in the Lower Yellow River. In addition, both the averaged carrying capacity at JHT and suspended sediment carried by larger discharge (higher than 5800 m^3/s) at Gaocun decreased dramatically. Furthermore, the dramatic decrease in suspended sediment concentration in the reach Huayuankou–Gaocun is mainly caused by overflooding around the cross-section of Jiahetan. This finding may be beneficial to the management of the river channel.

1 INTRODUCTION

The Suspended Sediment Concentration (SSC) of hyper-concentrated floods in the Lower Yellow River (reaches between the Xiaolangdi Hydrology station and the Lijin Hydrology station) shows a dramatic decrease along the river channel. For the averaged SSC of hyper-concentrated flood during the flood season, the averaged SSC at Aishan (AS) decreases by about 81% compared with that at Huayuankou (Chen 2014). The reason for this kind of dramatic decrease differs between studies. The SSC of flood that occurred in September 7, 1977 was about 911 kg/m^3, which is the maximum concentration reported. So, the hyper-concentrated flood event that occurred in 1977 is taken as a typical example to investigate the reason for the dramatic decrease in SSC along the river reach.

2 STUDY REACH

The Lower Yellow River (LYR) runs from Taohuayu to Lijin (LJ), and the length is approximately 786 km, as shown in Figure 1 (He et al. 2012a). The LYR can be divided into three reaches with distinctly different geomorphologies (Wu et al. 2005). The first reach is characterized with a typical braided pattern and runs from Taohuayu to Gaocun (GC). The second reach lies between GC and Aishan (AS), which is a transition reach between the braided pattern and the meandering pattern (Wu et al. 2005, Xia et al. 2010). The third part runs from AS to LJ, which is characterized by a meandering pattern. The lengths of the three reaches are about 207 km, 165 km, and 300 km, respectively. The sinuosities of the three reaches are about 1.15, 1.33 and 1.21, respectively. The slope of the channel bed in the braided reach decreases from 0.000265 to 0.000172. The slope of the river channel in the transition reach is about 0.000115, which is smaller than that of the braided reach. The slope of the meander reach is about 0.0001, which is the smallest among the three reaches. The distances between the levees of the three reaches are about 5–20 km, 1.4–8.5 km, and 0.45–5 km, respectively.

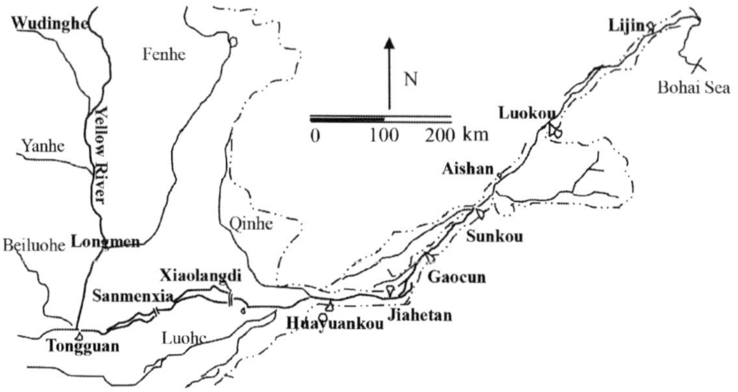

Figure 1. Sketch map of the Lower Yellow River.

3 ANALYSIS

There are two hyper-concentrated flood events that occurred in the LYR in 1977. The first flood event occurred during 7th to 14th, July. The second flood event occurred during 4th to 12th, August. The discharge-concentration relationships and carrying capacity are analyzed by discharge-concentration relationships and numerical simulation.

3.1 *Analysis of discharge-concentration relationships*

The discharge-concentration relationships of the first flood event at the eight hydrology stations are constructed. As the different channel formation of reaches in the LYR, relationships at HYK, JHT, GC, and LJ are shown in Figure 2. The daily SSCs are relatively high in the upper reach (maximum daily SSC at HYK is higher than 450 kg/m^3), which decreases from the upper stream to the downstream (maximum daily SSC at GC decreases to 250 kg/m^3). For the first flood event, the maximum daily SSC decreases from approximately 481 kg/m^3 at XLD to roughly 181 kg/m^3 at LJ. In addition, it reveals a dramatic decreased maximum daily SSC in the reach of HYK-GC, as the maximum daily SSCs at HYK and GC are 465 kg/m^3 and 230 kg/m^3, respectively (Fig. 2). The distances of HYK and GC from the Xiaolangdi reservoir are 132 km and 309 km, respectively. So, the maximum daily SSCs at GC decrease by about 51% compared with the SSC at HYK, and the length of the river reach of HYK-GC accounts for about 28% of the length of HYK-LJ. For the second flood event, the maximum daily SSCs at GC decrease by about 34% compared with the SSC at HYK. So, for these two flood events, the decrease in the reach of HYK-GC accounts for about 77% of the total decrease of SSC along the river reach of XLD-LJ.

For individual hydrological events, the discharge-concentration relationship can be divided into five possible forms by comparing the ratio of concentration over discharge on the increasing and decreasing stages for the same discharge (Williams 1989). Moreover, the typologies of the discharge-concentration relationship that occurred in 1977 among these eight stations are in the counterclockwise loop (Fig. 2). As shown in Figure 2, the typologies of the discharge-concentration relationship can be divided into three stages; constant concentration with increasing discharge, constant discharge with dramatically increased concentration, and decreased discharge and concentration. The pattern of concentration over discharge in the HYK-JHT reach is different from that of the GC-LJ reach. According to the decrease in concentration during the falling limb, the typologies of the discharge-concentration relationship among these eight stations can be divided into two groups. The first group of discharge-concentration typology can be found at the HYK-JHT reach, whereas the second group of topology may be displayed at the GC-LJ reach. The difference between the first group (such as HYK) and the second group (such as LJ) is that the concentration increase in the first

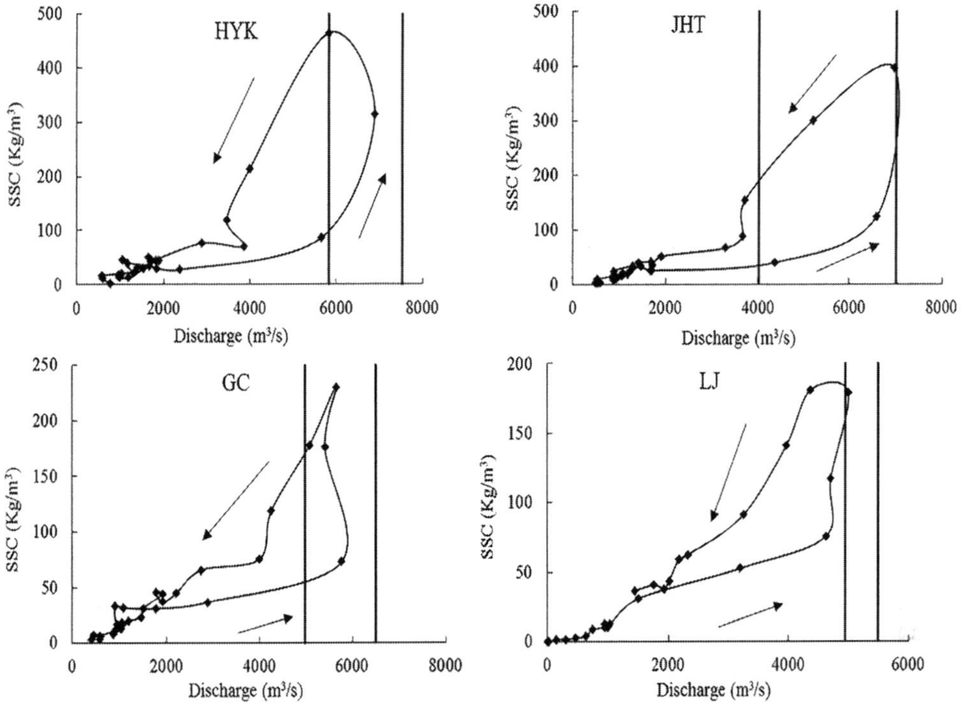

Figure 2. Typical hysteresis effect observed in suspended sediment during the flood events that occurred from June 25 to July 22, 1977 in the LYR. The arrows indicate the increase in the discharge during the flood events.

group is more dramatic. The counterclockwise loop of the discharge-concentration relationship indicates that the bed material is abundant and could be eroded by large floods. Thus, large discharge leads to the possibility of a large amount of sediment transport rates.

The ratio of concentration over discharge at GC is much less than that of the upper reach (Fig. 3). It means that, the decreasing SSC at large discharge is more apparent at GC. For example, for the first flood events, the SSC at GC is much less than that of the XLD, HYK, and JHT, when the discharge is approximately more than 5800 m³/s. For XLD, the SSC at the rising limb is larger than that at HYK, which is influenced by the flood output of joint operation of the Xiaolangdi reservoir and the Sanmenxia reservoir. The vertical lines shown in Figure 3 represent the maximum and minimum values of bankfull discharge at each hydrology station (He et al. 2015), and the averaged values are viewed as the measured bankfull discharge. The significantly increased wetted area and decreased flow velocity in over flooding flow may lead to reduced flow transport capacity and deposition, as a result, sediment concentration may show a declining trend. According to the measured bankfull discharge, the over-flooding at JHT is obvious. The hydrography of SSC at the upper reach of HYK is delayed compared with the flood hydrography, and the delay is about one day. However, the delay diminished at JHT. It means that, the SSC of flood occurred at JHT increases to its maximum value when the discharge reaches the maximum value.

Deposition in the HYK-GC reach accounts for about 38% of the total deposition in the LYR. More importantly, there are serious deposition on the floodplain and significant scouring in the main channel, and the volume of deposition on the floodplain is greater than that of bed scouring in the main channel (Xia et al., 2010). As for the serious deposition in the HYK-GC reach, the discharge-concentration relationships of the second flood event are more complex. For GC, the SSCs at both low and high discharge are smaller than that of cross-sections in the upper reach, e.g. XLD (Fig. 3).

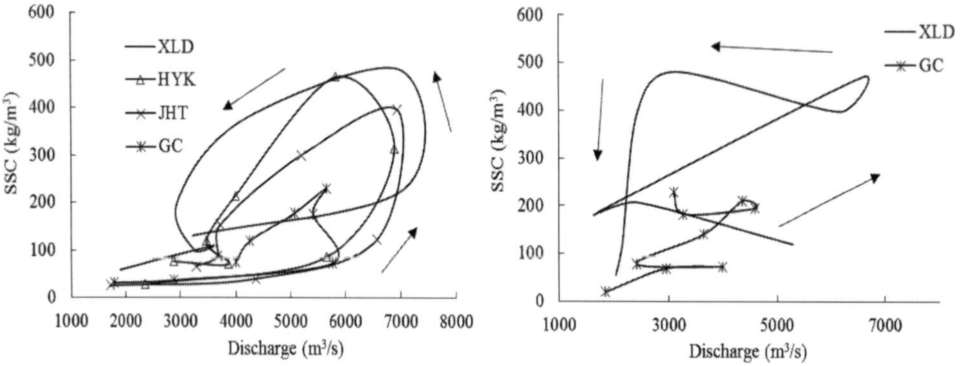

Figure 3. Typical hysteresis effect observed in suspended sediment during the two flood events that occurred in 1977 in the LYR. The arrows indicate the increase in the discharge during the flood events.

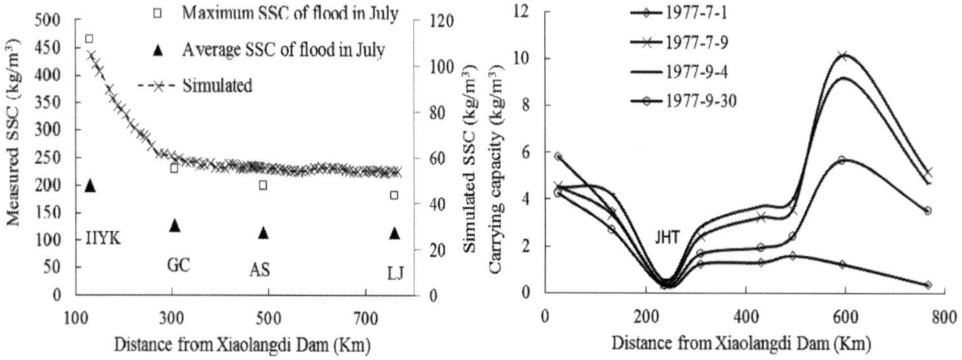

Figure 4. Longitudinal variation of SSC and carrying capacity in the Lower Yellow River.

3.2 *Analysis of carrying capacity*

To analyze the carrying capacity, the averaged value of the whole flood season is more reasonable. So, a one-dimensional model for hyper-concentrated flood is used to simulate the flood propagation of flood that occurred in 1977 (He et al., 2012b). The model simulated the flood prorogation at the reach between XLD and LJ. It simulated the flood that occurred during July and August, 1977. The simulated SSC, shown in Figure 4, is the averaged concentration of the whole simulated period. For the simulated averaged value, the SSC decreases from 104.8 kg/m^3 (HYK) to 54 kg/m^3 (LJ), and the ratio of decrease is about 48.5%. Compared with simulated SSC at HYK, the averaged SSC at GC decreased by 44%. So, the decreasing SSC at the reach of HYK-GC is more apparent. For simulated average SSC, the decrease in the HYK-GC reach accounts for 91% of the total decrease of SSC along the river reach of XLD-LJ.

According to the measured data, the averaged SSC of the first flood event decreased from 199 kg/m^3 (HYK) to 115 kg/m^3 (LJ), and the averaged SSC at LJ decreased by 42% compared with that at HYK (Chen 2014). For the second flood event, the averaged SSC at LJ decreased by 45% compared with that at HYK (Chen 2014). For the measured average SSC, the decrease in the HYK-GC reach accounts for about 71% of the total decrease of SSC along the river reach of XLD-LJ. So, the simulated averaged carrying capacity can be adopted to analyze the variation of averaged carrying capacity along the river reach.

In the LYR, high sediment concentration will change the topography of cross-sections dramatically and quickly, characterized by serious erosion or deposition in the main channel by peak discharge (Shu & Fei 2008). The typical cross-section at JHT is shown in Figure 5. As to the serious deposition in the XLD-GC reach, the averaged carrying capacity decreases

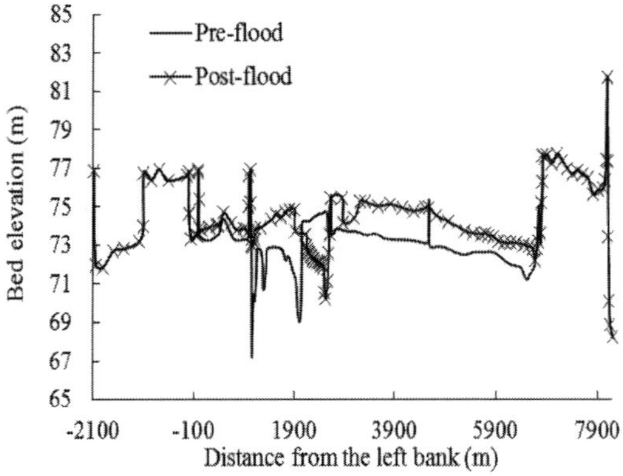

Figure 5. Cross-sectional geometries of pre-flood and post-flood at JHT in the Lower Yellow River.

during the reach of XLD-JHT, and then it increases in the reach down the hydrology station of JHT (Fig. 4). Comparing with the averaged carrying capacity at XLD, the averaged carrying capacity at JHT decreased by more than 90%.

So, the overflooding that occurred at JHT leads to the reduced carrying capacity at JHT and serious deposition around the cross-section of JHT. Most of the sediments carried are deposited, which leads to dramatically decreased concentration at GC when the discharge reaches the bankfull discharge.

4 CONCLUSIONS

Based on the measured data of hyper-concentrated flood that occurred in 1977, the SSC along the LYR are analyzed by typologies of discharge-concentration relationships and numerical simulation. The analysis shows that, the decrease in SSC in the reach of HYK-GC accounts for more than 70% of the total decrease in the LYR. The main reason is overflooding around the cross section of JHT, which leads to serious deposition and dramatically decreased carrying capacity at JHT. It also leads to dramatically decreased SSC carried by flood with a discharge more than 5800 m^3/s at GC. This finding can provide information for the management of the related river channel.

ACKNOWLEDGMENTS

This research was supported by (1) the "Twelfth Five Year" National Science and Technology Support Program (No. 2012BAB02B02), (2) the National Program on Key Basic Research Project (973 Program) (No. 2011CB403305), and (3) the National Natural Science Foundation of China (Nos. 51579230 and 51109198).

REFERENCES

Chen, A.J. 2014. Study of sediment concentration variation of hyper-concentration floods in Lower Yellow River *Tsinghua University* (in Chinese).

He, L., Wang, G.Q., Zhang, C. 2012a. Application of Loosely Coupled Watershed Model and Channel Model in Yellow River, China. *Journal of Environmental Informatics* 19(1): 30–37.

He, L., Duan, J.G., Wang, G.Q., Fu, X.D. 2012b, Numerical Simulation of Unsteady Hyperconcentrated Sediment-Laden Flow in the Yellow River. *Journal of Hydraulic Engineering* 138(11): 1–12.

He, L., Yan, Y.X., Yan, M. 2015. Analysis on the definition of bankfull stage by geometric criterion. *Journal of Hydroelectric Engineering* 34(5): 114–118 (in Chinese).

Shu, A.P., & Fei, X.J. 2008. Sediment transport capacity of hyperconcentrated flow. *Science in China Series G: Physics Mechanics and Astronomy* 51(8): 961–975.
Williams, G.P. 1989. Sediment concentration versus water discharge during single hydrologic events in rivers. *Journal of Hydrology* 111(1–4): 89–106.
Wu, B.S., Wang, G.Q., Ma, J.M., Zhang, R. 2005. Case study: River training and its effects on fluvial processes in the Lower Yellow River, China. *Journal of Hydraulic Engineering* 131(2): 85–96.
Xia, J.Q., Wu, B.S., Wang, G.Q., Wang, Y.P. 2010. Estimation of bankfull discharge in the Lower Yellow River using different approaches. *Geomorphology* 117(s1–2): 66–77.

Advanced Engineering and Technology III – Xie (Ed.)
 ISBN 978-1-138-03275-0

Projection pursuit evaluation for the water ecological civilization

Xianfeng Huang, Yongle Jia & Guohua Fang
College of Water Conservancy and Hydropower Engineering, Hohai University, Nanjing, China

ABSTRACT: Based on the notions and requirements of constructing water ecological civilization city, the evaluation index system of water ecological civilization composed of 29 indicators is constructed based on five aspects, namely water security, water ecology, water management, water landscape, and water cultures. According to the economic and social development level and regional characteristics, five evaluation grade standards are established including ideal, harmonious, basically harmonious, inharmonious, and extremely inharmonious. Furthermore, a grade evaluation model is built based on the accelerating genetic algorithm (RAGA) Projection Pursuit (PP). Then, water ecological civilization of Maanshan City in Anhui province, China is taken as an example. The result indicates that the water ecological civilization of Maanshan City in 2012 is in the basic state of harmony, which is consistent with the reality.

1 INTRODUCTION

The notion of ecological civilization evolved in the 1960s (Liu 2013). Since then, many scholars around the world have carried out a lot of research work. As an important part, water ecological civilization is an extension of ecological civilization from the perspective of literal meaning and connotation (Zuo 2013). The Ministry of Water Resources of the People's Republic of China issued "Views on Accelerating Water Ecological Civilization Construction" to promote the construction of ecological civilization water in January 2013. However, there still exit many problems in China, such as, relative lack of water resources, single function of water ecosystem, few robust institutions for water management, and low public welfare of water landscape. The persistence of these problems will, to a certain extent, result in the construction of water ecological civilization in China, being more arduous and worthwhile.

Water ecological civilization refers to the sum of material and spiritual wealth created by humans in the process of protecting the water ecosystem and living in harmony with water (Tang 2013). It also refers to the harmonious coexistence between the human society and the water resource system. Modern cities cannot be well developed without water, which rely greatly on the coordinated use of urban facilities for water supply, waste water discharge, sewage treatment, flood control, and drainage. In the early stage of urban ecological planning, people mainly focused on the adjustment of urban structure, urban configuration, and art design. Until the early 1990s, research gradually focused on the problems of urban water and water circulation (Wang 2003). At present, domestic and foreign research on urban water ecological civilization is still in the initial stage. The main issues include an unclear definition of the substance of urban water ecological civilization, the lack of a coordinated water ecological civilization construction mode involving human, water, and nature, and the absence of a systematic and comprehensive index system for evaluating urban water ecological civilization (Hu 2001).

In this paper, an evaluation index system is constructed based on the analysis of connotations and characteristics and the basic requirements of constructing ecological civilization cities. In view of the complexity of the water ecological civilization system (Fu 2016), a grade evaluation model is built using the projection pursuit method and the RAGA to obtain the optimal projection vector. This model can be used to evaluate the state of urban water ecological civilization comprehensively and objectively, as well as to provide some appropriate recommendations on how to improve the water ecological civilization grade.

2 EVALUATION INDEX SYSTEM OF CITY WATER ECOLOGICAL CIVILIZATION

The evaluation index system method first selects a series of representative indices constituting the evaluation index system by considering some principles and combining with the specific situation of the region evaluated. Then, it chooses an appropriate mathematical analysis method to process the index system data and to obtain a comprehensive parameter. The comparison of the comprehensive parameters and related standards can reflect the current situation or the threshold value of the evaluation area (Ma 2012).

In recent years, with the increasing emphasis and investment on ecological civilization construction, the corresponding research is unceasingly thorough, especially on the evaluation index system. Hua Gao divided the whole system which containing 23 evaluation indices into

Table 1. Evaluation index system of urban water ecological civilization.

System layer A	State layer B	Index layer C
Water security	Flood control and drainage	Standard-reaching rate of urban flood control (*C1*)
		Standard-reaching rate of city drainage (*C2*)
	Water quality	Compliance rate of water quality in urban water function areas (*C3*)
Water ecology	Regional water environment	Sustainability of ecological water requirement (*C4*)
		Degree of maintaining water environment (*C5*)
		Percentage of urban water surface (*C6*)
	Animals and plants living in rivers and lakes	Richness degree of aquatic species (*C7*)
		Rationality of plant disposition (*C8*)
	Water and soil conservation	Rate of land restoration (*C9*)
		Rate of water loss and soil erosion control (*C10*)
		Grass and forest coverage (*C11*)
Water management	Water resources management	Water consumption per 10,000 GDP (*C12*)
		Water consumption of large-scale industrial added value (*C13*)
		Discharge standard rate of effluent (*C14*)
		Reuse rate of urban sewage after treatment (*C15*)
		Leakage rate of water supply pipe network (*C16*)
		Popularization of water-saving society (*C17*)
		Effect of water source protection (*C18*)
	Engineering management	Realization of engineering achieving flood control and water supply standards (*C19*)
		Intact degree of engineering facilities (*C20*)
		Implementation of "three simultaneous" in water and soil conservation projects (*C21*)
	Planning compilation	Amount of planning presented (*C22*)
	Governmental functions	Qualified rate of managing water conservancy project (*C23*)
		Matching laws and regulations (*C24*)
Water landscape	Natural water landscape	Natural landscape (*C25*)
		Water conservancy scenery (*C26*)
	Urban water landscape	Length of pro waterfront (*C27*)
Water culture	Concrete performance	Cultural characteristics of waters and scenery (*C28*)
	Propaganda and education of water culture	Popularization of water culture (*C29*)

the water resources system, water ecological system, water landscape system, water engineering system, and water management system combined with research on an urban water ecological civilization evaluation system in Shandong Province (Gao 2013). Kejian Chu constructed an evaluation indicator system of water ecological civilization composed of three classes and 26 indices for reservoir and river network regions in the hilly areas of lower Yangtze River, based on four aspects including water resource security, water ecological environment, water cultures, and water management (Chu 2015). Haijiao Liu built an evaluation index system of water ecological civilization based on four aspects, namely water resources development and utilization, water ecological environment protection, water landscape construction and water management (Liu 2013); Fuqiang Tian developed an evaluation system of urban water ecological civilization consisting of 20 indices based on six aspects, besides the strictest water resources management and the optimal allocation of water resources, and applied it to assess the current construction situation of urban water ecological civilization in Zhengzhou (Wang 2015).

On the basis of existing research, according to the "Notice for Building Water Ecological Civilization Cities and Countryside with Beautiful Water Scenery Pilot" issued by the Water Conservancy Department of Anhui Province, China, following the systematic, operational, hierarchical, combination of quantitative and qualitative principles (Wang 2013) and focusing on water security, water ecology, water management, water landscape, and water culture, this paper establishes a unique evaluation index system of urban ecological civilization. The indicators are detailed in Table 1.

3 ESTABLISHMENT OF THE MODEL FOR EVALUATING URBAN WATER ECOLOGICAL CIVILIZATION

Grade evaluation of urban water ecological civilization is based on representative index values that comprehensively evaluate the level of a region with the mathematical model established and provide the policy-making basis for the region's water ecological civilization construction. As the indices used in the actual operation are multi-dimensional, non-linear, and incommensurable, the traditional data analysis methods are limited by too many variables, leading to difficulty in finding the inherent law of the data. Projection pursuit is an effective exploratory analysis method that can be used to process and analyze these data, especially non-normality population distribution and high-dimensional data. Therefore, this study uses the projection pursuit grade evaluation model based on the RAGA to assess the harmonious degree of urban water ecological civilization evaluation.

3.1 *The basic idea of the projection pursuit method*

Projection pursuit method provided by the sample data is put forward to avoid the formalization and mathematization of the comprehensive evaluation method of a conventional system. This method projects the high-dimensional data onto a lower-dimensional subspace through a given combination, and determines these projected values that can reflect the structure or feature of high-dimensional data by maximizing or minimizing the projection indices. Thus, it allows us to study high-dimensional data easily by simply analyzing the data from a low-dimensional space (Fang 2011). Meanwhile, this method can eliminate the effects of irrelevant projection direction, detect how the single index influences the composite score in the data analysis process, and avoid the uncertainty and subjectivity of artificial determining index weights.

3.2 *Calculation steps of the projection pursuit method*

The modeling process of the projection pursuit model is divided into four main steps, i.e. normalizing the original data, constructing a projection index function, optimizing the projection index function and grade evaluation, and particularly obtaining a projection vector that can reveal the characteristics of data further by optimizing the projection index function.

1. Normalization processing of the sample evaluation index set
   For the benefit indices:

$$x(i,j) = \frac{x^*(i,j) - x_{min}(j)}{x_{max}(j) - x_{min}(j)} \quad (1)$$

   For the cost indices:

$$x(i,j) = \frac{x_{max}(j) - x^*(i,j)}{x_{max}(j) - x_{min}(j)} \quad (2)$$

   where $x(i, j)$ indicates the normalized value of the jth index of the ith sample; $x^*(i, j)$ indicates the original value of the jth index from the ith sample; and $x_{max}(j)$ and $x_{min}(j)$ indicate the maximum and minimum values of the jth index, respectively.

2. Construction of the index function
   The projection pursuit method is used to project the p-dimension data onto a one-dimensional projection vector $a = (a(1), a(2), \ldots, a(p))$ based on projection values, i.e.

$$z(j) = \sum_{j=1}^{p} a(j)x(i, j), \quad j = 1, 2, \ldots, n \quad (3)$$

   where a is a unit vector, that is $\sum_{j=1}^{p} a^2(j) = 1$.

   When constructing the projection index, it is required that projected value characteristics should be locally dense as much as possible, preferably agglomerating into several point groups, but, on the whole, the projected point groups scatter. Therefore, the projection index function can be expressed as follows:

$$Q(a) = S_z D_z \quad (4)$$

   where S_z indicates the standard deviation of $z(i)$ and D_z indicates the local density of the projection value $z(i)$, i.e.

$$S_z = \sqrt{\frac{\sum_{i=1}^{n}(z(i) - E(z))^2}{n-1}} \quad (5)$$

$$D_z = \sum_{i=1}^{n}\sum_{j=1}^{n}(R - r(i, j))u(R - r(i, j)) \quad (6)$$

   where $E(z)$ is the average of the projected value sequence. R is the window radius of the local density, whose value can be determined according to the experiment, and the range of the parameter can be determined as $r_{max} + p/2 \le R \le 2p$. In practical applications, 10% of the projection sample variance is generally taken as the value of R to make the projection index deviate from the normal distribution of the highest degree. Here $r(i, j)$ indicates the distance between two samples, with $r(i, j) = |z(i) - z(j)|$ and $u(t)$ indicates the unit step function, i.e. $u(t) = 1$ when $t \ge 0$, and $u(t) = 0$ when $t < 0$.

3. Optimization of the projection index function
   From the formula for calculating the projection index function $Q(a)$, it is easy to know that $Q(a)$ is only related to the projection direction a when the evaluation index and the sample set are given. Therefore, the optimal projection direction can be estimated by solving the maximum of $Q(a)$, i.e.

$$\max\ Q(a) = S_z D_z$$
$$s.t.\ \sum_{j=1}^{p} a^2(j) = 1 \tag{7}$$

This is a complex nonlinear optimization problem, which can be solved by using the acceleration genetic algorithm.

4. Grade evaluation
   By substituting the best projection direction solved in the third step into formula (3), the evaluation level of sample points projected values $z^*(i)$ can be obtained. The projection pursuit grade evaluation model $y^* = f(z)$ can be established according to the level of samples and the corresponding projection value. Then, by calculating the projection value of the sample to be evaluated, the grade of the sample can be obtained by substituting it into $y^* = f(z)$.

4 CASE STUDY

Based on the projection pursuit method, the grade evaluation model is applied to the water ecological civilization of Maanshan City in 2012. Maanshan City (31°24′~32°02′N and 117°53′~118°52′E) is located in the easternmost tip of Anhui province, China. Its total land area is 4049 km^2. It has rich river networks and three lakes with ecological function, accounting for 15.9%.

4.1 *Classification of urban water ecological civilization*

At present, there is no clear standard for the classification of urban water ecological civilization. Thus, this paper classifies urban water ecological civilization into five grades, namely ideal, harmonious, and basically harmonious, inharmonious, and extremely inharmonious, ranging from I to V based on the domestic and foreign economic and social levels and combined with the natural environment of Maanshan City, as listed in Table 2.

With the wide application of research, for quantitative indicators, the status quo and grade standard adopt the specific value. In contrast, for qualitative indicators, the value of the status quo adopts the realization degree to achieve the target value in 2016. The realization degree can be calculated using equation (8) as follows:

$$f = \frac{C_i}{G_i}, \quad i = 1, 2, \ldots, p \tag{8}$$

where C_i indicates the status quo value of the ith index in 2012 and G_i indicates the target value of the ith index in 2016.

4.2 *Application of the comprehensive grade evaluation model*

The projection pursuit model can be programmed using Matlab by following the previous steps. The optimal projection vector can be expressed as a^* = (0.1790, 0.1712, 0.1796, 0.1927, 0.1944, 0.1923, 0.1841, 0.1976, 0.1990, 0.1825, 0.1756, 0.1808, 0.1833, 0.1726, 0.1963, 0.1829, 0.1860, 0.1709, 0.1877, 0.1860, 0.1989, 0.1894, 0.1887, 0.1875, 0.1895, 0.1836, 0.1955, 0.1827, 0.1695). Meanwhile, the corresponding projection values of I, II, III, IV, and V are 5.3797, 4.6840, 3.5896, 2.5043, and 0, respectively. The status projection value of Maanshan City in 2012 is 4.6255.

According to the best projection characteristic value of each grade, a scatter diagram of the projection characteristic value can be presented (Figure 1), and the evaluation model of urban water ecological civilization can be established as follows:

$$y^* = -0.001(z^*)^3 - 0.1042(z^*)^2 - 0.1477z^* + 5.0028, \quad R^2 = 0.9984 \tag{9}$$

Table 2. Status value of Maanshan City in 2012 and evaluation index grade.

Evaluation index		Status value	Standard values of different grades				
			I	II	III	IV	V
(*C1*)	Standard-reaching rate of urban flood control (%)	80	100	80	65	50	0
(*C2*)	Standard-reaching rate of city drainage (%)	60	100	80	65	50	0
(*C3*)	Compliance rate of water quality in urban water function areas (%)	70	95	80	60	50	0
(*C4*)	Sustainability of ecological water requirement (%)	83	90	80	60	40	0
(*C5*)	Degree of maintaining water environment (%)	85	90	80	60	40	0
(*C6*)	Percentage of urban water surface (%)	12	25	20	10	5	0
(*C7*)	Richness degree of aquatic species (%)	75	90	80	60	40	0
(*C8*)	Rationality of plant disposition (%)	80	90	80	60	40	0
(*C9*)	Rate of land restoration (%)	84	90	80	60	40	0
(*C10*)	Rate of water loss and soil erosion control (%)	80	85	80	70	60	0
(*C11*)	Grass and forest coverage (%)	42.5	50	40	30	15	0
(*C12*)	Water consumption per 10,000 GDP (m^3)	135	100	200	300	400	1000
(*C13*)	Water consumption of large-scale industrial added value (m^3)	67	5	10	20	40	200
(*C14*)	Discharge standard rate of effluent (%)	87.5	95	80	60	40	0
(*C15*)	Reuse rate of urban sewage after treatment (%)	12	50	40	30	20	0
(*C16*)	Leakage rate water supply pipe network (%)	21.2	10	15	20	25	30
(*C17*)	Popularization of water-saving society (%)	80	90	80	60	40	0
(*C18*)	Effect of water source protection (%)	85	90	80	60	40	0
(*C19*)	Realization of engineering achieving flood control and water supply standards (%)	80	90	80	60	40	0
(*C20*)	Intact degree of engineering facilities (%)	83.4	85	75	65	50	0
(*C21*)	Implementation of "three simultaneous" in water and soil conservation projects (%)	78	95	85	70	50	0
(*C22*)	Amount of planning presented (%)	0	90	80	60	40	0
(*C23*)	Qualified rate of managing water conservancy project (%)	80	90	80	60	40	0
(*C24*)	Matching laws and regulations (%)	80	90	80	60	40	0
(*C25*)	Natural landscape (%)	90	90	80	60	40	0
(*C26*)	Water conservancy scenery (%)	65	90	80	60	40	0
(*C27*)	Length of pro waterfront (%)	80	90	80	60	40	0
(*C28*)	Cultural characteristics of waters and scenery (%)	85	90	80	60	40	0
(*C29*)	Popularization of water culture (%)	90	90	80	60	40	0

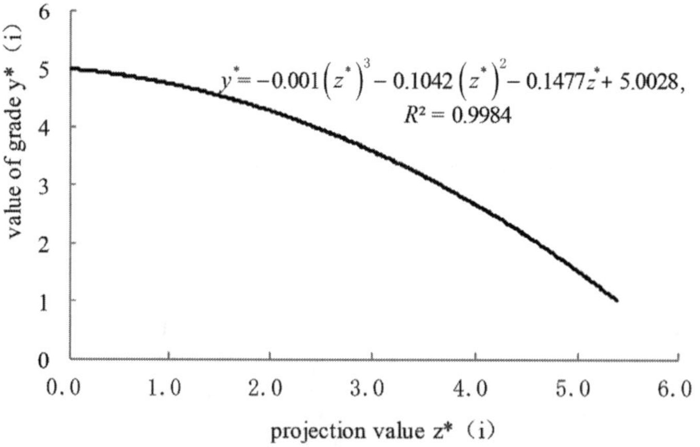

Figure 1. The relationship between the projection eigenvalue and the water ecological civilization level of the city.

Table 3. The error analysis of the projection pursuit grade assessment model.

Empirical values	1	2	3	4	5
Calculated value	1.0368	1.9221	3.0837	3.9637	5.0028
Absolute error	0.0368	0.0779	0.0837	−0.0363	0.0028
Relative error	3.68%	−3.89%	2.79%	−0.91%	0.06%

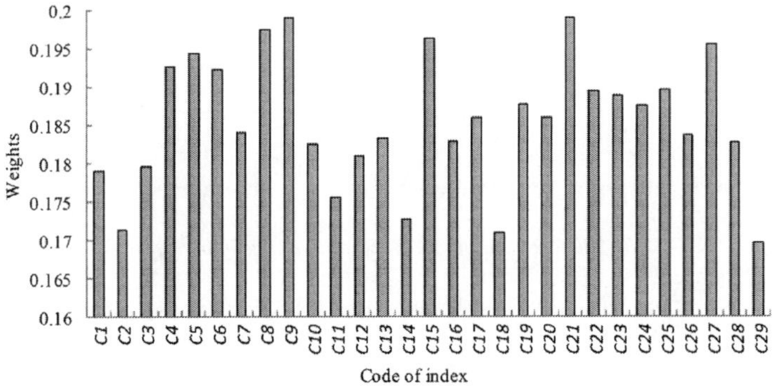

Figure 2. The histogram of single index weights.

Then, the error of the evaluation model established can be analyzed. The results are summarized in Table 3.

The mean absolute error of the evaluation model is 0.0475 and the average relative error is 2.27%, as given in Table 3. Thus, the grade evaluation model established has high accuracy and can be used to evaluate the level of urban water ecological civilization. By substituting the status projection values of Maanshan City into equation (9), it can be shown that the corresponding water ecological civilization is in the state of harmony. It also shows that the model is convenient for evaluating urban water ecological civilization.

In addition, each component size of the best projection direction reflects the degree to which the evaluation index significantly influences the level of urban ecological water civilization. The weight histogram is shown in Figure 2.

According to the evaluation results, the overall view of water ecological civilization in Maanshan City achieved the basic harmony. However, there is great room for improvement because some indicators of the individual level are still very low.

5 CONCLUSION

Based on the notions and basic requirements of urban water ecological civilization, an evaluation index system of urban water ecological civilization composed of 29 indicators was constructed based on aspects such as security of water resources, water ecology, water management, water landscape, and water culture. A total of five an evaluation grade standards were formulated based on the national planning and domestic and foreign economic and social development levels.

The projection pursuit grade evaluation model was constructed. The theoretical basis of this evaluation method was solid and the result was accurate, objective, and universal. The best projection direction calculated by the evaluation standard value sample showed the influence degree of each individual index on the overall state of the water ecological civilization, having a certain reference value and guiding significance to the actual work. By evaluating the status of water ecological civilization of Maanshan City using the evaluation model, the results obtained are found to be consistent with the reality. All these findings prove that the proposed model is feasible and can be applied to evaluate the status of water ecological civilization of similar cities.

ACKNOWLEDGMENTS

This work was funded by the Natural Science Foundation of Jiangsu Province (BK20130849) and the Priority Academic Program Development of Jiangsu Higher Education Institutions (PAPD).

REFERENCES

Chu KJ, Chou KF, Jia YZ. 2015. Evaluation indicator system of city water ecological civilization in-reservoir and river network region in hilly areas of the lower Yangtze River. *Sichuan Environment* 34:44–51.

Fang GH, Huang XF. 2011. *Theories, methodologies and applications for multi-objective decision making*. Beijing: Science Press.

Fu XM, Fang GH, Huang XF. 2016. Coordination degree evaluation of city aquatic ecological civilization: the case of Maanshan city. *South to North Water Transfers and Water Science & Technology* 01:21–25.

Gao H, Cao XY, Cai BG. 2013. Study on Evaluation System for Water Eco-Civilization City in Shandong Province. *China Water Resources* 10:08–10.

Hu C, Chu JY, He SM. 2001. A review of water ecological civilization connotation and index system construction. *Yellow River* 12:74–76.

Liu W. 2013. The basic theory of ecological civilization and the research status commentary. *Ecological Economy* 02: 34–37.

Liu HJ, Huang JW, Shi YZ. 2013. Water ecological civilization evaluation on the typical urban of the lower yellow river. *Yellow River* 12:64–67.

Ma F, Wang Q, Wang GL. 2012. Evaluation of Water Resources Carrying Capacity Based on Index System with Parameter Projection Pursuit Model. *South to North Water Transfers and Water Science & Technology* 10:62–66.

Tang KW. 2013. Discussion on concept and assessment system of aquatic ecological civilization. *Water Resources Protection* 04:1–4.

Wang JH, Hu P. 2013. Studies on evaluation system of water ecological civilization. *China Water Resources* 15:39–42.

Wang FQ, Wang L, Wei HB. 2015. Assessment of current construction situation of urban water ecological civilization in Zhengzhou. *South to North Water Transfers and Water Science & Technology* 13:639–642.

Wang PF, Wang C, Feng J. 2003. Advances in research of urban water ecosystem construction mode. *Journal of Hohai University Natural Sciences* 05:485–489.

Zuo QT. 2013. Discussions on key issues of water ecological civilization construction. *China Water Resources* 04:2–3.

Advanced Engineering and Technology III – Xie (Ed.)
© 2017 Taylor & Francis Group, London, ISBN 978-1-138-03275-0

Influence of hydroelectric projects of Mekong River in Laos based on text analysis

L.X. Li & Y.L. Zhao
Business School, Hohai University, Nanjing, Jiangsu, China

ABSTRACT: According to the classification and screening of all news about the Mekong River basin in recent years, this paper analyzed the background of the hydroelectric projects of Mekong River in Laos, including the cause of dispute over the dams, and the pros and cons of the construction. The construction of the dams in the basin of Mekong River is the key to all conflicts between Laos and other countries. It is worth considering how to coordinate the interests of all parties in order to realize both the development and utilization of water resources and protection of the ecological environment.

1 INTRODUCTION

The main stream of the Mekong River runs a total of 4,908 km. It is deemed as the most important cross-border river system. Mekong River originates from Zadoi County, Yushu Tibetan Autonomous Prefecture, Qinghai Province, and China. The river flows across Yunnan Province in China, Laos, Myanmar, Thailand, Cambodia, and Vietnam and runs into the South China Sea at Ho Chi Minh City, Vietnam.

As early as the 1950s, economic development in the Mekong River basin had attracted the attention of developed countries, especially France and America. In 1957, by taking the opportunity provided by the USA to conduct cooperation in the basin of Mekong River, four downstream countries, namely Laos, Thailand, Cambodia, and Vietnam, co-founded the Mekong River Downstream Coordination Committee. The Committee was responsible for planning the economic development in the downstream basin and coordinating the cooperation among these countries (China Economic Net, 2011).

In the 1960s, advocated by the World Bank, the four countries established the "Mekong River Sub-regional Project Development Plan", which aimed to coordinate the development in this region. However, in the following two decades, the development in the Mekong River region was suspended as a result of frequent wars in the sub-region of Mekong River.

In 1992, promoted by the Asian Development Bank, the "Greater Mekong River Sub-regional Economic Cooperation Mechanism" was set up by China, Thailand, Myanmar, Vietnam, and Cambodia, aiming to boost sub-regional economic and social development by strengthening economic ties among their members.

In December 1995, the ASEAN Summit proposed a new idea: the cooperative development in the Mekong River basin. Meanwhile, the ASEAN Summit invited Japan, Korea, Asian Development Bank, and World Bank to take part in the investment. Then, based on the development idea, China, Vietnam, Thailand, Cambodia, and Laos planned to build 12 dams in the Mekong River for hydroelectric generation and agricultural irrigation.

Laos is a Mekong River riparian country with a territory area of 236.8 thousand km^2. It adjoins China in the north, Cambodia in the south, Vietnam in the east, Myanmar in the northwest, and Thailand in the southwest. The north part of Laos is higher than the south and borders with the Western Yunnan Plateau of China. The plateau lies in the east, constituted from Truong Son Ra between Laos and Vietnam. Mekong valleys, basins, and small plateaus are situated along the Mekong River, whose branches lies in the west. The Mekong

River is the biggest river in Laos, which runs 1900 km across the west of Laos and its capital Vientiane. The Mekong River runs 234 km as the border between Laos and Myanmar and 976.3 km as the border between Laos and Thailand. It is also a key river that runs as the border between Laos and other Southeast Asian countries.

Of the 12 dams, Xayaburi Dam in Laos is the first one planned to be built in the Mekong River. On November 7, 2012, the construction of Xayaburi Dam began, which is planned to be completed at the end of 2019. The total investment for constructing the dam is 3.8 billion dollars. It is designed with a generation capacity of 1,260 MW, which is the first large-scale hydroelectric project on the mainstream of the Mekong River.

Don Sahong Dam is the second one, which is 30–32 m (98–105 ft) high. It has a capacity of 260 MW, most of which is exported to Thailand and Cambodia. In 2006, the Laos government signed a cooperation agreement with Malaysia Mega First Co. Ltd to begin the research on the feasibility of Don Sahong Dam. In 2013, as the construction of the dam violated the provisions of the Mekong River Agreement signed in 1995, it was suspended under the pressure of the surrounding countries. However, its operation will be resumed in 2018.

2 DISPUTES OVER HYDROELECTRIC PROJECTS IN THE MEKONG RIVER

2.1 *Cause of conflicts over hydroelectric projects in the Mekong River*

This paper analyzed 254 pieces of news or journal articles about water events during 2012 January to 2014 August, of which 201 were related to which referred to water conservancy and hydropower projects, economic development, and environmental deterioration in the Mekong River basin, 48 were related to conflicts over Laos dams, and especially five described the hydroelectric projects of Thailand, Vietnam, China, and Cambodia.

In the early 2012, the construction of Laos Xayaburi Hydropower Station caused huge pressure on countries in the sub-region of Mekong River. Then, the conflicts over the Mekong River basin began to occur. The harm caused by the construction of the hydropower station was beyond the expectation of water organizations and scientific research institutes. Both agriculture and traffic were affected, and the ecosystem was also seriously destroyed.

Of all the disputes analyzed, one event brought about great dissatisfaction among the surrounding countries: in April 2012, a Thailand Company signed a cooperation agreement on the Xayaburi hydropower project with Xayaburi Power Company, which was co-invested by Laos and Thailand. Initially, Cambodia and Thailand strongly opposed the Xayaburi hydropower project of Laos. Cambodia even threatened that it would file a lawsuit against Laos in the World Court. Some scholars also believed that the project undertaken by Laos to construct dams in the Mekong River would cause irreversible and destructive catastrophe to the river.

In May 2012, Laos stopped the construction under the pressure from many countries. Although Laos promised to shut down the Xayaburi project, it can be seen from various angles that it still pursues the early construction. In July 2012, Laos officials announced to terminate the project in the basin of Mekong River, which was highly commended by all countries. However, a month later, Laos decided to restart the construction of the Xayaburi Hydropower Project, which is expected to complete by 2019. In November 2012, the construction of Xayaburi Dam restarted despite the resolute opposition expressed by the neighboring countries. At the same time, other countries, including Cambodia, started other hydroelectric projects, which were also questioned.

In 2013, reports about these disputes suddenly subsided, while the ecological destruction still existed in the Mekong River basin. Since March 2014, media reports about the protests from the neighboring countries against the Lao dam appeared to have an obvious increasing effect.

2.2 *Cause of the intensifying conflicts*

Based on the news and journal articles about water events since 2014 August, it can be found that 50 reports were related to activities of water organizations and 27 reports were related to conflicts in the Mekong River basin.

Among the 27 reports mentioned above, there was only one report concerning conflicts over the hydroelectric project of Vietnam and Cambodia, and the rest related to dissatisfactions on Laos hydroelectric projects.

Over the past year, the intensifying conflicts in the Mekong River have been caused mainly by the construction of Don Sahong Dam and Xayaburi Dam in the second half of 2014.

In August 2014, the Mekong River Commission began to deliberate on the issue of Don Sahong Dam, and then Laos decided to suspend the construction of the dam. In August 2014, despite the strong disagreement on Don Sahong Dam, the developer of Xayaburi Dam (the first hydroelectric project in the main stream of Mekong River) re-submitted a design plan regardless of mass opposition. Since then, a lot of news have reported the protest from the governments and citizens of the surrounding countries. Related ecological environment organizations started to criticize the action of Laos act as it would destroy the ecological environment. The relationship between Laos and neighboring countries deteriorated. However, Laos still ignored all the protests and continued the construction of Xayaburi Dam.

3 ADVANTAGES AND DISADVANTAGES OF HYDROPOWER DEVELOPMENT

3.1 *Influence on fishery*

In the recent year, there were totally 62 pieces of news referring to water organizations and events in the Mekong River, most of which pointed out that dam construction of Laos in the Mekong River would ruin the environment. Among these reports, nine pieces of news described the specific reasons for why the dam could not be built in the Mekong River, and the main reason reported was fishery. In four pieces of news, the Mekong River Commission, together with other organizations, believed that Mekong River dams would cause damage to fishery. This may be the main reason why the construction of Mekong River dams attracted so much attention. In addition, three pieces of news reported that Mekong River dams pose a threat to nearby biological species, such as mangrove forest and dolphin at the Mekong River Delta. Another two pieces of news reported that construction of Laos dam would threaten the residents living in the Delta.

Although the views of all residents living in the basin of Mekong River could not be collected from the news reports in the recent year, it can be concluded that fishery is a very important industry in the basin and most residents rely on this for their survival. Therefore, the influence of Mekong River hydroelectric projects on fishery will affect the common interests of all countries in this basin.

Mekong River is rich in freshwater fishery resources with an annual fishery production of up to 206 tons. There are approximately 200 dams completed, under construction or planned in the basin of Mekong River. Many dams have caused serious damage to the fish living environment and fishery output (Wang, 2012). In the reports issued by the World Bank and the International Union for Conservation of Nature (IUCN), it is indicated that Pak Mun Dam in Thailand (located at 3 km upstream the junction of Mekong River and Mun River) has blocked the fish breeding migration in the river. The number of fish species in Mun River sharply decreased from 265 to 96, and the fish catch dropped by 80% (Wang, 2001).

3.2 *Influence on biology*

As early as 2005, the Mekong giant catfish attracted the attention of the public due to its giant size. In the following years, animal protection organizations in Thailand were trying to carry out artificial reproduction for this species. As stated in a report issued by the World Wildlife Fund (WWF) in July 2010, the Mekong giant catfish may become extinct in the near future. In a few years after this, the fish appeared frequently on various scientific exploration magazines. In March 2015, the Mekong giant catfish was close to extinction, with only hundreds of the fish being left. The WWF gave a serious warning against the extinction of this endangered species.

The Mekong giant catfish is not the only species on the edge of extinction. Mekong River dolphin is also endangered. The dolphin was not a new endangered species. As early as 2009, the WWF reported that water pollution in the Mekong River had pushed Irrawaddy Dolphin to the edge of extinction. Many people were worried that Mekong River dams would continually reduce the number of Irrawaddy Dolphin.

In related news, the Mekong River Commission put forward two possible reasons that affected fishing. First, water pollution has damaged the environment, resulting in a large number of fish mortality. Second, the Mekong River Basin is one of the world's largest inland fishery areas, which is undergoing a huge hydropower construction. Because dams block critical fish migration routes between the river's downstream floodplains and upstream tributaries, fishery suffered serious setbacks (Guy ziv, 2012).

3.3 *Immigrant problem*

All large hydroelectric stations face the problem of migration. Migration can be divided into three types: environmentally forced migration, climatic migration and economic migration. There is no exception for the situation of Mekong River basin. Mekong River hydroelectric stations have so great influence that they will create massive irregular migration.

Migrants from hydroelectric stations in Laos will first flow into domestic cities. However, employments provided by cities of Laos will be rather limited as a result of the lower level of urbanization and poor industry and tertiary sector. Eventually, migration will result in the increasing pressure on the local government. Although a new place would create an improved natural environment for migrants, the sense of belonging will reduce and most migrants will not increase their happiness. Hence, the conflict between migrants and local people will be amplified, even leading to crime.

3.4 *Improvement of flood control capacity*

Flood control is a key function of dams. In summer, frequent flood occurs in the Mekong River. The combined impact on the flooding in Vietnam's Mekong River delta is contributed by local man-made structures, sea level rise, and dams upstream in the river catchment (Thi Viet Hoa Le, 2007). Mekong River dams in Laos play a role in preventing flood.

As a country in the basin of Mekong River, Laos has not suffered from much flooding. Since 2013, there have been 68 news articles about the flood in the Mekong River, of which 44 clearly mentioned the flooding in the Mekong River. Of these, 20 referred to the flood in Cambodia, 14 referred to Vietnam, nine referred to Thailand, one referred to China, and none referred to Laos.

The majority of these articles mentioned about the dissatisfaction of other countries over Laos Xayaburi Dam. Undeniably, dam is a modern technology and the most effective measure for flood control. Dams in Laos greatly reduced the flood harm while other counties in the Mekong River basin are still affected by serious flooding.

3.5 *Increase in power generation capacity*

Xayaburi Dam was designed with a generating capacity of 1260 MW. One of the purposes of Laos Mekong River dams is to build "a battery of Southeast Asia" by hydroelectric generation.

Electrical resource in the Mekong River basin is rather rare. Previously, there was no non-pollution hydroelectric engineering by utilizing the natural terrain fall. Laos Mekong River dams not only supply power for national use, but also function as key power plants in the Mekong River. The completion of Xayaburi Dam and Don Sahong Dam will release the power shortage in Southeast Asian countries.

3.6 *Promotion of economic development*

Laos is an undeveloped country living on agriculture. Since the construction of Mekong River water conservancy and hydropower projects in 2012, the water resource has been developed and made full use of it. This has boosted the economic development in Laos. In 2013, the economic growth rate in Laos reached 8%, and the GDP broke through 10 billion dollars for the first time. In 2014, Lao economic growth rate was up to 7.6%. Its social causes made new progress, such as economy and poverty alleviation, culture, education, science and technology and health service (Chen, 2014).

Despite the prevailing slow economy, the progress of Laos has been quite evident in recent years. Thailand promised to purchase 95% generating capacity of Xayaburi Dam, which will be a great economic support for Laos. The economic growth in Laos has been more likely associated with the newly increased water conservancy industry. It can be said that Mekong River hydroelectric projects saved Laos economy.

With the improvement in Laos economy, Laos international prestige rose to some extent in recent two years. In 2014, Laos put priority on the development of friendly relationship with neighboring countries. On this basis, it positively hosted ASEAN meetings, actively participated in the construction of ASEAN Economic Community, continued to strengthen friendly cooperation with relevant countries and strived for the assistance of foreign government and the support and cooperation of international organizations (Chen, 2015).

4 CONCLUSIONS AND PROSPECTS

In recent few years, the construction of Xayaburi Dam and Don Sahong Dam by Laos in the Mekong River has been deemed as the cause of all conflicts between Laos and other countries in this basin. On the basis of news event data at domestic and overseas in past few years, the background, demand and counterview of the construction of dams in Laos were sorted; the adverse effect of the construction on Mekong River fishery, typical biology and residents was analyzed; the role that the construction played in developing hydropower resources, improving flood control capacity, and promoting economic development was stated. It may be referred to by hydropower stations in rivers, especially international rivers.

In addition, in the construction of the 12 dams on the Mekong River, more attention will be paid to the appeal in the upstream and downstream areas and the interests of each county shall be coordinated. The damage caused by dam construction to the ecological environment shall be taken into account.

ACKNOWLEDGMENTS

This work was supported by the National Natural Science Foundation of China, No. 71503068; the Fundamental Research Funds in Key Research Areas for the Central Universities, No. 2015B09614; and the Fundamental Research Funds for the Central Universities, No. 2014B15114.

REFERENCES

Bakker K. 1999. The politics of hydropower: Developing the Mekong. *Political Geography.*

Chen Dinghui. 2014. Laos: A Review of 2013 Development & Outlook for 2014 (Chinese).

Chen Dinghui. 2015. Laos: A Review of 2014 & Outlook for 2015 (Chinese).

China Economic Net. 2011. Xayaburi Dam was halted and "Greater Mekong Development Plan" was stranded (Chinese).

Guy Ziv; Eric Baranb; So Nam; Ignacio Rodríguez-Iturbed; Simon A. Levin. 2012. Trading-off fish biodiversity, food security, and hydropower in the Mekong River Basin. *National Acad Sciences.*

Shapiro G; Markoff J. 1997. A matter of definition. In: Roberts, C.W. (Ed.) Text Analysis for the Social Sciences. Mahwah, NJ: Lawrence Erlbaum.

Thi Viet Hoa Le; Huu Nhan Nguyen; Eric Wolanski; Thanh Cong Tran; Shigeko Haruyama. 2007. The combined impact on the flooding in Vietnam's Mekong River delta of local man-made structures, sea level rise, and dams upstream in the river catchment. *Estuarine, Coastal and Shelf Science.*

Wang Xiaomin. 2001. Effect of Mekong River Hydropower Development on Environment (Chinese).

Wang Xiaomin. 2012. Mekong River Hydropower Development and Fishery Development: The revelation in Columbia River (Chinese).

Advanced Engineering and Technology III – Xie (Ed.)

Brief review of development, applications, and research of PCCP

Shuangling Zheng
Department of Hydraulic and Hydropower Engineering, State Key Laboratory of Hydroscience and Engineering, Tsinghua University, Beijing, China

ABSTRACT: Prestressed Concrete Cylinder Pipe (PCCP) has excellent characteristics and has been widely used in domestic and international water transfer projects. This paper introduces the performance, development, and applications of PCCP, and presents a brief review of some related problems occurring in designs and engineering. With the rapid development of economy in China, and more attention paid to the safety of water transfer, environmental protection, and sand-saving, PCCP will have a brighter prospect and more widely used in the future. Furthermore, PCCP with a diameter of 4000 mm was developed and applied in the South-to-North water diversion middle route project, which will be useful in improving the improvement of the research and further application of PCCP.

1 INTRODUCTION

Prestressed Concrete Cylinder Pipes (PCCPs) are widely used in long-distance water transmission, city water supply, industrial pressured water delivery, and sewage disposal (Hu and Hao, 1999; Qiao, 2003; Rong, et al, 2000; Shen, 2005; Sun and Wang, 2006; Sun and Zhou, 2006; Wen, 2005; Yu and Liu, 2005; Zhang, 1999; Zhang, 2007; Zhou and Sun, 2001). PCCP water delivery is used in emergency water supply (Section Beijing) of the Beijing–Shijiazhuang section of the middle line within China's South-to-North Water Diversion Project with the parallel laying of DN4000 (Beijing Institute of Water, et al, 2010; LIU and Wang, 2005; Wang, et al, 2009; Yang, et al, 2009; Yao, et al, 2009). This is the first application of super-large diameter PCCP in China, which provides a significant value for the promotion of the theoretical research and engineering application of super-large diameter PCCP in China.

PCCP is a kind of composite pipe composed of steel cylinder, steel wire and concrete. The basic structure can be used to enwind circumferential prestressed wire on the high-strength concrete core with steel cylinder and smooth internal wall, and then to make a solid and durable protection layer with cement mortar on the external side of the pipe with steel socket joints welded at both ends and grooves to accommodate waterproof rubber ring (Hu and Hao, 1999; Qiao, 2003; Rong, et al, 2000; Shen, 2005; Sun and Wang, 2006; Sun and Zhou, 2006; Wen, 2005; Yu and Liu, 2005; Zhang, 1999; Zhang, 2007; Zhou and Sun, 2001).

PCCPs can be classified into lining-type (PCCP-L) and embedded-type (PCCP-E) according to the different locations of the steel cylinder within the pipe core (Hu and Hao, 1999; Sun and Wang, 2006; Sun and Zhou, 2006; Wen, 2005). As for PPCP-L, the concrete pipe core is covered with steel cylinder with the circumferential prestressed wire wound on the steel cylinder, and then sprayed with a mortar protection layer. PCCP-L is typically molded by the centrifuge process with a small diameter ranging between 400 and 1200 mm. As for PCCP-E, the steel cylinder is embedded at the location of 1/3 of the internal wall to the concrete core with the winding of circumferential prestressed wire on the concrete core, and then roller sprayed to form a mortar protection layer. Typically, it is molded with a vertical vibration process with a large diameter of about 1200 mm and currently a maximum diameter of 7600 mm can be achieved by this process (Hu and Hao, 1999; Wen, 2005).

Due to its unique structure, PCCP combines the tensile strength and anti-permeability of steel pipe with the compressive strength and abrasion resistance of concrete to bear large internal and external loads. PCCP can offer very low head loss during the service with high water flux due to its smooth internal wall and low roughness (Liu and Wang, 2005; Tian, et al, 2005; Yuan, 1995; Yu, et al., 2009). The sealing property of PCCP joints is good and the resistance to differential settlement by the steel flexible sockets is high with good adaptation to the foundation. Additionally, PCCP is durable with a long service time and low maintenance costs (Beijing Institute of Water, et al, 2010).

2 OVERVIEW OF THE DEVELOPMENT OF PCCP IN FOREIGN COUNTRIES

The history of the production and application of PCCP is more than 70 years. In 1893, Bonna, a French engineer, designed and fabricated a steel cylinder concrete pipe with a diameter of 1800 mm. The pipes were laid within the inlet network in Colombes, Paris with a length of 1.5 km and an internal pressure of 0.35 MPa. This project provided the foundation for the following development and application of PCCP. In 1939, Bonna Company developed and fabricated PCCP, with the pipes laid in the suburb of Paris. In 1942, Lock Joint Company in the US introduced the French concept and manufactured PCCP successfully. A series of research in the US promoted a further development of PCCP technology in the US (Beijing Institute of Water, et al, 2010; Hu and Hao, 1999; Qiao, 2003; Rong, et al, 2000; Shen, 2005; Wang, et al, 2009; Wen, 2005; YAO, et al, 2009; Yu and Liu, 2005; Zhang, 1999; Zhang, 2007; Zhou and Sun, 2001).

The research and application of PCCP are widely conducted in West Europe, North America, Middle East, and North Africa, in which the US and Canada are the countries with the largest application and promotion. PCCPs are used within the infrastructure facilities such as water, power and civil engineering in more than nearly 90% of the large and medium cities in the US with the total length of about 45,000 km (Hu and Hao, 1999).

In 1976, the large pipes were fabricated by Company A for the water transmission project in Arizona, US with the length of 7.24 km and the internal pressure of 0.98 MPa. The depth of the earth covering the pipes is 10.5 m. The super-large diameter PCCPs were also used in Project Castaic in the US with the diameter of 5100 mm and the internal pressure of 0.95 MPa. The total length is 9.5 km with the depth of earth covering the pipes being 13.5 m. The largest PCCP water transmission project around the world is Project Great Man-Made River (GMR) in Libya, North Africa. It was built by Price Company of US and East Asia international consortium of South Korea in November, 1983. There were four stages in this project. In Stage 1, the total pipe length was 1872 km, in which 80% of the pipes were DN 4000 mm and the rest were DN1600~2800 mm with the working pressure of 0.6~1.8 MPa; in Stage 2, the total length was 1731 km with the working pressure of 2.0~2.6 MPa, and the double-rubber rings were applied as the sealing rings for pipe joints. Stage 3 and Stage 4 started in 1998 with the new pipeline construction of 825 km and the network connection of 1720 km in Stages 1 and 2. The design value for the coefficient of roughness was 0.0116 in this project, which was rather low in terms of hydraulics.

The service history of PCCP projects in the US is long. There were a few pipe failure accidents that occurred in recent years. The research focus of PCCP has been changed from the fabrication and development of the pipes to the maintenance and inspection of the pipes, safety enforcement, earthquake prevention, and disaster reduction.

3 RESEARCH, DEVELOPMENT AND APPLICATION OF PCCP IN CHINA

The research and application of PCCP in China started late with a history of only 30 years. In 1984, PCCP with the internal diameter of 600 mm was developed by Suzhou and a company in Liaoning, and its pilot application in water transmission pipeline succeeded. In 1985, PCCP with the internal diameter of 600 mm made with self-stress concrete was used for trial on the water supply pipeline for the power plant and the municipal city respectively by the cooperation between the Jiangsu Province Nanjing Cement Pipe Plant and the Beijing Civil Engineering Research Institute. The pipe diameters of the PCCPs developed in the early stage in China were small with the limitations of the technique and process equipment.

In order to meet the requirements from the large coal-fired power plants and water transmission projects, a large pipe introduction working group was established in 1985 by the previous Water Resources and Power Ministry to start the development of large PCCP and to import the relative techniques from the US. In 1988, Shandong Electricity Pipe Engineering Company introduced PCCP manufacturing process techniques and related key equipment such as steel cylinder spiral welding machine, equipment used to fabricate socket steel rim, pipe mold and vibrator (Ameron, US). A production line was established with the supporting equipment developed by the Hangzhou Mechanical Design Institute of National Electricity Company. The first batch of PCCPs with DN 2600 mm was produced in 1990 and successfully applied in a power plant water circulation project in Shandong. Currently, there are about 50 PCCP manufacturers in China with 85 production lines. The diameter of the PCCP produced ranges between DN 400 and 4000 mm with the internal pressure of 0.4~2.0 MPa. The PCCP with the largest diameter is the 4 m PCCP developed in 2005 for the South-to-North Water Diversion Project by Beijing Hanjian Heshan Pipe Co, Ltd, Shandong Electricity Pipe Engineering Company, and Xinjiang Guotong Pipe Co, Ltd, China.

PCCPs have been widely used in various applications such as water resources, power supply and municipal water supply, and drainage due to its unique excellent properties. Based on the incomplete statistical data, the total length of PCCPs installed in China is estimated to be more than 600 km, in which PCCPs with the pipe diameter above DN 1400 mm are laid at a length of more than 400 km.

Currently, among all the water diversion projects finished in China, the emergency water supply (Section Beijing) of the Beijing–Shijiazhuang section within the South-to-North Water Diversion Project is a long-distance water transmission project with the largest diameter PCCP with the total length of the main canal being 80 km. Within this sectional project, double-rubber-ring embedded-type PCCP (PCCP-E) with DN 4000 mm was used with parallel laying in the Huinanzhuang–Daning section. The total length of this section is 56.479 km with the design water transmission capacity of 50 m^3/s and the maximum capacity of 60 m^3/s. The design working pressure is 0.4~0.8 MPa. It is the first application of large-diameter PCCP in China with high technical requirement and high implementation difficulty in the main line project in the South-to-North Water Diversion Project.

Previously, the most famous large PCCP water transmission project in China was the Wanjiazhai Yellow River Diversion Project in Shanxi Province (Chen, 2005; Chen, 2010; Guo, 2003; Zhang, 2007). This project is a large water diversion project covering multiple drainage basins. The water is supplied to three energy bases, i.e., Taiyuan, Datong and Pingshuo, respectively with the water taken from the Wanjiazhai reservoir on the main stream of Yellow river. The total length of the connection section with the Yellow river diversion project is 139.35 km, in which the total length of PCCPs is 43.2 km with DN 3000 mm. The water transmission project of Ertix River-to-Urumqi in Xinjiang Province finished in 2006 is one of the PCCP projects with the highest degree of difficulty. The location of this project is in the Junggar Basin, which is at the edge of the desert with bad construction conditions. PCCPs with DN 2800 mm are used with parallel laying in the inverted siphon section of this project. The single length is 7394.25 m, with the working pressure of 1.6 MPa, and the depth of earth covering the pipes is 4 m.

As for the application of power supply, PCCP is mainly used in the make-up water, circulation water, and hydraulic conveying systems in the power plants, such as Shandong Huaneng Dezhou Power Plant's circulation water project and Jiangsu Huaneng Taicang Power Plant's cooling water pipe project. As for municipal city water supply, there are Shenzhen city water transmission project and Harbin Mopanshan water supply project (mainly for water supply, with the functions such as downstream flood control and agriculture irrigation). In the application of municipal water drainage, the largest project is Stage 2 project of wastewater control in Shanghai.

4 SOME RELATED ISSUES IN THE RESEARCH AND APPLICATION OF PCCP

4.1 *The industrial standard for PCCP*

American Water Works Association (AWWA) issued C301, the first design for PCCP in 1949 (ANSI/AWWA C301-99; ANSI/AWWA C304-99; GB/T 1965–2005). The production and design were separated with the revisions, and C301 (Manufacturing standard) and C304 (Design Standard) were issued as the standards for PCCP by the American National Standards Institute and AWWA in 1992. Both standards were revised in 1999. The research and application of PCCP started late in China, and the development of the industrial standard also started late. The *Industrial Standard for the Production of Prestressed Concrete Cylinder Pipe* was issued by the National Construction Materials Bureau in 1996. The *Design Code for the Pipe Structure of Prestressed Concrete Cylinder Pipe* was issued in 2002. Standard JC625 was modified in 2005 and upgraded to a national standard, i.e., *Prestressed Concrete Cylinder Pipe*. However, both standards are the reference and revision to the US standards. Currently, some parts of the production and design of PCCP in China are still using the ANSI standards from the US, such as the connection section in the Wanjiazhai Yellow River Diversion project in Shanxi Province and the Beijing section of the middle line in the South-to-North Water Diversion project.

4.2 *Summary of the studies on the critical issues in the production, design, and construction of super-large diameter PCCP*

There will be more development, production, and application of super-large diameter pipes with a diameter of 4 m PCCP in the Beijing section of the middle line in the South-to-North Water Diversion Project. Compared with the small-diameter PCCP, the technical difficulty in the production, design, transportation, and installation of super-large diameter PCCP is high due to its high weight (up to 77 t for a single part with the length of 5 m) and deep laying depth. There are a series of critical challenges for its technology and higher requirements for the safe operation of these pipelines. The cooperation on the summary of the studies will be carried out between the related design institutes, construction, and R&D organizations, which will facilitate the further development of PCCP technology in China.

4.3 *The monitoring of prototype hydraulic parameters*

The choice of pipe roughness is very critical in the design of the long-distance water transmission project, especially in the middle line of the South-to-North Water Diversion Project with a low head loss. The total investment on the project will be increased with a big roughness value; however, the normal requirements for the water transmission may not be satisfied with a very low value. Therefore, both the total amount of investment and the normal operation of the project are involved with the proper and rational selection of the roughness value. Currently, the roughness value of about 0.014 is chosen for PCCP in the projects undertaken in China. However, the roughness value recommended by foreign PCCP manufacturers is 0.0115, indicating a big difference. Therefore, the reinforcement of the monitoring on the hydraulic parameters for the established PCCP projects will provide the reference data for the engineering design in the future.

4.4 *Anti-corrosion and inspection/maintenance issues*

The actual service life of PCCP is the point of concern due to potential corrosion within the environment. Currently, the main measure used is cathodic protection. Additionally, there may be engineering problems caused by cracks with a long-term operation. The research focus of PCCP in the US has been changed to the inspection, maintenance, and earthquake prevention. The inspection/maintenance of PCCP is becoming a new topic with more number of ongoing PCCP projects in China.

4.5 *Summary of PCCP failure accidents and emergency warning*

The production and application of PCCP started in the US in 1942, and failure accidents of PCCP have been occurring since 1970s. According to the statistical data in 2003, there were more than 500 accidents with PCCP failure within the past 30 years. There have already been more than 40 failures on the 3000 km pipelines constructed since 1983 in the PCCP project of Libya GMR. There is still no PCCP failure accident in China due to the short application time of PCCP. Only one accident occurred in November, 2008, in which there was serious water leakage on the PCCP because the wall of the PCCP was drilled through accidently by a highway geological exploration and construction staff in the PCCP water supply pipe of connection section of the Yellow River Diversion project in Shanxi Province. Therefore, it is suggested to summarize the explosion and failure accidents of PCCP in foreign countries with the Yellow River Diversion Project accident as the reference to establish the emergency treatment plan for accidents and to improve the capability of handling the emergency cases.

5 CONCLUSIONS

As a kind of pipeline material with excellent performance, there are multiple applications of PCCPs in water conservancy, power supply, water supply, and drainage. There is a proven history with wide applications in advanced foreign countries. The application time in China is short with rapid development. The development and application of PCCP of diameter 4 m in the Beijing section of the middle line in the South-to-North Water Diversion Project initiated the application of super-large diameter PCCP in China. This will significantly facilitate the development of PCCP in China, which is proposing higher requirements on the design, production, transportation, construction, and maintenance.

REFERENCES

ANSI/AWWA C301-99. Prestressed Concrete Cylinder Pipe, Steel-Cylinder Type, for Water and Other Liquids [S].

ANSI/AWWA C304-99. Standard for Design of Prestressed Concrete Cylinder Pipe [S].

Beijing Institute of Water, et al. National key technology R&D program of 'the eleventh-five year plan' (2006BAB04 A04): Research of structure safety and quality control of PCCP with large diameter [R]. 2010.

Chen Qing-fang. Application of PCCP in the Linkage Sections of Yellow River Diversion Project [J]. Shanxi Water Resources, 2005, (5):51–52.

Chen Qing-fang. The Accident Treatment of PCCP Water-supply Pipelines of the Linkage Sections of YRDP (Yellow River Diversion Project) [J]. Sci-Tech Information Development & Economy, 2010, 20(12):219–220.

GB/T 19685–2005. PCCP [S].

Guo Yong-feng. Practice of PCCP in Shanxi Wanjiazhai Yellow River Diversion project [J]. Special Structure, 2003, 20(4):40–41, 47.

Hu Lian-sheng, Hao Man-cang. Prestressed concrete cylinder pipe (PCCP) and its development [J]. Shanxi Hydrotechnics, 1999, (6):84–85.

Liu Jing, Wang Dong-li. Calculation and analysis on friction head loss of PCCP of Mid-route of South-to-North Water Transfer Project [A]. The 2nd youth science and technology forum organized by Chinese hydraulic engineering society [C]. Xi'an: 2005. 255–261.

Qiao Xiao-ping. Application of PCCP in water supply [J]. Shanxi Water Resources, 2003, (4):32–33.

Rong Zhi-jun, Hao Man-cang, Jia Bao-ping, et al. Comparison on PCCP, PCP and Steel Pipe [J]. Shanxi Hydrotechnics, 2000, (4):48–50.

Shen Zhi-ji. Development and discussion of PCCP in China [J]. Water & Wastewater Engineering, 2005, (4):1–4.

Sun Hai-liang, Zhou Guan-yang. Manufacture principle, industrial flow and typical lay of PCCP [J]. Electric Power Machinery, 2006, 27(4):78–82.

Sun Shao-ping, Wang Guan-ming. Characteristic of prestressed concrete cylinder pipe [J]. Municipal Engineering Technology, 2006, 24(2):121–123, 125.

Tian Shi-yang, Peng Xin-min, Li Dong-li, et al. Analysis on test of water conveyance resistance for prestressed concrete cylinder pipe [J]. Water Resources and Hydropower Engineering, 2005, (10):74–77.
Wang Dong-li, Liu Jin, Shi Wei-xin, et al. Study on key design techniques of PCCP for South-to-North Water Transfer Project (Mid-route) [J]. Water Resources and Hydropower Engineering, 2009, 40(11):33–39.
Wen Kailiang. Development and application of PCCP [J]. Shanxi Hydrotechnics, 2005, (2):86–87.
Yang Jin-xin, Hu Hui-qing, Wang Dong-li. Study Oil safety monitoring scheme for PCCP of Mid-route of South-to-North Water Transfer Project [J]. Water Resources and Hydropower Engineering, 2009, 40(7):55–61.
Yao Xuan-de, Shi Wei-xin, Wang Dong-1i. Analysis on leakage risk of large diameter PCCP pipeline for South-to-North Water Transfer Project [J]. Water Resources and Hydropower Engineering, 2009, 40(7):66–69.
Yu Jing-yang, Yuan Yi-xing, Li Wen-qiu, et al. Discussion on friction head loss calculation for large diameter PCCP [J]. Water & Wastewater Engineering, 2009, 35(7):108–110.
Yu Lin, Liu Qun-chang. Review of PCCP with lager diameter and FRPM [J]. China Rural Water and Hydropower, 2005, (7):108–109.
Yuan Qian-sheng. Test study on water resistance of concrete pipe with lager diameter [J]. Yellow River, 1995, (7):29–30.
Zhang Shu-kai. Review of development of PCCP in China [J]. China Building Materials, 2007, (4):37–40.
Zhang Ya-ping. Application of Prestressed Cylinder Concrete Pipe [J]. Water & Wastewater Engineering, 1999, 25(5):54–57.
Zhou Guan-yang, Sun Hai-liang. Research and application of PCCP in China [J]. Refrigeration Air Conditioning & Electric Power Machinery, 2001, (1):2.

Advanced Engineering and Technology III – Xie (Ed.)
 ISBN 978-1-138-03275-0

Evaluation of regional cultivated land ecological security of Zhejiang province based on the entropy weight method and the PSR model

Keying Yu, Le Sun, Ying Hu, Lei Chen, Ying He & Panpan Sun
Hangzhou University of Commerce, Zhejiang, China

ABSTRACT: The evaluation of cultivated land ecological security is an important basis for improving the security situation of the farmland ecosystem and promoting the sustainable utilization of cultivated land. In order to better understand the cultivated land ecological state of Zhejiang Province, this paper dissects the connotation of cultivated land ecology, and constructs a theory system based on the ecological security that consists of the law of land use development, the principle of sustainability, and the idea of ecological thinking, so as to take appropriate measures on the cultivated land ecological security of Zhejiang Province. This paper adopts the Pressure—State—Response (PSR) framework and selects 29 indices to construct the evaluation index system of cultivated land ecological security in Zhejiang Province. In addition, it establishes an integrated evaluation model based on the entropy weight method and the integrated evaluation method, to carry out the regional study on the cultivated land ecological security of all regions in Zhejiang Province. The grade difference of cultivated land ecological security of different regions in Zhejiang Province is not very large, basically at the general grade. The cultivated land ecological security in Taizhou, Zhoushan is in the poor state, and the cultivated land ecological security in Hangzhou, Ningbo, Jiaxing, Wenzhou, Lishui, and Jinhua is in the general state, and the cultivated land ecological security in Huzhou, Shaoxing, and Quzhou is in the good state. On the basis of the above conclusions, this paper explores the rational ways to solve the production and development and puts forward the corresponding countermeasures and suggestions according to the state and the actual situation of cultivated land ecological security in all regions.

1 INTRODUCTION

Cultivated land resource is a very important agricultural means of production, with a variety of functions, such as food production, space carrying, and ecological service. The cultivated land ecosystem gradually evolves into a composite system of society, economy, and ecology with high coupling after a long-term intervention of mankind. With an increased use of cultivated land resources, both in depth and extent, the scarcity of cultivated land is enhanced, and the ecological problems of using the cultivated land are gradually highlighted. According to the results of the second national land survey, the land of Zhejiang takes up 1.10% of the area of the country, which is one of the smallest provinces in China. With the increased process of urbanization and industrialization, the scale of non-agricultural land has been expanding, and the cultivated land has become one of the scarcest resources for the economic development in Zhejiang. Therefore, carrying out the research on the cultivated land ecological security and optimizing the improvement path of cultivated land ecological security have great theoretical and practical significance in easing the contradiction between people and land, guaranteeing the food safety, and promoting the economic and social sustainable development.

At the end of the 1980s, the Organization for Economic Cooperation and Development (OECD) and the United Nations Environment Programme (UNEP) proposed the concept of the PSR model, namely Pressure–State–Response (Zhuo Fengli, 2012), which has become the most generally used method in the evaluation on current sustainable development and of ecological security due to its advantages of integrity, flexibility, and clear causal relationship (Zuo Wei, Zhou Huizhen & Wang Qiao, 2003; Liu Li, Qiu Daochi & Su Hui, 2004). Liu Yong and co-workers (Liu Yong, Liu Youzhao & Xu Ping, 2004), taking the ecological security of the land resource in Jiaxing, Zhejiang as the evaluation goal, constructed an evaluation index system by selecting 24 indices from three aspects, namely nature, economy, and social. They carried out an integrated evaluation of the ecological security state of land resources in Jiaxing City in 1991 and 1997. Mao Zhang-yuan and co-workers (Mao Zhang-yuan, Li Zhan & Xue Jibin, 2015), on the basis of the evaluation index system and examples of the regional land ecological security, established a land ecological security evaluation index system composed of 25 indices with Yiwu characteristics, and evaluated and classified the situations of land ecological security in Yiwu from 2008 to 2012.

All of the above studies are focused on the dynamic evaluations of land ecological security; however, research on the evaluation of space sub-regions is relatively few. In order to fully understand the ecological state of cultivated land in Zhejiang Province, and take targeted measures on the problems of cultivated land ecological security in different cities of Zhejiang Province, as well as to ensure the sustainable utilization of cultivated land resources, we adopted the PSR model and selected 29 indices from three aspects, namely pressure, state, and response, which affect the cultivated land ecological security. Furthermore, we established an evaluation index system for the cultivated land ecological security of Zhejiang Province, and calculated the weight of each index by using the entropy weight method so as to conduct the integrated evaluation of cultivated land ecological security in the cities of Zhejiang Province. We used this model to analyze the situations of cultivated land ecological security of 11 cities in 2014 in Zhejiang Province.

2 RESEARCH METHOD AND DATA SOURCE

2.1 *Research method*

2.1.1 *Constructing evaluation index system*

The key to evaluating the cultivated land ecological security is to establish a scientific evaluation index system and determine the weight of each index. The determination of the weight of each index is an important foundation to obtain the objective and accurate evaluation results. As an important part of the research on the cultivated land ecological security, its index is selected by not only considering the current state of land ecology, but also reflecting the potential impacts and the impacts of human activities on land ecological security. The cultivated land ecological security is characterized by systematization and complexity, and its evaluation index system is a complex system that is integrated with population, economy, resources, environment, and society. By the application of the PSR model, this paper comprehensively analyzes the influence factors on cultivated land ecological security in Zhejiang Province, and constructs a cultivated land ecological security evaluation index system for Zhejiang Province containing four layers, namely target, criterion, factor, and index, and 29 indices on the basis of selecting the index under the principle of scientific, operable, comparable, and regional systems.

2.1.2 *Standardization management of index*

By using the method of range standardization, the original data are processed by the dimensionless method, and the original data matrix is constituted by n evaluation indices and m cities in Zhejiang Province, $X = (x_{ij})_{m\times n}$, $(i = 1,2,...m, j = 1,2,...n)$, *where* x_{ij} is the value for the j index of the i area.

For the positive index (the bigger the value, the more the safety) adopting

$$\text{Positive index } X'_{ij} = \frac{X_{ij} - \min\{X_j\}}{\max\{X_j\} - \min\{X_j\}} \tag{1}$$

For negative indicators (the smaller the value, the more the safety) adopting

$$\text{Negative index } X'_{ij} = \frac{\max\{X_j\} - X_{ij}}{\max\{X_j\} - \min\{X_j\}} \tag{2}$$

The standardized evaluation matrix can be obtained from the calculation $A = (y_{ij})_{m\times n}$.

2.1.3 *Determination of index weight*

The entropy weight method is a mathematical method used for calculating the integrated index based on the comprehensive consideration of the information provided by each index. It has the advantage of obtaining the weight value of each index objectively and accurately. In order to get the weight of each index, first, the information entropy of each index should be calculated according to the standardized value of original data. The calculation formula is given by:

$$e_j = -k\sum_{i=1}^{m} f_{ij} \ln f_{ij} \left(\text{when } f_{ij} = 0, \textit{let } f_{ij} \ln f_{ij} = 0\right) \tag{3}$$

where $K = \frac{1}{\ln m}$, $f_{ij} = \frac{y_{ij}}{\sum_{i=1}^{m} y_{ij}}$. Then, the weight of each index can be determined based on the entropy value obtained according to the formula:

$$W_j = \frac{1 - e_j}{\sum_{i=1}^{m} (1 - e_j)} \left(0 \le W_j \le 1, \sum_{j=1}^{m} w_j = 1\right) \tag{4}$$

where W_j is the entropy (weight) value of the j index and e_j is the entropy value of the j index.

Thus, the greater the index, the smaller the weight value; on the contrary, the smaller the entropy value, the greater the weight value. The weight of each index can be obtained according to formulas (1)–(4).

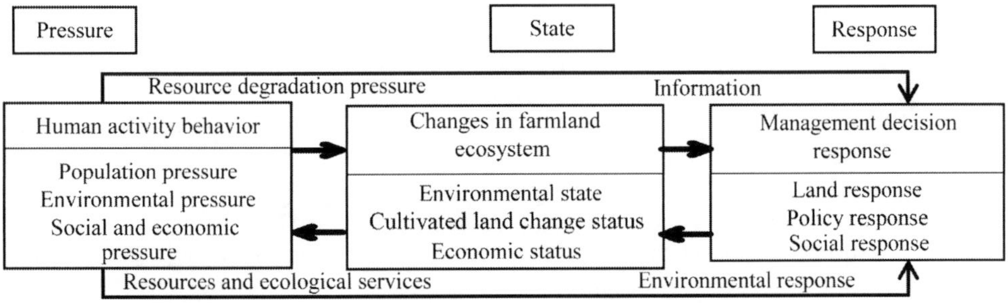

Figure 1. PSR framework model for cultivated land ecological security evaluation.

Table 1. Integrated evaluation index system of cultivated land ecological security in Zhejiang province.

Target layer	Standard layer	Factor layer	Index layer	Security tendency	Weight
	Pressure	Population pressure	Population density (person/km^2)	–	0.02547114
			Natural population growth rate (%)	–	0.031644027
		Environmental pressure	Grain sown area (10.4 hm^2)	+	0.020351687
			Cultivated land area per capita (hm^2/person)	+	0.056729622
			Unit agricultural fertilizer use (kg)	–	0.016713681
			Unit agricultural pesticide use (kg)	–	0.029991952
			Proportion of agricultural affected	–	0.020542408
		Social and economic pressure	Proportion of cultivated land (%)	–	0.017671027
			GDP annual growth rate (%)	+	0.034058028
			Per capita disposable income of rural residents (yuan)	+	0.030254755
			First industry accounts of GDP (%)	–	0.029163055
Ecological security of cultivated land index	State	Environment state	Soil and water loss rate (%)	–	0.02120633
			Protected area accounts of jurisdiction area (%)	+	0.081757381
			Crops disaster area (10.4 hm^2)	–	0.023774929
		Cultivated land change state	Reduce the area of cultivated land (hm^2)	–	0.020568399
			Increase the area of cultivated land (hm^2)	+	0.025320327
			Land reclamationrate (%)	+	0.052975197
		Economic state	Per capita grain output (kg/person)	+	0.045446805
			Agricultural output value (100 million yuan)	–	0.036613407
	Response	Land response	Paddy field area (hm^2)	+	0.020713833
			Dry land area (hm^2)	–	0.021471871
			Effective irrigated area ratio (%)	+	0.02438386
		Policy response	Handling capacity of industrial wastewater disposal facilities (ton/day)	+	0.091760672
			Utilization rate of industrial solid waste (%)	+	0.01674677
			Total power of agricultural machinery (10,000 kW)	+	0.044530966
		Social response	Water and soil erosion control rate (%)	+	0.043024153
			Ratio of environmental pollution control investment accounts for GDP (%)	+	0.071376679
			Residential area holding by house holds in agricultural production and operation (m^2/family)	–	0.024811204
			Water logging control situation (%)	+	0.020925836

2.1.4 *Calculation of the integrated evaluation index of cultivated land ecological security*
The cultivated land ecological security index is an index used to evaluate the degree (between 0 and 1) of cultivated land ecological security. The situation of cultivated land regional ecological security can be judged by using the integrated evaluation method according to the standardized value and the weight of each index. The specific formula is given by:

$$A_i = \sum_{j=1}^{n} w_j y_{ij}$$

where A_i is the integrated evaluation value of the i evaluation area and w_j is the weight value of the j evaluation index, the standardized value of the j index in the i year. The more the A_i tends to 1, representing the higher degree of cultivated land ecological security, the more the A_i tends to 0, representing the lower degree of cultivated land ecological security.

2.1.5 *Grading criteria of cultivated land ecological security*

An integrated evaluation table (Table 2) is set up for the cultivated land ecological security in Zhejiang Province according to the comprehensive safety value obtained from the calculation, referring to the relevant literature (Li Ling, Hou Shutao, Zhao Yue, Zheng Xuling, 2014; Pei Tingting, Chen Ying, Zhao Yan'an, Wang Daojun, Liu Shu'an, 2014; Yuan Shao-feng, Shi Wei-wei, Sun Le, 2011) and combining with the actual situation in Zhejiang Province.

Table 2. Integrated evaluation criteria of regional land ecological security.

Rank	Ecological security value interval	Ecological security degree	System characteristics
I	<0.3	Adverse	The service function of the land ecosystem closes to collapse, and the ecological process is very difficult to reverse; the ecological environment is severely damaged, and the ecosystem structure is incomplete along with loss of function, and ecological restoration and reconstruction is very difficult. A lot of ecological environment problems have been occurred and evolved into the ecological disaster, and the service function of the land ecosystem has been degenerated adversely;
II	0.3–0.4	Poor	The ecological environment has been destroyed greatly, the ecosystem structure is destroyed greatly, and the function is degenerated and incomplete. The environment is difficult to recover, and the ecological problem is huge with more ecological disasters;
III	0.4–0.5	General	The service function of the land ecosystem has been degenerated; the ecological environment is destroyed, and the ecosystem structure has been changed, but the basic functions can still be maintained. It can easily deteriorate after the disturbance, and the ecological problem is significant, and sometime ecological disasters occur;
IV	0.5–0.5	Good	The service function of the land ecosystem is perfect; the ecological environment is less damaged, the ecosystem structure is still perfect, and the function is still good. It can recover after the disturbance, and the ecological problem is not obvious with few ecological disasters;
V	>0.6	Ideal	The service function of the land ecosystem is basically perfect and the ecological environment is not destroyed; the ecosystem structure is perfect with strong functions and the regeneration and recovering ability of the ecosystem is strong, and the ecology question is not remarkable with less ecological disasters.

2.2 *Research method*

The original data source for this research is mainly obtained from the Statistical Yearbook of Zhejiang Province, and referred to the authoritative information on the land resource use data that are published in the Zhejiang Statistical Yearbook on Natural Resources and Environment and the Second Agricultural Census Data Compiling Agricultural Volume of Zhejiang Province, as well as on the relevant government websites.

3 RESULTS AND ANALYSIS

3.1 *Integrated evaluation on cultivated land ecological security in Zhejiang province*

As can be seen from Table 3, the integrated index of cultivated land ecological security in 11 cities of Zhejiang Province is 0.3766–0.5763. Based on the criteria of Table 2, it is in the state of poor–general–good, which indicates that the overall level of cultivated land ecological security in Zhejiang Province is not ideal, and the factors that affect the cultivated land ecological security in the areas are different.

3.1.1 *Pressure safety evaluation on cultivated land*

As shown in Table 3, the ecological security pressure of cultivated land shows good performance (more than 0.2) in Huzhou and Shaoxing, which is characterized by a larger per capital area of cultivated land, a smaller number of farmland disaster areas, with smaller ecological pressure in cultivated land. The ecological security pressure of cultivated land does not perform well in Ningbo and Taizhou, which is characterized by more pesticides and chemical fertilizers for unit area of cultivated land and larger environment pressure on the agricultural land. As for Ningbo with a higher level of urbanization, the rapid urbanization has resulted in a huge demand for land resources, so as to cause a larger space pressure on the cultivated land.

3.1.2 *State safety evaluation on cultivated land*

From Table 3, it can also be seen that the state indices of cultivated land in Ningbo, Jiaxing, Huzhou, and Quzhou are higher (more than 0.15). Among them, the state of the cultivated land in Jiaxing is better, especially in high rate of land reclamation and less reduction of cultivated land in recent years. The economic state of cultivated land in Quzhou and Huzhou is better with a bigger grain output per capita and a higher agricultural output. The environmental state of cultivated land in Ningbo is better, which is related to the high ratio of the protected area to the jurisdiction area. The state index of cultivated land in Hangzhou and Taizhou is lower. As can be seen from the results, the integrated evaluation values of

Table 3. Pressures, state, influence, and integrated index of the cultivated land ecological security in the cities of Zhejiang.

City	Pressure	State	Response	Integrated index	Safety grade
Hangzhou	0.185190751	0.079865553	0.232333057	0.497389361	III
Ningbo	0.134008305	0.152122525	0.176798967	0.462929797	III
Jiaxing	0.171247371	0.170181333	0.136077552	0.477506256	III
Huzhou	0.220826952	0.152629659	0.202866858	0.576323469	IV
Shaoxing	0.212191594	0.111568856	0.198124718	0.521885167	IV
Zhoushan	0.153263267	0.101517633	0.121845743	0.376626643	II
Wenzhou	0.156558367	0.148538981	0.178728433	0.483825781	III
Jinhua	0.18242916	0.123610688	0.14201691	0.448056758	III
Quzhou	0.190319527	0.15598839	0.18686543	0.533173348	IV
Taizhou	0.148748796	0.097808864	0.134270099	0.38082776	II
Lishui	0.197132121	0.129994708	0.157215575	0.484342404	III

the reduced area and the agricultural output in Hangzhou both are 0, which concentrates on a relatively large area of land for construction in Hangzhou and higher unit GDP energy consumption, namely the safety value of resource and economic state is lower. Particularly, Taizhou shows a high rate of soil erosion, and a relatively low ratio of the protected area to the jurisdiction area, so the integrated value of the safe state of its cultivated land is lower.

3.1.3 *Response safety evaluation on cultivated land*

Human activities can significantly affect the level of the integrated evaluation on the cultivated land responses, in which the policy response and the social response are particularly important. As can be seen from Table 3, the response integrated value of cultivated land in Hangzhou and Huzhou is higher. The higher handling capacity of the industrial wastewater disposal facilities in Hangzhou indicates that it responds to the government policy actively; the rate of soil erosion control and environmental pollution control investment of Huzhou accounts for a higher proportion of GDP, indicating a good social response. The response integrated value of cultivated land in Zhoushan and Taizhou is low. The integrated evaluation values of the paddy field area and the handling capacity of the industrial wastewater disposal facilities in Zhoushan are 0, indicating that the cultivated land response and the policy response are inactive; the proportion of environmental pollution control investment accounts for GDP and water logging control conditions in Taizhou are not ideal. It indicates that the economic development level and the optimization of the industrial structure have a great impact on the improvement of cultivated land ecological security; thus, the measures to improve the cultivated land ecological security will be taken in the future.

3.2 *Integrated evaluation and classification of cultivated land ecological security in Zhejiang province*

From the figure, the following can be summarized:

1. The cultivated land ecological security in Huzhou, Quzhou and Shaoxing is at the good grade. Among them, the pressure state of cultivated land in Shaoxing and Huzhou is better, which is characterized by smaller population density, more areas of per capita cultivated land, higher per capita disposable income for rural residents and higher annual growth rate of GDP. The safety condition of cultivated land in Quzhou is relatively good, characterized by smaller rate of soil erosion, smaller ratio of the protected area to the jurisdiction area, and smaller number of crop disaster areas.
2. The cultivated land ecological security in Jiaxing, Hangzhou, Ningbo, Jinhua, Lishui, and Wenzhou is at the general grade. The aforementioned areas are mostly the developed areas of Zhejiang Province, as well as the main areas for producing grains, with a larger resource demand by the industry and agriculture compared with the other areas, which specifically indicate less per capita water resources, and larger use of chemical fertilizers and pesticides. In addition, in the six regions, except Lishui, the population density is larger and the growth rate of natural population is higher; thus, the population pressure will inevitably lead to an increase in pressure on resources, especially to a great impact on the cultivated land ecological security. The areas mentioned above belong to those whose ecological environment of cultivated land is required to be improved.
3. The ecological security of cultivated land in Zhoushan and Taizhou is at the worst grade. The per capita cultivated land resources and the grain sown area are less. The number of disaster areas of agricultural land is larger and the proportion of the first industry in GDP is low in Taizhou, affecting the cultivated land ecological security in this area. Ecological pressure safety of cultivated land is poor, indicating larger use of chemical fertilizers and pesticides for unit area of cultivated land and causing greater pressure on the environment of agricultural land. At the same time, the proportion of the investment on environmental pollution control in GDP and the water logging control situation is not ideal, producing greater pressure on the cultivated land ecological security. These two areas should belong to the key regions for controlling the ecological environment of cultivated land.

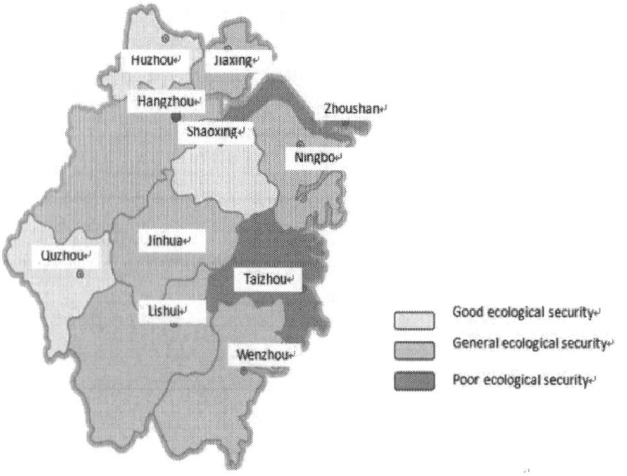

Figure 2. Ecological security grade division of cultivated land in Zhejiang Province.

4 CONCLUSION AND DISCUSSION

4.1 *Conclusion*

The evaluation results show that the cultivated land ecological security in Zhoushan and Taizhou is in the poor state. For Zhoushan City, the main factor for the ecological security problem of cultivated land is the state index that includes three categories: environmental state, change state of cultivated land and economic state. In recent years, the increase in the area of cultivated land and the sown area of grain have been low in Zhoushan; thus, the per capita area of cultivated land and the per capita output of grain are also low. For Taizhou, the number of crop disaster areas is larger, and the ratio of the protected area to the jurisdiction area is relatively low, which proves that the infrastructure construction of cultivated land is not perfect, and that the government needs to develop policies to better protect the cultivated land ecological security.

The cultivated land ecological security in Hangzhou, Ningbo, Jiaxing, Wenzhou, Jinhua, and Lishui is in the general state. These cities should accelerate the transformation of rural economic structure and the upgrading of the industrial structure, as well as improve the proportion of the third industry, so as to gradually improve the level of urbanization by these means. Moreover, they should increase the fund investment for environmental protection and governance from the government, and improve the discharge rate of industrial wastewater and the integrated disposal rate of solid wastes. They should also gradually increase the coverage rate of forests, alleviate the ecological pressure of cultivated land, and improve the ecological security.

The cultivated land ecological security in Huzhou, Shaoxing and Quzhou is in the good state. In the future, we should focus on reducing emissions from industrial wastes, controlling population growth and soil erosion, and reducing the rate of land pollution by deducting the applied quantity of chemicals so as to further improve the cultivated land ecological security in this area.

4.2 *Discussion*

This paper constructs the cultivated land ecological security evaluation zoning in Zhejiang Province based on the PSR model. The evaluation of the cultivated land ecological security can make a quantitative analysis on the cultivated land ecological security in Zhejiang Province. Thus, this evaluation is a complex system engineering, which can well reflect the safety ecological state of cultivated land in all regions. At the same time, the entropy weight method is adopted to determine the weight of each index, which overcomes the subjectivity of the Delphi entropy weight determination method. This research constructed a cultivated land ecological security evaluation index system for Zhejiang Province containing four layers, namely target, criterion, factor, and index, and 29 indices on the basis of selecting the index under the principle of scientific, operable, comparable, and regional systems.

REFERENCES

Li Ling, Hou Shutao, Zhao Yue, Zheng Xuling. Evaluation and Forecast of Land Ecology Security of He'nan Province Based on P-S-R Model [J]. Research of Soil and Water Conservation. 2014, (1):188–192.

Liu Li, Qiu Daochi, Su Hui, et al. Exploration for Evaluation of Land Intensive Use in Metropolis [J]. Journal of Southwest China Normal University (Natural Science Edition), 2004, 29 (5):887–890.

Liu Yong, Liu Youzhao, Xu Ping. Evaluation on Ecological Security of Regional Land Resources—A Case Study of Jiaxing City, Zhejiang Province [J]. Land Management College of Nanjing Agricultural University, Resources Science, 2004, (3):69–75.

Mao Zhangyuan, Li Zhan, Xue Jibin. Evaluation of Land Ecological Security Based on PCA—A Case Study of Yiwu City, Zhejiang province [J]. Journal of Anhui Agricultural Sciences, 2015, (20):197–200.

Pei Tingting, Chen Ying, Zhao Yan'an, Wang Daojun, Liu Shu'an. Evaluation on the Ecological Security of Baiyin City Based on P-S-R Model [J]. Chinese Agricultural Science Bulletin. 2014, (2):215–221.

Yuan Shaofeng, Shi Wei-wei, Sun Le. Study on regional land ecological security evolvement: A case of Hangzhou. [J]. Shanghai Land and Resources. 2011, (2):25–29.

Zhuo Fengli. Land Ecological Security Evaluation Based on Entropy Weight and Set Pair Analysis: A Case of Hebei Province [J]. Areal Research and Development, 2012, 31 (6):111–114.

Zuo Wei, Zhou Huizhen, Wang Qiao. Conceptual Framework for Selection of An Indicator System for Assessment of Regional Ecological safety [J]. Soils, 2003, 24 (1):2–7.

Advanced Engineering and Technology III – Xie (Ed.)
© 2017 Taylor & Francis Group, London, ISBN 978-1-138-03275-0

Uptake of uranium by red mud from aqueous solution

Wanying Wu
Department of Environmental Science and Engineering, Guangzhou University, Guangzhou, Guangdong, China

Nan Chen
Guangdong Provincial Key Laboratory of Radionuclides Pollution Control and Resources, Guangzhou, Guangdong, China

Ziming Feng
Department of Environmental Science and Engineering, Guangzhou University, Guangzhou, Guangdong, China

Jinwen Li, Diyun Chen & Dongmei Li
Guangdong Provincial Key Laboratory of Radionuclides Pollution Control and Resources, Guangzhou, Guangdong, China

ABSTRACT: This paper studied the removal of uranium by red mud from aqueous solution. It performed dynamic tests to investigate variable factors such as reaction time, temperature and initial concentration. The test results indicated that the optimal conditions was 3 g/L of red mud, pH = 4.2, 24 hours of adsorption time, and 10 mg/L of initial uranium concentration at 25°C. Under these conditions, the removal rates reached up to 73.25%. The results also indicated that the adsorption data fitted both the pseudo-second-order kinetic model and Langmuir equations.

1 INTRODUCTION

Red mud is a waste residue generated by the Bayer process, which is widely used to produce alumina. For every ton of alumina produced, approximately 1 to 2 tons of alumina residues are generated (Cao et al, 2007). According to different production processes, red mud can be classified into three types: red mud from the sintering process, Bayer red mud, and red mud from the combined process. As the fourth largest alumina production country in the world, China still lacks a suitable method to store or apply large quantities of red mud (Gu et al, 2011). The traditional storing method of red mud leads to the occupation of vast areas of land. At the same time, dry red mud produces dust that results in great damage to the environment. Alkalinity and metals present in red mud will percolate the soil, causing soil salinization and groundwater pollution. Meanwhile, red mud contains a significant quantity of uranium that has radiological and chemical toxicity effects. Many studies have reviewed the reuse of red mud, such as adsorption of phosphorus, Cd^2+, Cr^6+, Pb^2+, and Ni+ from aqueous solution (Li et al, 2006; Ren et al, 2008; Kang et al, 2011; Wang et al, 2013; Zeng et al, 2013).

Uranium is produced as a fuel for nuclear power, which can easily cause water pollution and lead to ecological damage. The main nuclear wastewater treatment techniques include the chemical precipitation method, the adsorption method, and the ion exchange method (Wang et al, 2013; Zhu et al, 2013). In recent years, some studies have used natural clay minerals as adsorbents in nuclear wastewater treatment and made great progress. Some minerals, such as attapulgite, vermiculite, montmorillonite, illite, and pyrophyllite, are also widely used as absorbents (Deng et al, 2011; Liu et al, 2012; Xiao et al, 2014; Cui et al, 2015; Chen et al, 2015; Zhang et al, 2015). However, only a few studies have used red mud as an adsorbent in nuclear wastewater treatment.

To determine an environmentally acceptable method for industrial waste reuse and nuclear wastewater treatment, this paper used red mud as an absorbent and studied its adsorption property on variable factors such as adsorption time, reaction temperature, and initial concentration of uranium. The aim of the study was to determine the optimal conditions of adsorption and provide a theoretical basis for industrial wastewater treatment using red mud.

2 MATERIALS AND METHODS

2.1 *Reagent and instrument*

U_3O_8 is a stable reagent and the experimental water used is double-deionized water.

The main instruments include an SFG-02.400 series heating and drying oven, a PHSJ-4A pH meter, a TDL-5-A centrifuge, a 721G VIS spectrophotometer, an SHA-82A water-bath constant temperature vibrator.

2.2 *Experimental materials*

Red mud was obtained from Pingguo Aluminum Co. Ltd, Guangxi Province, China.

It was pre-dried to a constant weight below a temperature of 105°C. Dry red mud was ground in the laboratory mortar grinder and sorted through a 250 μm mesh sieve to maintain a uniform particle size. Then, the red mud sample was preserved in a sample bag.

2.3 *Preparation of standard uranium solution*

U_3O_8 (1.1792 g) was placed in a 100 ml beaker, in which 10 ml hydrochloric acid, 3 ml H_2O_2, and two drops of 1 mol/L nitric acid were dissolved. Then, the solution was transferred to a 1 L volumetric flask, diluted it with secondary deionized water, the concentration of which was 1 g/L. Different concentration of uranium solution were diluted from the uranium standard solution.

2.4 *Experimental methods*

The single factor assessment method and the quantitative analysis method were performed to optimize the basic experimental conditions for adsorption.

The red mud sample (3 g/L) and uranium solution (40 ml) were placed in a centrifuge tube (50 ml) and shaken at a speed of 200 r/min. The process was repeated again at a speed of 4000r/min to separate red mud from uranium solution. Residual concentrations in the supernatant uranium solution were determined by atomic absorption spectrophotometry.

The removal rate and the uranium adsorption capacity are calculated as follows:

$$\psi_t = \frac{C_0 - C_t}{C_0} \times 100\% \tag{1}$$

$$Q_t = \frac{(C_0 - C_t)V}{m_p} \tag{2}$$

where C_0 is the initial uranium concentration (mg/L), C_t is the uranium concentration at any time of t (mg/L), V is the volume of the solution (L), and m_p is the dosage of dry red mud (g).

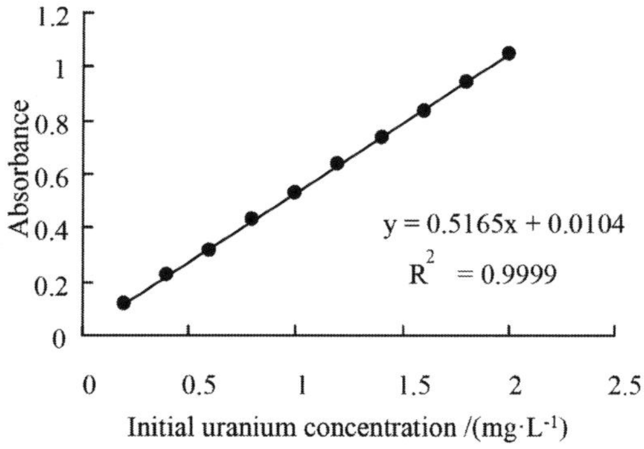

Figure 1. Standard curve of uranium.

Table 1. Feasibility study data.

Red mud dosage (g)	Concentration of remaining uranium solution (mg/L)	Removal rates (%)	Adsorption capacity (mg/g)
0.1202	3.8363	23.27	0.3873
0.1200	3.8845	22.31	0.3718
0.1201	3.8459	23.08	0.3844
0.1203	3.8652	22.70	0.3773

2.5 *Drawing the standard curve of uranium*

A total of 10 different initial uranium standard concentration solutions (from 0.2 mg/L to 2 mg/L) were prepared and their absorbance were determined to generate a standard curve. The result is shown in Figure 1. Through this standard curve, the absorbance can be calculated as uranium concentration.

3 RESULTS AND ANALYSIS

3.1 *Feasibility study*

The uranium adsorption ability was studied under the following conditions: 25°C; red mud dosage 3 g/L; pH = 4.2; initial concentration of the uranium solution 5 mg/L; and reaction time one hour. The result is summarized in Table 1.

From Table 1, it can be seen that red mud has a certain uranium adsorption capability. In order to study its optimal adsorption capability, different experimental conditions were investigated in the follow-up study.

3.2 *Effect of reaction temperature on adsorption*

The influence of reaction temperature on uranium adsorption was studied under the following conditions: red mud dosage 5 mg/L; pH = 4.2; initial concentration of the uranium solution 5 mg/L; adjusted reaction temperature 25, 35, 45, 55, and 65°C. The result is shown in Figure 2.

The removal rate of uranium increased with the temperature up to 65°C, which indicated that the reaction was an endothermic reaction, and the high temperature was beneficial to the adsorption. As the temperature increased from 25°C to 65°C, the curve is represented

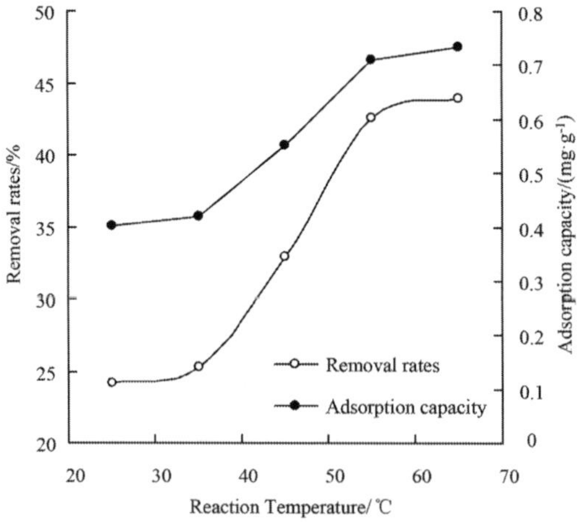

Figure 2. Effect of reaction temperature on adsorption.

by an "S" shape. The graph showed the quickly increased process from 25°C to 55°C. When the temperature reached 55°C, the increased removal rate declined gradually. Given that the higher temperature may cause quicker mechanical consumption, the reaction temperature was set as 25°C.

3.3 *Effect of reaction time on adsorption*

The effect of reaction time on uranium adsorption was studied under the following conditions: 25°C; red mud dosage 3 g/L; pH = 4.2; initial concentration of the uranium solution 5 mg/L; adjusted oscillating reaction time 1/6, 2/6, 0.5, 1, 2, 4, 8, and 24 h. The result is shown in Figure 3.

As shown in Figure 3, within the first 4 h, the removal rate of uranium was increased with increasing reaction time. It rapidly reached up to 30% when the reaction time was 4 h. However, when the reaction time increased from 4 h to 8 h, the removal rates increased by less than 5%. Furthermore, when the reaction time was 24 h, the removal rates reached up to 46.22%. This phenomenon indicated that the concentration of uranium ion was reduced with increasing reaction time. It was hypothesized that the reaction would not reach the basic adsorption equilibrium within 24 hours, so the removal rate continued to increase gradually. It also indicated that the reaction was slow and would require 72 hours to reach equilibrium. In order to save the industrial reaction time, the reaction time was set within 24 hours.

3.4 *Effect of initial concentration on adsorption*

Uranium solution at concentrations ranging from 2 to 50 mg/L was adsorbed by red mud (3 g/L) at 25°C and pH = 4.2. After placing the sample in a thermostatic bath with a stirrer for 24 hours, the concentration of uranium solution was determined and the effects of initial uranium concentrations on adsorption were studied. The adsorption data are shown in Figure 4.

Initially, the removal rate increased rapidly with the increase in the initial uranium concentration. As the initial concentration of uranium increased to 10 mg/L, the increase in the removal rate stopped and began to reduce. Meanwhile, uranium adsorption capacity kept increasing. This phenomenon indicated that the amount of the absorbent's active site is limited. When the absorbent's dosage and the pH of the solution were settled, the removal rates were reduced with increasing amounts of target ions. The adsorption data showed that, when the initial concentration reached to 50 mg/L, the removal rate was basically the same as the initial concentration was 5 mg/L. It indicated that the adsorption

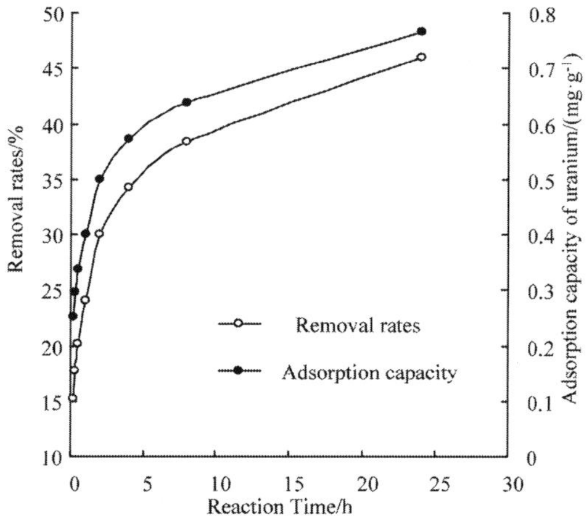

Figure 3. Effect of reaction time on the adsorption of uranium onto red mud.

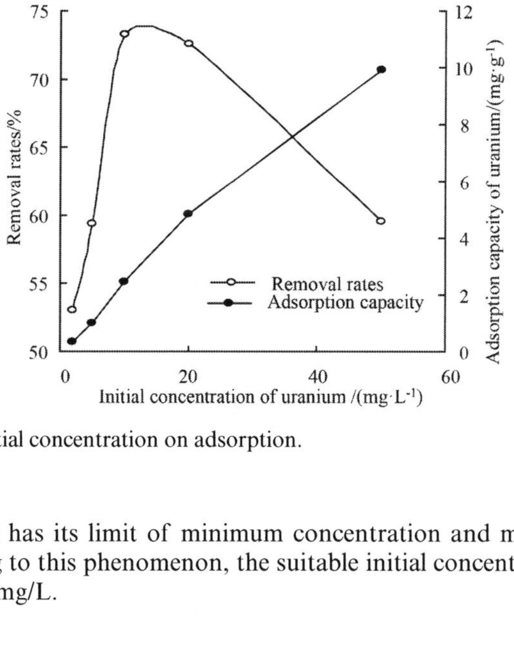

Figure 4. Effect of initial concentration on adsorption.

capacity of red mud has its limit of minimum concentration and maximum adsorption capability. According to this phenomenon, the suitable initial concentration of uranium is determined to be 10 mg/L.

3.5 *Sorption isotherm model for uranium onto red mud*

Supposing that the isothermal equilibrium adsorption experimental data fitted the Langmuir and Freundlich adsorption model, it was found that the Langmuir thermodynamic model was more suitable for the adsorption process (Figure 5, Figure 6 and Table 2). The Langmuir thermodynamic model indicated that the homogeneous adsorbate was present on the adsorbent's surface, and red mud was suitable for low adsorption concentration conditions.

The Langmuir adsorption isotherm equation is expressed as follows:

$$\frac{1}{Q_e} = \frac{1}{Q_m} + \frac{1}{K_1 Q_m C_e} \quad (3)$$

where Q_e is the adsorbed amount at any time e (mg/L), Q_m is the optimal adsorption capacity (mg/g), and K_1 is the Langmuir equilibrium constants.

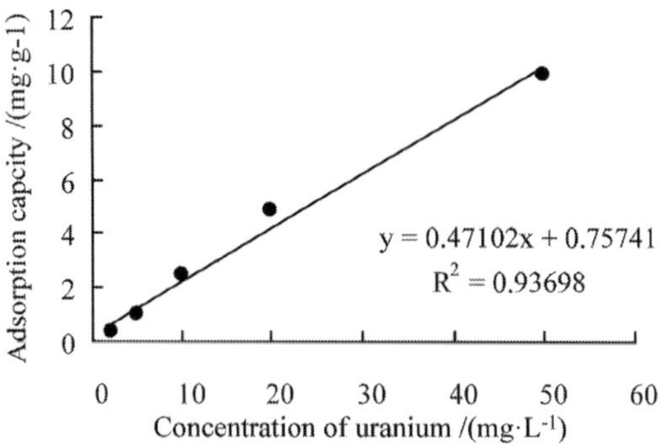

Figure 5. Langmuir adsorption isotherm for uranium.

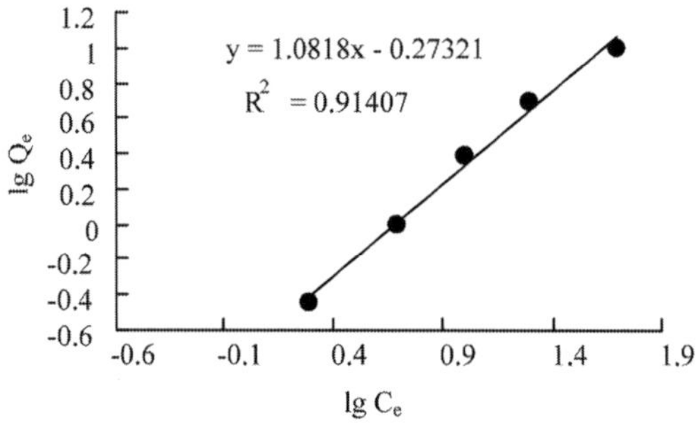

Figure 6. Freundlich adsorption isotherm for uranium.

Table 2. Parameters of the sorption isotherm model for uranium onto red mud.

	Freundlich isotherm			Langmuir isotherm		
Metalions	K_1	n	R^2	K_L	Q_m (mg/g)	R^2
Uranium	1.3142	0.9244	0.9141	1.6080	1.3203	0.9370

The Freundlich adsorption isotherm equation can be written as follows:

$$\lg Q_t = \lg K_L + \frac{1}{n}\lg C_t \tag{4}$$

where Q_t is the adsorbed amount at any time t (mg/g), C_t is the ion concentration at any time of t (mg/L), and n and K_L are the Freundlich characteristic constants that are related to the adsorption intensity and the adsorption capacity.

From Figure 5, Figure 6 and Table 2, it can be seen that the Langmuir and Freundlich sorption isotherm could fit the curve well, but the Langmuir model was more suitable. In general, it indicated that the monolayer adsorption on the surface of red mud played a leading role in the adsorption process. When the adsorption reached equilibrium, its adsorption rate was equal to the desorption rate, with no intermolecular interactions between the molecules on the surface.

3.6 *Adsorption dynamic model for uranium onto red mud*

Supposing that the adsorption data at different times fitted to a kinetic model, it was found that the adsorption of uranium onto red mud could fit the pseudo-second-order kinetic equations well (Figure 7, Table 3).

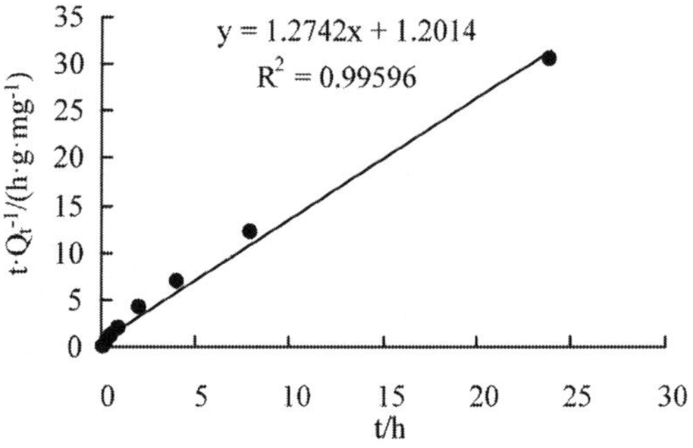

Figure 7. Pseudo-second-order kinetic equations of uranium adsorption onto red mud.

Table 3. Kinetic sorption parameters for uranium adsorption onto red mud.

Kinetic sorption parameters	Q_e (mg/L)	K_f (g/mg · min)	R^2
Red mud	0.7848	1.3514	0.9960

The pseudo-second-order kinetic equation can be expressed as follows:

$$\frac{1}{Q_e} = \frac{1}{Q_m} + \frac{1}{K_1 Q_m C_e} \tag{5}$$

where Q_t is the adsorption capacity at any time t (mg/g), Q_e is the equilibrium adsorption capacity (mg/g), K_f is the reaction rate constant, and t is the adsorption time (h).

From Figure 7 and Table 3, it can be seen that the adsorption process of uranium was suitable to the pseudo-second-order kinetic equation.

4 CONCLUSION

1. Red mud has a good adsorption efficiency on uranium. When its adsorbent dosage was 3 g/L, pH = 4.2, and the reaction temperature was 25°C, the uranium adsorption capacity reached up to 73.25%.
2. Bayer red mud showed an efficient adsorption capability from an acid solution, especially the wastewater containing uranium with concentrations below 10 mg/L. Therefore, red mud is suitable for removing low concentrations of uranium solution.
3. The adsorption dynamics of uranium onto red mud fitted the pseudo-second-order kinetic equations and Langmuir adsorption isotherms well.
4. Because the composition of red mud is very complex, this study could not cover all the aspects. However, the preliminary study shows a novel idea to deal with nuclear wastewater and reuse industrial residues. Also, it provides a theoretical basis for the application of red mud in wastewater treatment.

REFERENCES

Cao, Y., Li, W.D. & Liu, Y.G. 2007. Properties of Red Mud and Current Situation of Its Utilization. *Bulletin of the Chinese Ceramic Society* 26(1): 143–145.

Chen, Y., Cheng, H.F. & Deng, Y.T. et al. 2015. Progress in Uranium Adsorption by Clay Minerals. *Geology of Chemical Minerals* 02: 93–98.
Cui, R.P., Li, Y.L. & Jing, C. 2015. Adsorption of Uranium from Aqueous Solution on Illite. *Environmental Chemistry* 02: 314–320.
Deng, J.X., Meng, J. & Cheng, W. et al. 2011. Waste-water Treatment Technique for a Uranium Tailings Pond. *Uranium Mining and Metallurgy* 30(2): 100–103.
Gu, H.N., Wang, N. & Zhang, N.C. et al. 2011. Study on Radioactivity Level of Red Mud and Rediological Constraints of Usability as Building Materials. *Light Metal* 05: 19–21.
Kang, J.W., Hu, X.B. & Tian, J.M. et al. 2011. Study on the Treatment of Zinc Smelting Waste-water using Red Mud. *Journal of Taiyuan University of Echnology* 04: 375–378.
Li, Y.Z., Liu, C. & Yi, Z.K. et al. 2006. Phosphate Removal from Aqueous Solution using Activated Red Mud. *Acta Scientiae Circumstantiae* 26(11): 1775–1779.
Liu, J., Chen, D.Y. & Zhang, J. et al. 2012. Adsorption Characteristics and Mechanism of Uranium on Attapulgite. *Environmental Science* 08: 2889–2894.
Ren, S.J. & Wang, Y.Q. 2008. Research on Adsorbing Cd^{2+} in Water by Granular Red Mud. *Environmental Engineering* 02(5): 70–73.
Wang, J., Li, H.C. & Liu, J. et al. 2013. Progress on Radioactive Pollution and Related Problems caused by Uranium Mining. *Environ Health* 11:1033–1036.
Wang, X.J., Yue, Q.Y. & Zhao, P. et al. 2013. Research on the Adsorption Capability of Red Mud Granular Sobent for the Removal of Cd(II) and Pb(II). *Industrial Water Treatment* 05: 61–64.
Xiao, Y.Q., Liu, W.J. & Zhou, Y.T. et al. 2014. Adsorption Behavior of U(VI) and Mechanism Analysis by Organically Modified Vermiculite. *Atomic Energy Science and Technology* 12: 2187–2194.
Zeng, J.J., Wang, D.B. & Feng, Q.G. et al. 2013. Adsorption pf Hexavalent Chromium by using Modified Red Mud. *Journal of Guangxi University: Nat Sci Ed* 03: 673–678.
Zhang, X.F., Chen, D.Y. & Peng, Y. 2015. Adsorption Behavior of Uranium by Zeolite Loaded with P-calix[4]arene Acetate. *China Environmental Science* 06: 166–169.
Zhu, L., Wang, J. & Liu, J. et al. 2013. Preliminary Study on Uranium, Thorium and Some Other Metals Leached from Uranium Tailing under Simulated Natural Environmental Conditions. *Environmental Chemistry* 04: 678–685.

Advanced Engineering and Technology III – Xie (Ed.)
© 2017 Taylor & Francis Group, London, ISBN 978-1-138-03275-0

Research on the evaluation index of heliport noise

Bin Shao, Dong Li, Guan-hu Wang & Shao-feng Guo
Air Force Engineering University, Xi'an, Shanxi, China

Yu-hui Zhang & Jia-ling Li
88th Team, 94126 Army, Xi'an, Shanxi, China

ABSTRACT: The current heliport noise evaluation index cannot properly reflect the influence of military helicopter noise on susceptible targets around the heliport. In this regard, the characteristic of military helicopter noise is discussed, and the content and links of the existing airport noise evaluation index are analyzed. To elucidate the effects of the total noise level and a single burst noise simultaneously, a combined evaluation method using the Maximum Noise Level A (L_{Amax}) and the Weighted Equivalent Continuous Perceive Noise Level (L_{WCEPN}) is proposed for the evaluation of military heliport noise. The validity of the method is tested through an actual case.

1 INTRODUCTION

In recent years, with the increase in the number of military helicopters and training tasks, as well as the rapid development of urbanization, the impact of the heliport on the surrounding environment has become an important issue that cannot be ignored. Due to the huge noise of military helicopters, and with the enhancement of the surrounding people's environmental consciousness and awareness of rights, the complaints about military heliport noise also increase year by year. More attention has been paid to the relocation of the original heliport and the site selection of a new heliport. The prediction and evaluation of heliport noise has been one of the essential factors that affect the site selection.

At present, the evaluation methods of airport noise at home and abroad are mainly used for fixed wing aircraft, and research on helicopter noise has been mainly focused on the noise generation and propagation mechanism, noise detection, feature extraction and recognition, etc. (Zhong, 2008; Duan, 2009; Wang, 2010; Yamada, 2010).

There are only a few studies in the literature on the methods of helicopter noise evaluation. Currently, the airport noise evaluation index adopted by countries can scientifically reflect the noise effect of fixed wing aircraft; however, the noise effect of military helicopters on the surrounding environment cannot be accurately reflected due to the characteristics of military heliport.

There is a similar problem in the *Aircraft Noise Standards for Airport Surrounding Areas (Draft)* (2014) and other current standards in China, which recommend the use of the Maximum Sound Level (L_{max}) to evaluate helicopter noise. The aforementioned factors require a further study on the noise evaluation indices and evaluation methods of military heliport noise.

2 THE CHARACTERISTICS OF HELIPORT NOISE

Military heliports are important infrastructures of national defense. In recent years, with the rapid increase in military helicopters and the enhancement of training intensity, and with the increase in flying sorties of helicopters and trained subjects change, the problems of military heliport noise present new characteristics (Wang, 2003; Shinohara, 2009; Liu, 2012; Zhang, 2014).

1. Wide affected area and long duration
Compared with the civil airport, military heliport contains more helicopters and more flying sorties. The impact of civil airport noise on the surrounding environment mainly involves in both ends of the runway and the take-off and landing phases. However, the military helicopter flight program is more complex. The noise of military helicopter itself is very loud, and formation flying and other flight training subjects make the impact of military helicopter noise wider. Compared with other fixed wing aircraft, helicopters need low altitude hover, hover on the ground, flight idle, and other subjects, which make the effect of low- and medium-level helicopter noise on the surrounding environment last for a longer period of time.

2. More obvious burst noise
Civil aircraft fly based on a strict flight plan. The noise can be forecasted and controlled to a certain extent. However, military helicopters need training in all-weather and different subjects, as well as a case of emergency is quickly dispatched. So, the noise occurs randomly, which requires that the evaluation method must consider both the continuous effect of noise and the effect of burst noise.

3. Significant impact on people
At present, evaluation methods of aircraft noise are mainly based on the evaluation of noise energy. In fact, the impact of different frequencies of sound on the human body is also different. The experimental results indicate that the frequency of helicopter noise is relatively low. Therefore, under the same level condition, helicopter noise is easier to make people irritable than fixed wing aircraft noise.

3 EXISTING EVALUATION INDEX AND ANALYSIS OF ITS CHARACTERISTICS

3.1 *The maximum sound level A* (L_{Amax})

Aircraft noise is mixed noise that is composed of a series of different frequency combinations of sounds; however, the sensitivity of the human ear to different frequencies of noise is different. So, in order to make the auditory subjective feelings and the evaluation results consistent, the undesired sound pressure level of different frequencies should be weighted properly in measuring instruments. The sound levels measured by the weight method are called weighted sound pressure levels. A different weighted sound pressure level has a different correction factor. The A-weighted Sound Pressure Level (level A, L_A) is based on loudness, which is designed by simulating the curve $40L_N$ of equal-loudness contours. This method filters most of the low-frequency noise while the high-frequency noise will retain the original value, making the evaluation results more consistent with human auditory characteristics (Du, 2011). L_A can be calculated as follows:

$$L_A = 10\lg\left[\sum_{i=0}^{24} 10^{0.1(L_{pi}+\Delta_i)}\right] \tag{1}$$

Here, L_{pi} represents the sound pressure level of each frequency band, and Δ_i indicates the A-weighted correction factor of the i-th frequency band.

In a prescriptive measurement period or for a separate noise event, the maximum value of the measured A-weighted Sound Pressure Level is called as the Maximum Sound Level A, denoted as L_{Amax} (units: dB(A)) (Figure 1). This index is commonly used to evaluate individual noise events.

3.2 *Effective perceived noise level,* L_{EPN}

In the evaluation of the impact of aircraft noise on the human body, not only the energy and frequency characteristics of noise, but also the characteristics of continuous noise time need should be considered for the introduction of Effective Perceived Noise Level, L_{EPN}. The

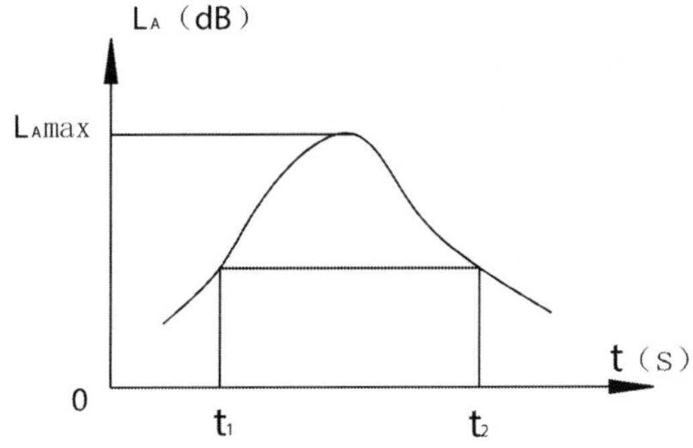

Figure 1. The maximum sound level A.

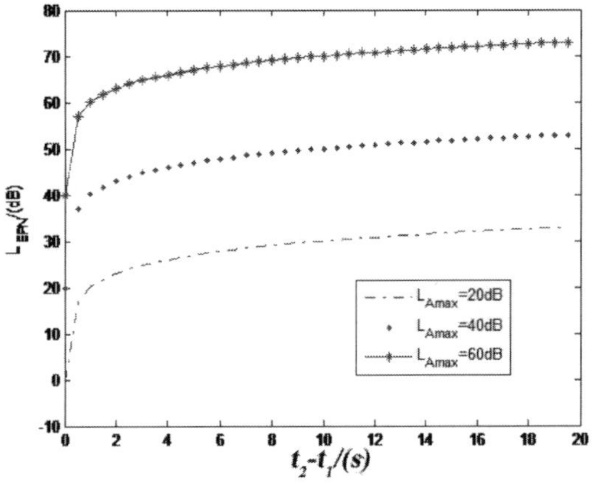

Figure 2. Relationship between L_{EPN} and $t_2 - t_1$.

International Civil Aviation Organization (ICAO) takes the L_{EPN} as a measure of aircraft noise. Generally, the sound level meter is used to measure the whole process noise level A or level D, and then the L_{EPN} can be approximately calculated by the following formula:

$$L_{\mathrm{EPN}} = L_{\mathrm{Amax}} + 10 1_{\mathrm{g}} \frac{t_2 - t_1}{20} + 13 = L_{\mathrm{Dmax}} + 10 1_{g} \frac{t_2 - t_1}{20} + 7\,(\mathrm{dB}) \tag{2}$$

Here, L_{Amax} and L_{Dmax} represent the Maximum Sound Level A and Maximum Sound Level D, respectively, and $t_2 - t_1$ indicates the descent time required by the maximum value of level A or level D of 10 dB (unit: s).

The approximate relationship between the L_{EPN} and the descent time $t_2 - t_1$ under different L_{Amax} conditions is shown in Figure 2.

Figure 2 shows that the value of L_{EPN} depends on the maximum noise level and the descent time: when $t_2 - t_1 < 1.0$ s, $L_{\mathrm{Amax}} > L_{\mathrm{EPN}}$; when $t_2 - t_1 = 1.0$ s, $L_{\mathrm{Amax}} = L_{\mathrm{EPN}}$; when $t_2 - t_1 > 1.0$ s, $L_{\mathrm{Amax}} < L_{\mathrm{EPN}}$. This also reflects the effect of noise duration on the perceived noise.

3.3 *Weighted Equivalent Continuous Perceived Noise Level, L_{WECPN}*

Mainland China and Brazil mainly use the Weighted Equivalent Continuous Perceived Noise Level (L_{WECPN}) to measure the airport noise of fixed wing aircrafts. The calculation formula L_{WECPN} is given by:

$$L_{\mathrm{WECPN}} = \overline{L}_{\mathrm{EPN}} + 10\lg\left(N_1 + 3N_2 + 10N_3\right) - 39.4 \tag{3}$$

Here, $\bar{L}_{EPN}$ indicates the average energy value of N flights' Effective Perceived Noise Level (dB); N_1, N_2, N_3 indicates the sorties of day, evening, and night, respectively.

According to the above formula, $\bar{L}_{EPN}$ assumes 50 dB, and the effects of different time periods are calculated separately. The relationship between L_{WECPN} and the flight numbers N_1, N_2, and N_3 is shown in Figure 3.

As shown in Fig. 3, L_{WECPN} takes both the single average perceived noise level and the influence of sorties of different times into account, and its essence is the shock of all the day's aircraft noise averaged per second (perceived noise level). There are large differences in the number of flights per day, so it has a great impact on the final results depending on whether the maximum number of flights or the average number of flights is considered.

3.4 *Day–Night Average Sound Level*

America mainly uses the Day-Night Average Sound Level (L_{dn}) to measure the airport noise, which is calculated as follows:

$$L_{dn} = 10\lg\left[\frac{1}{24}\left(15\times10^{L_d/10} + 9\times10^{(L_n+10)/10}\right)\right] \tag{4}$$

Here, L_d represents the daytime equivalent noise level (dB) and L_n represents the equivalent noise level at night (dB). L_{dn} also represents the impact of noise shock averaged per second.

In the literature on the Aircraft Noise Standards for Airport Surrounding Areas (Draft), the conversion relationship between L_{WECPN} and L_{dn} is proposed as follows:

$$L_{WECPN} = L_{dn} + 13 \tag{5}$$

In addition, some other noise evaluation indices include the British Noise and Number Index (*NNI*) and the Canadian Noise and Exposure Forecast (*NEF*). However, these indices also have some problems in evaluating heliport noise. For example, the *NNI* is calculated by a sensory noise peak, but it does not consider the total noise; the *NEF* considers the type of aircraft and flying sorties, but it does not consider the effect of a single noise.

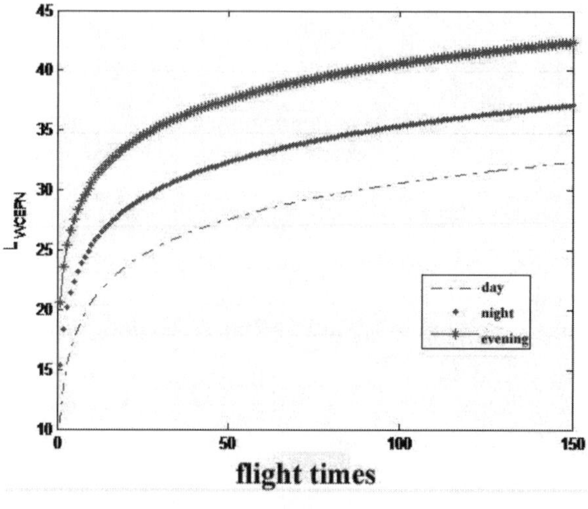

Figure 3. Relationship between L_{WECPN} and flight times.

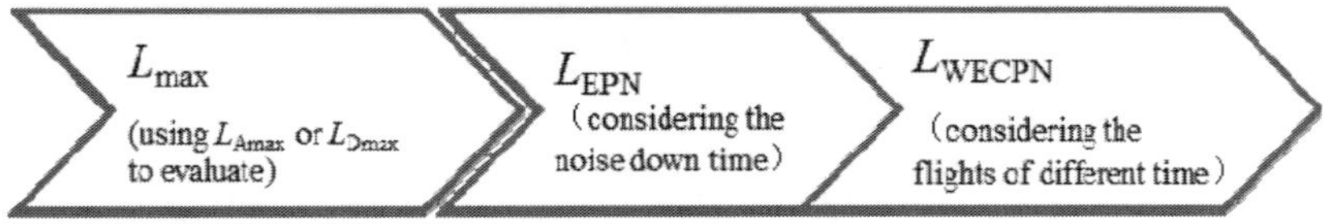

Figure 4. Relationship between the evaluation parameters.

4 SELECTION OF JOINT EVALUATION INDEX

In the last section, several noise evaluation indices are introduced. Our current standards propose L_{Amax} for the evaluation of helicopter noise. The standard references the Hong Kong Civil Airport Standard *NEF25*, namely the L_{Amax} limit standards of residential area and office district, in the time period of 7:00–19:00 are 85 dB(A) and 90 dB (A) (Liu, 1994). In fact, L_{Amax} and L_{EPN} can only evaluate the single noise event, thus evaluating only the impact of a flight on the ground observation points. In order to consider the impact of flight times, the *Airport Ambient Aircraft Noise Environment Standards,* promulgated in 1988 in China, use L_{WECPN} to assess the total noise value of the whole day. The relationship between the three indices is shown in Figure 4.

L_{Amax} cannot reflect the influence of flight times and the duration of noise. L_{EPN} reflects the impact of noise duration on perceived noise, but the role of the maximum sound level and the duration of noise cannot be distinguished. Although the L_{WECPN} takes the effect of the duration of noise and flights into account, there is a larger difference in the result because military helicopters randomly take-off and land just for the tasks. Furthermore, the change of sorties at different time periods will also influence the evaluation result. The noise of the helicopter is more loud under the conditions of the same sound level, so the national standard usually reduces to 5–10 dB(A), relative to fixed wing aircraft, when based on the energy equivalent evaluation. In this paper, according to the standard of Taiwan and on the basis of conversion relationship, the limit value of L_{WECPN} is set as 67 dB(A), which is 8 dB(A) lower than the fixed wing aircraft.

In order to reflect the influence of total noise and a single burst high noise on the environment, the combination of both the evaluation indices L_{WECPN} and L_{Amax} is proposed for military heliport noise. This approach not only overcomes the disadvantages of using L_{Amax} alone, but also has the advantage of using L_{WECPN} to reflect the total noise. In order to make the evaluation more comprehensive and reasonable and develop the related noise standard to deal with the problems of large randomness of military helicopter noise, the noise value of biggest one-day flight sorties and average daily flight sorties should be considered.

5 INDEX VERIFICATION

The actual noise situation and the villagers' feeling about a military heliport was investigated and tested. The locations of airport noise measure points are shown in Figure 5. In the map, points A, B, C, and D represent the heliport and 1#, 2#, 3#, 4# represent the villages around the heliport. The monitoring results of measure points and the calculation results of evaluation indices are summarized in Table 1.

From Table 1, it can be seen that the monitoring results of L_{Amax} at both daytime and nighttime are over 79 dB, i.e., above both the noise limitation and the maximum noise limitation of burst noise at night in the acoustic environment functional area in category 2, which are stipulated in the *Environmental Quality Standard for Noise* (GB3096-2008). So, the monitoring results of L_{Amax} basically agree with the time at which the villagers reflect strongly. However, except points A and D, the values of the other measure points around the heliport are under the limit standard of L_{Amax}: 85 dB(A) in the residential area and 90 dB(A) in the office district. In addition, the limit value of L_{dn} is set as 57 dB(A) (Table 1). This index has the advan-

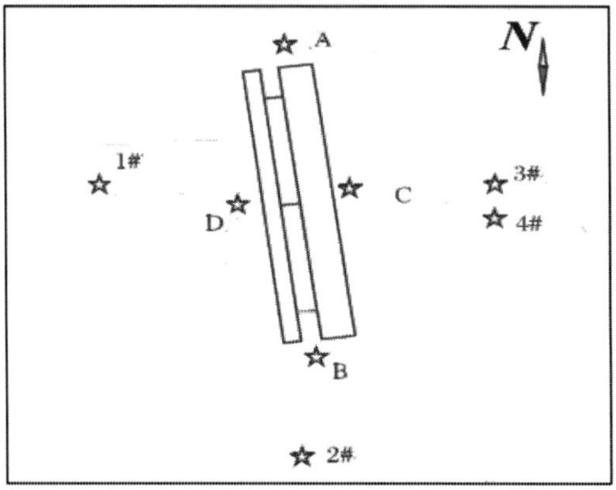

Figure 5. Sketch map of airport noise measure points.

Table 1. L_{EPNmax}, L_{EPNmin}, L_{WECPN}, L_{dn}, and L_{Amax} of each monitoring point (dB).

Monitoring point	A	B	C	D	1#	2#	3#	4#
Relative location of the runway	N, 400 m	S, 400 m	E, 200 m	W, 200 m	W, 1400 m	S, 2800 m	E, 1800 m	E, 1700 m
L_{Amax} (day)	92.1	88.5	92.8	87.8	79.9	80.2	79.8	79.8
L_{Amax} (night)	91.5	87.2	91.9	85.6	79.7	79.4	80.0	79.6
L_{EPNmax}	97.8	98.6	102.9	101.9	90.0	90.3	89.9	89.9
L_{EPNmin}	90.1	89.7	94.2	92.6	85.0	85.7	85.0	85.1
L_{WECPN}	78.1	78.8	83.4	82.6	71.2	71.6	71.3	71.1
L_{dn}	67.3	62.2	68.4	59.5	55.2	55.6	55.3	55.1

tage of reflecting the effect of cumulative noise, but, in fact, the calculation results are still below the limit value of 57 dB(A). So, L_{dn} cannot correctly reflect the impact of helicopter noise on the surrounding villages. According to the calculation results presented in Table 1, all the L_{WECPN} values of the villages are above 67 dB(A), indicating the impact of military heliport noise on the surrounding environment above the limits. This result is consistent with the practical investigation indicating that most villagers think that the helicopter noise has caused a greater impact on their lives.

So, according to the planning of guaranteeing 36 helicopters after the completion of the heliport, the future daily flight sorties must be greater than 100 times the test. With the increase in the number of flight sorties, L_{WECPN} will continue to increase, so L_{Amax} alone cannot play a good role in the evaluation. The influence of the total noise level and the single maximum noise level must be considered simultaneously. The combination of both indices L_{WECPN} and L_{Amax} can be a very good evaluation method of noise effects on the surrounding residents provided they meet the national standards. The rationality and validity of the proposed evaluation methods will also be verified.

6 CONCLUSION

Helicopter noise has the characteristics of suddenness, low frequency, and persistence, and has great effects on the sensitive area around the heliport. It has been an important factor affecting the development and construction of the heliport. However, there are many shortcomings in the current standards that propose L_{Amax} for the evaluation, whose scientificity and rationality need to be improved. On the basis of the actual situation of military helicopter flight training, this paper made an exploration on the characteristics of helicopter noise and related evaluation indices, and discussed the limit values of those indices. A method for using both L_{Amax} and L_{WECPN} was put forward to synthetically evaluate the noise of the helicopter, and a field noise testing and investigation of a military heliport was made. The result indicated that the current helicopter noise evaluation index L_{Amax} could not correctly evaluate the effects of military helicopter airport noise on the surrounding environment. However, the combination of both evaluation indices L_{Amax} and L_{WECPN} was more consistent with the actual survey result. This indicates that the combined evaluation method is reasonable, feasible, and more realistic, which has a certain reference for the study of helicopter noise.

REFERENCES

2014. *Aircraft Noise Standards for Airport Surrounding Areas (Draft)* (GB9660-201X).

2008. *Environmental quality standard for noise* (GB3096-2008).

2009. *Hong Kong Planning Standards and Guidelines.*

Duan Guangzhan & Chen Pingjian. 2009. Helicopter rotor noise calculation based on CFD. *Journal of Aerodynamics* 27(3): 314–319.

Du Jitao. 2011. Research on Airport Noise Prediction Model and its Application. Nanjing: Nanjing *University of Aeronautics and Astronautics.*

Ichiro Yamada. 2010. Airport noise model taking account of soundproofing embankment and aircraft ground operation. Sydney: *Proceedings of 20th International Congress on Acoustics, ICA.*

Liu Shaoyu & Du Junming. 1994. *Air noise in Hong Kong: current and status and future.* Beijing: *National Conference on Acoustics.*

Liu Zhou et al. 2012. A general model for aircraft noise calculation. *Journal of Vibration and Shock* 31(17): 124–128.

Shinohara & Tani. 2009. *Consideration to a new aircraft noise monitoring system at Narita Airport,* Internoise.

Wang Huaming et al. 2003. An experimental study of helicopter noise. *Acta Acustica* 28(2): 177–181.

Wang Yang et al. 2010. Analysis of the influence of helicopter flight parameters on the rotor blade vortex noise in the takeoff and landing. *Journal of Aerodynamics* 28(3): 322–327.

Zhang Yuhui et al. 2014. Research on evaluation approach of helicopter airport noise. *Journal of Air Force Engineering University (Natural Science Edition)* 15(4): 21–24.

Zhong Weigui & Zhang Yitao. 2008. Analysis of the helicopter tail-rotor and vortex interaction noise based on FW-H equation. *Helicopter Technique* 155: 31–34.

Food engineering and technology

Advanced Engineering and Technology III – Xie (Ed.)
© 2017 Taylor & Francis Group, London, ISBN 978-1-138-03275-0

Effect of different storage temperatures and packing methods on the quality of *Dictyophora indusiata*

S. Liu, X.M. Duan, L. Jia, X.J. Rao, Z.L. Zhang, Y.H. Xie & L. Li
Beijing Vegetable Research Center, Beijing Academy of Agriculture and Forestry Sciences, National Engineering Research Center for Vegetables, Beijing, China

L.B. Wang
USDA, ARS, U.S. Horticultural Research Laboratory, Fort Pierce, Florida, USA

ABSTRACT: *Dictyophora indusiata* (Vent. ex Pers.) are prone to postharvest deterioration during storage at high temperatures. To date, there are no reports available on the preservation technology of fresh *Dictyophora indusiata*. Fresh *Dictyophora indusiata* were packed using different methods, and physiological changes associated with postharvest deterioration were monitored during subsequent storage for 20 days at 0°C or 4°C and 70–80% relative humidity. The effect of cold storage and polyethylene (PE) film bag on the postharvest physiology and quality of *Dictyophora indusiata* was addressed. The application of 0°C+PE film bag packaging inhibited weight loss and thus improved the quality of *Dictyophora indusiata* as revealed by sensory evaluation, weight loss, vitamin C, total sugar, total phenolics, total flavonoid, and soluble protein content. Our results indicated that the use of the 0°C+PE film bag was an effective method to retain the postharvest physiology and quality of *Dictyophora indusiata*.

1 INTRODUCTION

Dictyophora indusiata (Vent. ex Pers.), known as the "queen of the mushrooms" due to the beautiful appearance, delicious taste, and health benefit, is an edible mushroom widely eaten in Asian countries, especially in China (Deng et al. 2013). They are not only appreciated as a highly tasty and nutritional, but also as an important source of biologically active compounds with a pronounced medicinal value (Ishiyama et al. 1999).

Storage temperature is one of the main factors that affect post-ripening and qualities such as respiration, transpiration, senescence, and other physiological processes (Singh et al. 2010). In general, low temperature can obviously reduce the respiration rate, growth of spoilage microorganisms, and postharvest physiological change, delay senescence, and maintain the nutrients of fruit and vegetables (Tian et al. 2013). Optimum storage temperature varies depending on the fruit or vegetable. Packaging technology is also a key to quality preservation, which is used to reduce the rate of respiration for mushrooms and limit overall quality deterioration. According to several authors, packaging films can delay mushroom spoilage with minimum changes in physiochemical and extend the shelf life of mushrooms (Ares et al. 2006, Taghizadeh et al. 2010). Therefore, packaging and storage at low temperatures are extensively applied to preserve vegetable quality.

Dictyophora indusiata are rich in nutrient components and water content, and are more likely to lose freshness at high temperatures due to the lack of appropriate postharvest technologies. The major changes that occur during its storage include loss in weight due to moisture loss, change in color due to browning, fungal decay, loss of nutritional value, and reduction in marketability. Therefore, the present work aimed to investigate the effect of cold storage and polyethylene (PE) film packaging on the quality of *Dictyophora indusiata*, and explore the suitable storage method to improve the overall quality and prolong the storage period of *Dictyophora indusiata*.

2 MATERIALS AND METHODS

2.1 *Raw material and treatments*

Eggs of *Dictyophora indusiata* were collected from Guiyang city of Guizhou Province, China, and harvested in the early morning. All the eggs were packed in sealed foam boxes (holding moisture) containing ice bottles (cooling), and immediately carried to the laboratory within 6h at ambient temperature. The eggs of *Dictyophora indusiata* were incubated at 27°C and RH 95% for 72 h. The incubation rate was about 75%. Then, fresh *Dictyophora indusiata* was obtained and randomly distributed into three groups: (1) mushrooms packed with PE film bags (0.06 mm in thickness) were encased into plastic boxes (PB) stored at 0 ± 0.5°C and 70–80% Relative Humidity (RH); (2) mushrooms packed with PE film bags (0.06 mm in thickness) were encased into plastic boxes stored at 4°C and RH 70–80%; (3) mushrooms without packaging were enclosed in Expanded Polystyrene (EPS) boxes stored at 0 ± 0.5°C, RH 70%–80% as Control Check (CK). All the treatments were performed in triplicate.

2.2 *Sensory evaluation*

Sensory quality of *Dictyophora indusiata* in terms of changes in visual appearance, color, decay, shrinkage, and acceptability was evaluated according to commercial practice and following 1–5 grades. (5, excellent = good quality, no browning rot, no shrinkage; 4, good = good quality, color slightly dim, without browning rot, slight shrinkage; 3, fair = extremely slight browning, slightly soft, no decay, slight shrinkage, edible; 2, poor = poor quality, partial browning, soft, slightly rotten, severe shrinkage, edible; 1, very poor = completely broken, severe browning rot, not edible). A score of 3.0 indicated minimum commercial acceptability. Quality evaluation was determined by three trained individuals.

2.3 *Weight loss*

Weight loss was determined by the difference in the weight of the sample before and after storage and is expressed on a wet weight basis (%).

2.4 *Determination of chemical quality attributes*

Vitamin C (Vc) content was determined by the molybdenum blue colorimetric method (Li, 2000). The content of soluble proteins was determined as described by Bradford (1976). The total phenolic content was determined according to Han et al. (2014). Total flavonoid content was measured according to Lin & Tang (2007). The anthrone-sulfuric acid method was used to measure the total sugar content (Laurentin & Edwards, 2003).

2.5 *Statistical analysis*

All experiments were carried in triplicate and the average values are reported in the analysis. Statistical analysis was performed using the SPSS 19.0 (SPSS Inc., Chicago, IL, USA) software with analysis of variance (ANOVA) and Duncan's new multiple range tests ($p < 0.05$). Differences at $P < 0.05$ were considered significant.

3 RESULTS AND DISCUSSION

3.1 *Effect of storage temperature and packing method on the sensory evaluation of* Dictyophora indusiata

Sensory quality of *Dictyophora indusiata* stored for 20 days is shown in Fig. 1. Throughout the storage time, the samples stored at 4°C showed a significant reduction with the sensory score varying from 5.0 to 2.9, while 0°C low temperature and PE packaging retarded this reduction

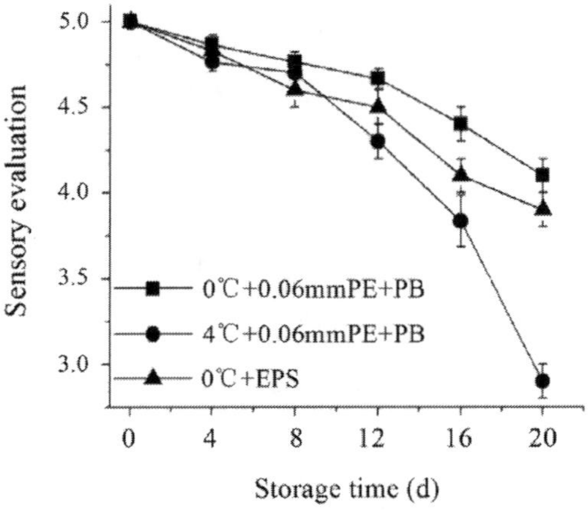

Figure 1. Effect of storage temperature and packing method on the sensory quality of *Dictyophora indusiata*.

and showed a score of 4.1 in the 0°C+PE+PB mushrooms and 3.9 in the 0°C+EPS mushrooms at the end of storage. At the 20th day, the samples stored at 4°C became unacceptable for the market (score above 3.0) with the sensory quality score of 2.9. The high quality of mushrooms stored at the 0°C+PE+PB conditions was preserved, and the difference between the mushrooms stored at 4°C and 0°C was significant ($P < 0.05$) at the end of storage.

3.2 *Effect of storage temperature and packing method on the weight loss of* Dictyophora indusiata

The weight loss in all the treatments markedly increased with the prolonging storage period (Fig. 2). However, mushrooms stored at 4°C showed a greater weight loss with an average rate of 0.90% per day, while the rate of weight loss in the 0°C+PE+PB samples was 0.65% per day and that in the 0°C+EPS samples was 0.71% per day during the storage. At the 20th day, the rate of weight loss in the 0°C+PE+PB, 4°C+PE+PB, 0°C+EPS samples were 12.90%, 17.94%, and 14.23%, respectively. Significant differences were found between the 0°C and 4°C treatments ($P < 0.05$). The PE film effectively reduced the weight loss of *Dictyophora indusiata* at the lower (0°C) temperature.

3.3 *Effect of storage temperature and packing method on the vitamin C, soluble protein and total sugar content of* Dictyophora indusiata

The changes in the content of vitamin C, total sugar, and soluble protein are shown in Fig. 3. The vitamin C content of mushrooms stored at 4°C showed a slight increase during the first 4 days, thereafter the content decreased rapidly in subsequent shelf-life storage. In contrast, the mushrooms stored at 0°C showed a slight increase during the first 8 days and experienced a gradual decrease from d 8 to d 20 (Fig. 3a). After 20 days of storage, the vitamin C content in the mushrooms stored at 4°C significantly decreased, its vitamin C content was 0.57 mg/g, while the vitamin C content in the 0°C+PE+PB and 0°C+EPS mushrooms was 1.01 mg/g and 0.78 mg/g, respectively. Among the three treatments, the 0°C+PE+PB treatment resulted in a higher vitamin C content when compared with the 0°C+EPS treatment from d 8 to d 20 of storage. Losses in vitamin C content in the 0°C-treated mushroom were slowed down and much delayed.

The total sugar content in all the mushrooms decreased greatly during 20 days of storage (Fig. 3b). However, the use of 0°C and PE film bag delayed this trend and significant difference was observed between the 0 and 4°C treatments at the end of storage ($P < 0.05$).

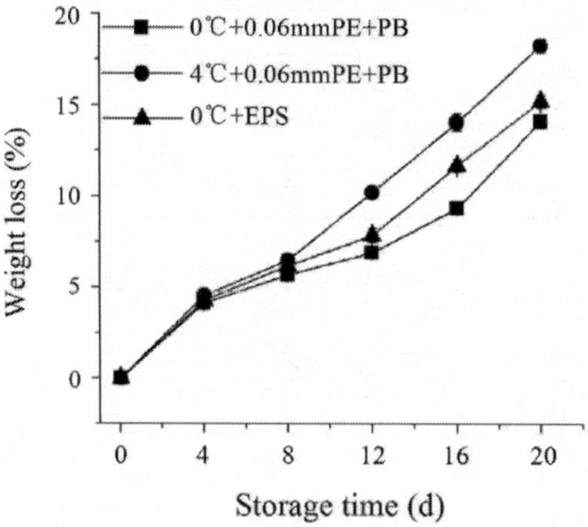

Figure 2. Effect of storage temperature and packing method on the weight loss of *Dictyophora indusiata*.

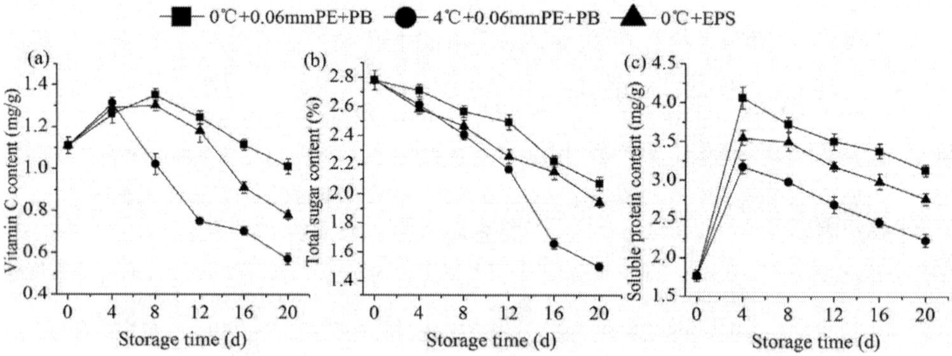

Figure 3. Effect of storage temperature and packing method on the vitamin C content (a), total sugar content (b) and soluble protein content (c) of *Dictyophora indusiata*.

Among the three treatments, the 4°C+PE+PB treatment had the lowest total sugar content from d 8 to d 20 of storage and the 0°C+PE+PB treatment had the highest total sugar content throughout the storage. The total sugar content of *Dictyophora indusiata* decreased ($P <$ 0.05) at 4°C storage, due to the higher respiration rate and the formation of polysaccharide (Guo et al. 2008).

The content of soluble protein increased slightly during the first 4 days, and then decreased markedly at the end of storage (Fig. 3c). The mushrooms stored at 4°C had a relatively low content of soluble protein. At early storage, the vitamin C and soluble protein contents of *Dictyophora indusiata* were slightly increased. This phenomenon could be attributed to the after-ripening of *Dictyophora indusiata* after incubation from the eggs.

3.4 *Effect of storage temperature and packing method on the total phenolic and total flavonoid contents of* Dictyophora indusiata

The content of total phenolics and total flavonoids in *Dictyophora indusiata* decreased gradually during the storage (Fig. 4 a and b). The total phenolic content of all treatments decreased during the remaining 20 d, which is due to polyphenols that cause browning during the storage (Ye et al. 2012). The samples stored at 4°C had a relatively low content of total phenolics and

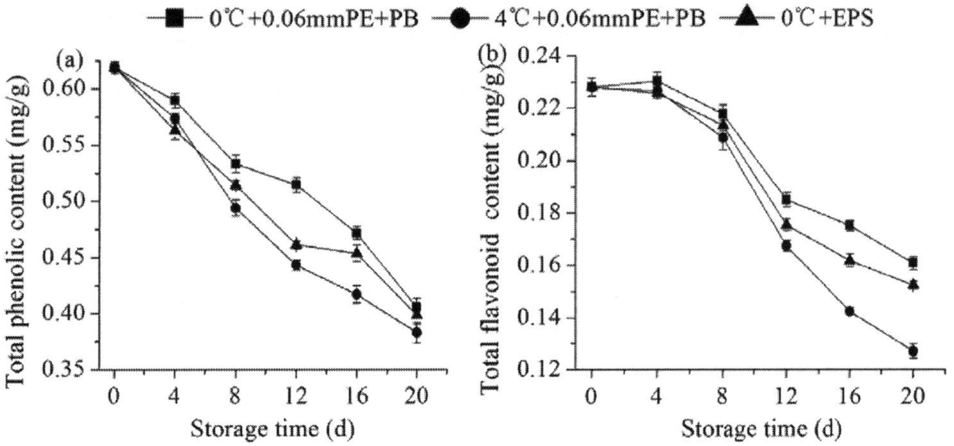

Figure 4. Effect of storage temperature and packing method on the total phenolic (a) and total flavonoid content (b) of *Dictyophora indusiata*.

total flavonoids and showed the most rapidly decreasing rates. At the end of storage, the total phenolic contents in the 0°C+PE+PB, 4°C+PE+PB, and 0°C+EPS samples were 0.405 mg/g, 0.383 mg/g, and 0.398 mg/g, respectively. The total flavonoid contents in the 0°C+PE+PB, 4°C+PE+PB, and 0°C+EPS samples were 0.161 mg/g, 0.127 mg/g, and 0.152 mg/g, respectively. The results indicated that low temperature (0°C) and PE film bag could effectively maintain the content of total phenolics and total flavonoids in *Dictyophora indusiata* during the storage.

4 CONCLUSIONS

In conclusion, the effect of cold storage and PE film bag on the postharvest physiology and quality of *Dictyophora indusiata* was addressed. According to the physiological change in *Dictyophora indusiata* during storage, we determined that the suitable storage temperature of *Dictyophora indusiata* was 0°C. Furthermore, we demonstrated that the application of 0°C+PE film bag packaging inhibited the increase in weight loss and the decrease in the sensory score and contents of vitamin C, soluble protein, total sugar, total phenolics, and total flavonoids, thereby improving the quality of *Dictyophora indusiata.* Our results indicated that the use of 0°C+PE film bag was an effective method to retain the postharvest physiology and quality of *Dictyophora indusiata.* This kind of preservation method extended the shelf life of *Dictyophora indusiata* up to 20 days and has a great potential in the practical application.

ACKNOWLEDGMENTS

The research was funded by the National Science and Technology Support Program (No. 2015BAD19B02) and the Science and Technology Innovation Special Construction Funded Program of Beijing Academy of Agriculture and Forestry Sciences (No. KJCX20140205). Corresponding author: S. Liu

REFERENCES

Ares, G., Parentelli, C., Gámbaro, A., Lareo, C. & Lema, P. 2006. Sensory shelf life of shiitake mushrooms stored under passive modified atmosphere. *Postharvest Biology and Technology* 41(2): 191–197.

Bradford, M.M. 1976. A rapid and sensitive method for the quantitation of microgram quantities of protein utilizing the principle of protein-dye binding. *Analytical Biochemistry* 72(1): 248–254.
Deng, C., Fu, H., Teng, L., Hu, Z., Xu, X., Chen, J. & Ren, T. 2013. Anti-tumor activity of the regenerated triple-helical polysaccharide from *Dictyophora indusiata. International Journal of Biological Macromolecules* 61: 453–458.
Guo, L., Ma, Y., Sun, D.W. & Wang, P. 2008. Effects of controlled freezing-point storage at 0°C on quality of green bean as compared with cold and room-temperature storages. *Journal of Food Engineering* 86(1): 25–29.
Han, C., Zuo, J., Wang, Q., Xu, L., Zhai, B., Wang, Z., Dong, H. & Gao, L. 2014. Effects of chitosan coating on postharvest quality and shelf life of sponge gourd (*Luffa cylindrica*) during storage. *Scientia Horticulturae* 166: 1–8.
Ishiyama, D., Fukushi, Y., Ohnishi-Kameyama, M., Nagata, T., Mori, H., Inakuma, T., Ishiguro, Y., Li, J. & Kawagishi, H. 1999. Monoterpene-alcohols from a mushroom *Dictyophora indusiata. Phytochemistry* 50(6): 1053–1056.
Laurentin, A. & Edwards, C.A. 2003. A microtiter modification of the anthrone-sulfuric acid colorimetric assay for glucose-based carbohydrates. *Analytical Biochemistry* 315(1): 143–145.
Li, J. 2000. Molybdenum blue colorimetric method to determine reduced vitamin C. *Food Science* 21(8): 42–45 (in Chinese).
Lin, J.Y. & Tang, C.Y. 2007. Determination of total phenolic and flavonoid contents in selected fruits and vegetables, as well as their stimulatory effects on mouse splenocyte proliferation. *Food Chemistry* 101(1): 140–147.
Singh, P., Langowski, H.C., Wani, A.A. & Saengerlaub, S. 2010. Recent advances in extending the shelf life of fresh *Agaricus* mushrooms: a review. *Journal of the Science of Food and Agriculture* 90(9): 1393–1402.
Taghizadeh, M., Gowen, A., Ward, P. & O'Donnell, C.P. 2010. Use of hyperspectral imaging for evaluation of the shelf-life of fresh white button mushrooms (*Agaricus bisporus*) stored in different packaging films. *Innovative Food Science & Emerging Technologies* 11(3): 423–431.
Tian, J.Q., Bae, Y.M. & Lee, S.Y. 2013. Survival of foodborne pathogens at different relative humidities and temperatures and the effect of sanitizers on apples with different surface conditions. *Food Microbiology* 35(1): 21–26.
Ye, J.J., Li, J.R., Han, X.X., Zhang, L., Jiang, T.J. & Miao, X. 2012. Effects of active modified atmosphere packaging on postharvest quality of shiitake mushrooms (*Lentinula edodes*) stored at cold storage. *Journal of Integrative Agriculture* 11(3): 474–482.

Advanced Engineering and Technology III – Xie (Ed.)

Extraction of soluble dietary fiber from soy sauce residue by microwave-ultrasonic assisted enzymatic hydrolysis and its antioxidant potential

T. Wang, W. Li, Y. Chu, M. Gao, Y. Dong, C. Zhang & T. Li
Jiangsu Key Construction Laboratory of Food Resource Development and Quality Safe, Xuzhou Institute of Technology, Xuzhou, China

ABSTRACT: Extraction of Soluble Dietary Fiber (SDF) from Soy Sauce Residue (SSR) was investigated by microwave-ultrasonic assisted enzymatic hydrolysis for the first time. The single factor experiment and the orthogonal experiment methodology were used to obtain the optimal condition for extracting SDF. DPPH scavenging activity and reducing power were measured to evaluate the antioxidant properties of SDF. The results revealed the optimal extraction conditions were as follows: ratio of solvent to material 15:1; microwave power 200 W; microwave-ultrasonic extraction time 200 s; enzymatic hydrolysis pH 5.5; and enzymatic hydrolysis time 50 min. Under these conditions, the maximum extraction rate of SDF reached 2.15%. Furthermore, the extracted SDF showed high DPPH scavenging activity and reducing power. The study confirms that SDF recovered from SSR has a good potential in the development of functional food.

1 INTRODUCTION

Soy sauce, as a famous seasoning in Asia, is mainly produced from soybean, wheat, and salt using a traditional fermentation technique. With the rapid development of soy sauce industry, a large number of Soy Sauce Residues (SSR) is retained, which contains a large amount of valuable components, such as protein, isoflavones, fibers, and oil (Hosoda et al., 2012). Unfortunately, most of the soy sauce residues are mainly disposed of using low-cost methods for feeds and fertilizers or discarded as industrial waste. Therefore, effective reuse of SSR can result in considerable economic value, and reduce the burden of the waste treatment of the soy sauce industry. However, to date, only a few related reports have been found in the literature.

In recent years, DF (Dietary Fiber) has received increased attention from the food industry, consumers, and researchers, due to its health benefits. The beneficial effects of DF include reducing caloric intake, decreasing total and Low Density Lipoprotein (LDL) levels, improving postprandial glucose response, and enhancing gastrointestinal immunity (Ma et al., 2015; Gunness & Gidley, 2010). Among those health benefits, Soluble Dietary Fiber (SDF) has been thought to play an important role. Soluble dietary fiber is carbohydrate polymers with 10 or more monomeric units, which are not hydrolyzed by endogenous enzymes in the small intestine of humans (Cui et al., 2011). It is mainly composed of pectins, β-glucans, mucilages, gum Arabic, konjac glucomannan, seaweed polysaccharides, microbial polysaccharides, non-digestible oligosaccharides, and inulin (Cui, Nie, & Roberts, 2011).

In the past years, studies have reported that fibers were mostly derived from cereals, fruits, vegetables, tubers, and algae (Kaushik, & Singh, 2011; Meyer, Dam, & Lærke, 2009; Raghavendra et al., 2004). Currently, interest is focused on alternative DF sources. Therefore, SSR, as a potential cheap biomass resource, is a good choice.

The extraction methods of SDF reported include: steam explosion and dilute acid soaking (Wang et al., 2015), shear emulsifying assisted enzymatic hydrolysis (Ma et al., 2015), and others. To our knowledge, there is little information on the extraction of SDF from SSR by the microwave-ultrasonic assisted enzymatic hydrolysis method. In this study, microwave-ultrasonic assisted enzymatic hydrolysis was chosen to extract SDF from soy sauce residue. The optimal extraction parameters were investigated with the single factor experiment and the orthogonal experiment. Additionally, the antioxidant properties of the extracted SDF, such as DPPH free radical-scavenging activity and reducing power, were examined. The findings of this study will provide a useful reference for the industrial extraction of SDF from SSR and the further utilization of SDF.

2 MATERIALS AND METHODS

2.1 *Materials*

Soy sauce residue was obtained from Xuzhou Hengshun Wantong food Brewing Co. Ltd. (Jiangsu, China), with a moisture content of 16.34 (w/w, %). It was dried at 60°C grounded into powder, and passed through a 60-mesh sieve. Deoiled residue was obtained as follows: 2 g of soy sauce residue powder was immersed in 10 ml light petroleum for 12h, which was duly covered to avoid solvent loss. After filtration, the residue was air-dried until a constant weight was obtained.

Papain (enzyme activity 3000 U/g) was purchased from Sinopharm Chemical Reagent Co., Ltd. (Beijing, China). Cellulase (enzyme activity 40000 U/g) was purchased from Yakult pharmaceutical Ind. Co., Ltd. All other chemicals used were of analytical grade.

2.2 *Extraction of SDF by microwave-ultrasonic assisted enzymatic hydrolysis*

A 2 g of deoiled residue was mixed with 20 ml deionized water in a 250 ml extraction flask, which was then placed in a microwave-ultrasonic synergistic extraction apparatus (CW-2000, Shanghai Tuoxin Analytical instrument Co., Ltd., China). The mixture was simultaneously irradiated with 100 W of microwave and ultrasonic power (600 W, unadjusted) for intervals of 200 s. After stopping the irradiation, the solution was first enzymatically hydrolyzed with 1% (w/w) papain (pH 6.5) at 65°C for 60 min to hydrolyze the protein. The resulting slurries were heated at 95°C for 5 min to terminate the enzymatic reaction. After cooling to room temperature, the slurries were subjected to sequential enzymatic digestion by 3% (w/w) cellulose (pH 4.5, 45°C for 60 min), and allowed to cool to room temperature and filtered. The filtrate obtained was concentrated at 70°C by rotary evaporation, and the solution was added into four volumes of 95% ethanol and kept at 25°C for 12 h. The precipitate was collected after centrifugation at 4000 r/min for 5 min and then freeze-dried. The extraction rate of SDF can be determined by using the following equation (1):

$$Exaction\ rate\,(\%) = \frac{m_1}{m_0} \times 100\% \tag{1}$$

where m_0 is the weight (g) of deoiled soy sauce residue and m_1 is the weight (g) of SDF after extraction.

2.3 *Optimization of the microwave-ultrasonic assisted enzymatic hydrolysis conditions*

2.3.1 *Effect of ratio of solvent to material (v / m) on the extraction rate of SDF*

The extraction parameters were as follows: ratio of water to SSR (v/m) 10:1; ultrasonic power 600 W; microwave power 200 W; microwave-ultrasonic extraction time 200 s; hydrolyzed by 0.3% (w/w) papain at 65°C for 60 min; and 1% (w/w) cellulose (pH 4.5) at 45°C for 60 min. The selected ratios of solvent to material (v/m) were 5:1, 10:1, 15:1, 20:1, 25:1, and 30:1. The other steps were the same as mentioned above.

2.3.2 *Effect of microwave power on the extraction rate of SDF*
Based on the optimal ratio of solvent to material (v/m), microwave power of 50 W, 100 W, 150 W, 200 W, 300 W, and 400 W was selected. The other steps were the same as mentioned above.

2.3.3 *Effect of extraction time on the extraction rate of SDF*
Based on the optimal ratio of raw material to solvent (m/v) and microwave power, the microwave-ultrasonic extraction time was set at 50 s, 100 s, 150 s, 200 s, 250 s, and 300 s. The other steps were the same as mentioned above.

2.3.4 *Effect of enzymatic hydrolysis pH and time by cellulose on the extraction rate of SDF*
Based on the optimal microwave-ultrasonic extraction conditions, the chosen pH of cellulose hydrolysis was 4, 4.5, 5, 5.5, 6, and 6.5, and the hydrolysis by cellulose time was 30 min, 40 min, 50 min, 60 min, 70 min, and 80 min. The other steps were the same as mentioned above.

2.3.5 *Orthogonal experiment*
On the basis of the single factor experimental results, and considering the five important parameters (ratio of solvent to material, microwave power, extraction time, pH, hydrolysis time) (Table 1), the L_{16} (4^5) orthogonal experiment was used to determine the optimal microwave-ultrasonic assisted enzymatic hydrolysis parameter to evaluate the combination effects of the five factors on the extraction rate of SDF.

2.4 *Antioxidant properties of SDF*

2.4.1 *DPPH (2,2-diphenyl-1-picrylhydrazyl)free radical scavenging capacity assay*
DPPH free radical-scavenging capacity of SDF was determined according to the method of Brand-Williams, Cuvelier, and Berset (1995) with some modifications. Briefly, a 2 mL SDF solution was mixed with 2 mL DPPH radical solution (0.2 mM). The absorbance of the sample was read at 515 nm after incubation at room temperature for 30 min (A_{sample}), while deionized water was used for the blank (A_{blank}). The scavenging activity of DPPH free radicals was calculated according to the following formula: DPPH scavenging activity (%) = $(1 - A_{sample}/A_{blank}) \times 100\%$. The IC50 value was determined by a nonlinear regression by plotting log (concentration) versus scavenging rate with Graph PadPrism 6.

2.4.2 *Reducing power assay*
The reducing power of SDF was determined based on the method described by Lin, Wei, and Chou (2006) with a few modifications. Briefly, a 0.5 ml SDF solution was mixed with 2.5 ml phosphate buffer (0.2 mmol/L, pH 6.6) and 2.5 ml potassium ferricyanide (1%, w/v), after incubation at 50°C for 20 min. Then, 0.5 ml trichloroacetic acid (10%, w/v) was added and the mixture was centrifuged at 1000 r/min for 10 min. The supernatant (2.5 ml) was mixed with 2.5 ml ferricchloride (0.1%, w/v) for 10 min at room temperature. Absorbance was measured at 700 nm.

Table 1. Factors and levels of the orthogonal experiment.

	Factor				
Level	A, ratio of solvent to material (mL/g)	B, microwave power (W)	C, extraction time (s)	D, pH	E, hydrolysis time (min)
1	10:1	50	150	4	40
2	15:1	100	200	4.5	50
3	20:1	150	250	5	60
4	25:1	200	300	5.5	70

2.5 *Statistical analysis*

Data are expressed as mean ± Standard Deviation (SD) (n = 3). Experimental design and data analyses were performed with Orthogonal Design Assistant V3.1 and SPSS 14.0 (SPSS, Inc., USA). Significance differences were evaluated by one-way analysis of variance (ANOVA). *P* values <0.05 and <0.01 were considered as statistically significant and highly significant, respectively.

3 RESULTS AND DISCUSSION

3.1 *Effect of ratio of solvent to material on the extraction rate of SDF*

The effect of the ratio of solvent to material (i.e., 5:1, 10:1, 15:1, 20:1, 25:1, and 30:1) on the extraction rate of SDF is shown in Figure 1A. It can be seen that as the ratio of solvent to material increased from 5:1 to 15:1 and the extraction rate increased remarkably from 0.648% to 1.725%. Then there was no significant difference. The main reason for this phenomenon could be that more ingredients are dissolved in the solvent with the increasing solvent to material ratio until they were fully extracted. Then, the large amounts of the solvent will have a little effect on the extraction rate of SDF, thereby leading to a waste of solvent and a higher processing cost. Therefore, 15:1 was selected as the optimal ratio of solvent to material.

3.2 *Effect of microwave power on the extraction rate of SDF*

The influence of microwave power on the extraction rate of SDF was examined by varying the microwave power levels at 50 W, 100 W, 150 W, 200 W, 300 W, and 400 W, with the other factors being fixed. The results shown in Figure 1B indicated that among the power levels tested, the highest SDF extraction rate was obtained at the power of 100 W, and then extraction rate decreased. This indicate that the increasing microwave power could help promote the dissolution of SDF, until it reached to 100 W, whereas the further increasing microwave power will increase the degradation of SDF. Thus, the optimal microwave power was found to be 100 W.

3.3 *Effect of microwave-ultrasonic extraction time on the extraction rate of SDF*

To find a suitable extraction time for the extraction of SDF, SSR was extracted for 50–300 s in the synergistic extraction apparatus. As shown in Figure 1C, the extraction rate increased rapidly when the extraction time varied from 50 s to 200 s, peaked at 200 s, and then there was a slow decrease with the increasing time. This was probably because the dissolved SDF was decomposed by the further increasing time. The results are consistent with a previous investigation which showed that the antioxidant ingredients increased first and then decreased slowly with the increasing microwave/ultrasonic extraction time (Wu et al., 2015). So, the optimal extraction time was 100 s.

3.4 *Effect of enzymatic hydrolysis pH on the extraction rate of SDF*

Cellulose activity was affected by the pH of the solvent (Puri et al., 2012), in order to investigate the effect of pH on the extraction rate of SDF, the solution was adjusted to different pH values (i.e., 4.0, 4.5, 5.0, 5.5, 6.0, and 6.5) when hydrolyzed by cellulose. As shown in Figure 1D, the extraction rate of SDF ranged from 1.342% to 1.118% when pH was varied from 4.0 to 6.5, and reached maximum at pH 4.5, indicating that cellulose could exert maximum activity at this pH value. Thus, the optimal pH value of cellulose hydrolysis was 4.5.

3.5 *Effect of enzymatic hydrolysis time on the extraction rate of SDF*

To determine cellulose hydrolysis time for the extraction of SDF, different hydrolysis times were selected from 30 min to 80 min. The results shown in Figure 1E indicated that as the

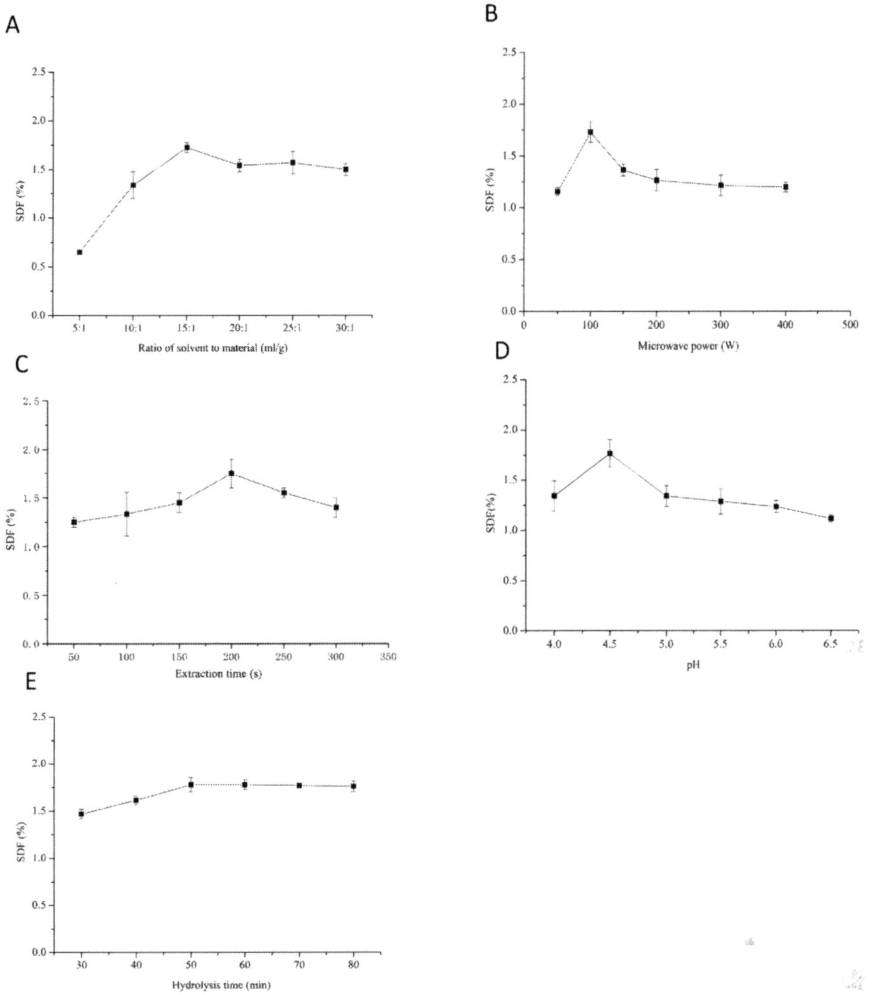

Figure 1. Effect of ratio of solvent to material (A), microwave power (B), microwave-ultrasonic extraction time (C), enzymatic hydrolysis pH (D), and enzymatic hydrolysis time (E) on the extraction rate of SDF.

hydrolysis time increased from 30 min to 50 min, the extraction rate of SDF increased from 1.47% to 1.78%, and then reached a plateau at 50 min, suggesting the completion of the enzymatic hydrolysis process. To avoid wasting time, 50 min was chosen as the optimal hydrolysis time.

3.6 *Optimization results by the orthogonal experiment*

The orthogonal experimental method is usually used to reduce the number of required experiments and obtain reasonable results (Tinsley & Brown, 2000). Thus, based on the results of single factor experiments, according to Table 1 (factors and levels of orthogonal design), 16 groups of experiments were carried out in the L_{16} (4^5) orthogonal experiment design. The experimental results and intuitive analysis are outlined in Table 2. The extraction rate of SDF ranged from 1.19% to 1.92%. Among them, the maximum extraction rate was obtained in the group $A_2B_3C_4D_1E_2$ (Run 7). However, according to the K value, the optimal formulation was $A_2B_4C_2D_4E_2$, not $A_2B_3C_4D_1E_2$, contained in the 16 groups. So, a verification experiment was required to examine the result of the optimal formulation of $A_2B_4C_2D_4E_2$. The results

Table 2. Results of the orthogonal experiment.

Number L_{16} (4^5)	A	B	C	D	E	SDF (%)
1	1	1	1	1	1	1.190
2	1	2	2	2	2	1.500
3	1	3	3	3	3	1.430
4	1	4	4	4	4	1.505
5	2	1	2	3	4	1.765
6	2	2	1	4	3	1.785
7	2	3	4	1	2	1.920
8	2	4	3	2	1	1.740
9	3	1	3	4	2	1.355
10	3	2	4	3	1	1.235
11	3	3	1	2	4	1.270
12	3	4	2	1	3	1.485
13	4	1	4	2	3	1.175
14	4	2	3	1	4	1.125
15	4	3	2	4	1	1.230
16	4	4	1	3	2	1.283
K1	1.406	1.371	1.382	1.430	1.349	
K2	1.802	1.411	1.495	1.421	1.514	
K3	1.336	1.462	1.413	1.428	1.469	
K4	1.203	1.503	1.459	1.469	1.416	
R	0.599	0.132	0.113	0.048	0.165	

Table 3. Results of variance analysis.

Factor	Square of deviance	Degree of freedom	F ratio	Critical value	Significance
A	0.797	3	22.771	4.760	*
B	0.040	3	1.143	4.760	
C	0.030	3	0.857	4.760	
D	0.006	3	0.171	4.760	
E	0.061	3	1.743	4.760	
Error	0.07	6			

* Represents significant difference ($p < 0.05$).

indicated that the optimal formulation led to a maximum extraction rate of 2.15%, which was higher than that of $A_2B_3C_4D_1E_2$ (1.92%) ($p < 0.05$). Therefore, the optimal condition for the extraction of SDF was at the ratio of solvent to material of 15:1, the microwave power of 200 W, the microwave-ultrasonic extraction time of 200 s, the enzymatic hydrolysis pH of 5.5, and the enzymatic hydrolysis time of 50 min.

The value of R (Max Dif) value indicates the significance of the factor's effect. Furthermore, a larger R value means that the factor has a bigger influence on the extraction rate. Thus, the order of effect of the five factors on the extraction rate of SDF could be determined according to the magnitude order of R. As shown in Table 2, the order of the effects of the five factors on the extraction rate of SDF was ratio of solvent to material (A) > enzymatic hydrolysis time (E) > microwave power (B) > extraction time (C) > enzymatic hydrolysis pH (D). It indicated that the ratio of solvent to material had the most significant effect on the extraction rate of SDF.

The corresponding ANOVA for the extraction of SDF is presented in Table 3. It indicated that the five factors affected the extraction rate of SDF in an interaction-dependent way with ratio of solvent to material considered as the most significant factor for the extraction of SDF ($p < 0.05$). This indicated that the effect of the ratio of solvent to material was more

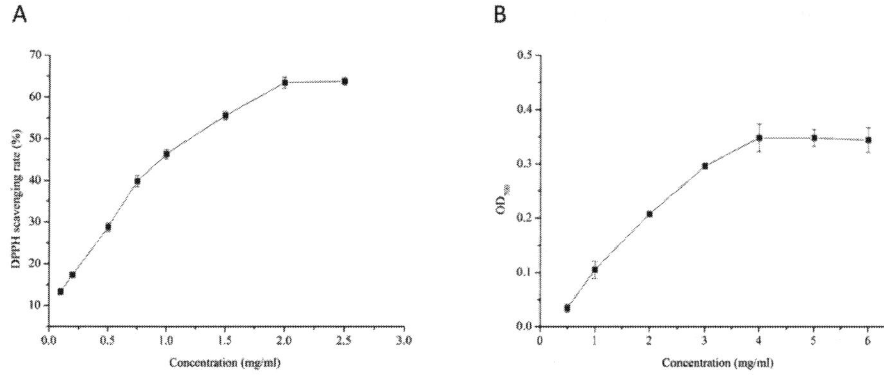

Figure 2. DHHP free radical-scavenging activity (A) and the reducing power (B) of SDF.

important than the other conditions. Therefore, operating at an appropriate ratio of solvent to material is preferable to SDF extraction.

3.7 *Antioxidant capacity of SDF*

3.7.1 *Evaluation of antioxidant capacity by the DPPH free radical-scavenging effect*

Antioxidant capacity is usually evaluated by the DPPH scavenging activity. DPPH, with characteristic absorption at 517 nm, is a stable organic nitrogen radical, and can be converted to 2,2-diphenyl-1-picrylhydrazine when reacted with antioxidants. The decrease in absorbance at 517 nm indicates the scavenging potential of the antioxidant (Von Gadow et al., 1997). As shown in Figure 2A, with the increase in SDF concentration, the DPPH scavenging activity increased, and reached a plateau at 2 mg/ml. The maximum scavenging rate of DPPH was 63.71%. Moreover, the IC_{50} value of SDF was 1.178 mg/ml. Thus, the SDF extracted from SSR had potential antioxidant capacities against free radical injury. The results are in agreement with those of Isita and MaHua (2015), who found that SDF extracted from defatted sesame husk, rice bran, and flax seed showed the DPPH scavenging ability that varied from 35% to 70% in the concentrations ranging from 0.05 to 0.2 mL/assay.

3.7.2 *Evaluation of antioxidant capacity by reducing power assay*

Previous reports have found that there is a direct correlation between reducing power and antioxidant activity (Duh, Du, & Yen, 1999). The reducing power is evaluated by absorbance at 700 nm; a higher absorbance indicates a higher reducing power. As shown in Figure 2B, the dose-dependent profile of reducing power was obvious, and the highest reducing power reached 0.348 at a SDF concentration of 4 mg/ml. Then, the value remained constant with the increasing concentration. The results indicate that SDF might serve as electron donors, which could terminate radical chain reactions by converting free radicals to more stable products when reacting with them.

4 CONCLUSIONS

In the present study, Soy Sauce Residue (SSR), a by-product from the soy sauce processing industry, was used for the preparation of Soluble Dietary Fiber (SDF). A novel technique, microwave-ultrasonic assisted enzymatic hydrolysis, was established for the extraction of SDF from SSR for the first time. The introduction of ultrasonic and microwave power could promote the dissolution of SDF, thus shortening the extraction time and enhancing the extraction rate. A single factor test and orthogonal experiment methodology were adopted to evaluate and optimize the extraction parameters, The maximum extraction rate of SDF reached 2.15% at the ratio of solvent to material of 15:1, the microwave power of 200 W, the microwave-ultrasonic extraction time of 200 s, the enzymatic hydrolysis pH of 5.5, and the enzymatic hydrolysis time of 50 min. Moreover, the results of *in vitro* antioxidant assays indicated that the extracted SDF had a good DPPH scavenging activity and reducing power. Based on the results, it was found that SSR was economical for the extraction of SDF, and the recovered SDF had a great potential in the development of functional food ingredients.

ACKNOWLEDGMENTS

This work was funded by the National Natural Science Foundation of China (No. 81273004, 31270577), Qing Lan Projec, the Open Project Program of Jiangsu Key Laboratory of Food Resource Development and Quality Safety, Xuzhou Institute of Technology (No. SPKF201311, SPKF201412) and the Science and Technology Planning Project of Xuzhou (KC14SM094).

REFERENCES

Brand-Williams, W., Cuvelier, M.E., & Berset, C. (1995). Use of a free radical method to evaluate antioxidant activity. *LWT—Food Science and Technology, 28*, 25–30.

Cui, S.W., Nie, S., & Roberts, K.T. (2011). Functional properties of dietary fiber. *Comprehensive Biotechnology (Second Edition), 4,* 517–525.

Cui, S.W., Wu, Y., & Ding, H. (2013). The range of dietary fibre ingredients and a comparison of their technical functionality. *Fibre-Rich and Wholegrain Foods,* 96–119.

Duh, P.D., Du, P.C., & Yen, G.C. (1999). Action of methanolic extract of mung bean hulls as inhibitors of lipid peroxidation and non-lipid oxidative damage. *Food and Chemical Toxicology, 37,* 1055–1061.

Gunness, P., & Gidley, M.J. (2010). Mechanisms underlying the cholesterol lowering properties of soluble dietary fibre polysaccharides. *Food & Function, 1(2),* 149–155.

Hosoda, K., Miyaji, M., Matsuyama, H., Imai, Y., & Nonaka, K. (2012). Digestibility, ruminal fermentation, nitrogen balance and methane production in holstein steers fed diets containing soy sauce cake at 10 or 20%. *Animal Science Journal, 83(3),* 220–226.

Isita, N., Mahua, G. (2015). Studies on functional and antioxidant property of dietary fibre extracted from defatted sesame husk, rice bran and flax seed. *Bioactive Carbohydrates and Dietary Fibre, 5,* 129–136.

Kaushik, A., & Singh, M. (2011). Isolation and characterization of cellulose nanofibrils from wheat straw using steam explosion coupled with high shear homogenization. *Carbohydrate Research, 1,* 76–85.

Lin, C.H., Wei, Y.T., & Chou, C.C. (2006). Enhanced antioxidative activity of soybean koji prepared with various filamentous fungi. *Food Microbiology, 23,* 628–633.

Ma, M., Mu, T., Sun, H., Miao, Z., Chen, J., & Yan, Z. (2015). Optimization of extraction efficiency by shear emulsifying assisted enzymatic hydrolysis and functional properties of dietary fiber from deoiled cumin (*Cuminumcyminum, L.*). *Food Chemistry, 179,* 270–277.

Meyer, A.S., Dam, B.P., & Lærke, H.N. (2009). Enzymatic solubilization of a pectinaceous dietary fiber fraction from potato pulp: Optimization of the fiber extraction process. *Biochemical Engineering Journal, 1,* 106–112.

Puri, M., Sharma, D., & Barrow, C.J. (2012). Enzyme-assisted extraction of bioactives from plants. *Trends in Biotechnology, 30(1),* 37–44.

Raghavendra, S.N., Rastogi, N.K., Raghavarao, K.S.M.S., & Tharanathan, R.N. (2004). Dietary fiber from coconut residue: Effects of different treatments and particle size on the hydration properties. *European Food Research and Technology, 6,* 563–567.

Tinsley, H.E.A., & Brown, S.D. (2000). Handbook of applied multivariate statistics and mathematical modeling. *Academic Press,* 183–208.

Von Gadow, A., Joubert, E., & Hannsman, C.F. (1997). Comparison of the antioxidant activity of aspalathin with that of other plant phenols of rooibos tea (*Aspalathuslinearis*), a-tocopherol, BHT, and BHA. *Journal of Agricultural and Food Chemistry, 45,* 632–638.

Wang, L., Xu, H., Yuan, F., Fan, R., & Gao, Y. (2015). Preparation and physicochemical properties of soluble dietary fiber from orange peel assisted by steam explosion and dilute acid soaking. *Food Chemistry, 185,* 90–98.

Wu, D., Gao, T., Yang, H., Du, Y., Li, C., Wei, L., Zhou, T., Lu, J. & Bi, H. (2015). Simultaneous microwave/ultrasonic-assisted enzymatic extraction of antioxidant ingredients from Nitraria tangutorun Bobr. Juice by-products. *Industrial Crops and Products, 66,* 229–238.

Advanced Engineering and Technology III – Xie (Ed.)
© 2017 Taylor & Francis Group, London, ISBN 978-1-138-03275-0

Degradation of AFB_1 by edible fungus

Meili Shao, Hongyang Sun, Ziyi Zhao, Hongyu Zhao & Liang Li
College of Food Science, Northeast Agriculture University, Harbin, China

ABSTRACT: In the present study, the ability of three strains of edible fungus (*Pleurotus ostreatus* No. 2, *Pleurotus citrinopileatus* No. 1, *Pleurotus edodes* No. 54) to degrade AFB_1 and the factors affecting the degradation of AFB_1 were assessed. The results indicated that the AFB_1-degradation rate of the edible fungus tested was strain specific, with the percent degradation ranging from 69.7% to 80.4% ($p < 0.05$). The degradation rate of AFB_1 was influenced significantly by incubation time, incubation temperature, pH of solution, volume of culture filter, and AFB_1 concentration. The three strains of edible fungus could be used as a biological agent for AFB_1 reduction.

1 INTRODUCTION

Aflatoxins are toxic metabolites produced by some *Aspergillus* species, especially *A. flavus* and *A. parasiticus* (Detroy R.W., Hesseltine C.W., 1968, Peltonen L., El-Nezami H., Haskard C., et al, 2001). These molecules are produced at pre-harvest and during the storage of cereal grains, particularly maize, rice, wheat, barley, oats, and sorghum (Prudente A.D., King J.M., 2002). They can also be found in nuts (e.g., almonds, peanuts, Brazil nuts, pistachio) and oil seeds (e.g., cottonseed). Among aflatoxins, aflatoxin B_1, which is more prevalent than any other analogue, has been considered the most potent toxin and carcinogen (Celyk K., Denly M., Savas T., 2003, Carolyn Haskard, Charlotte Binnion, Jorma Ahokas, 2000, A. Hernandez-Mendoza, H.S. Garcia, J.L. Steele, 2009, Adel Hamidi, Reza Mirnejad, Emad Yahaghi, 2013). Concerning the frequent presence of AFB_1 in cereals, the high intake of these products by humans and the toxicity and carcinogenicity of AFB_1, it is necessary to find strategies to remove, reduce, or degrade AFB_1.

In the past two decades, various physical and chemical biological methods have been used to detoxify aflatoxins from food and feed materials (Shetty P.H., Hald B., Jespersen L., 2007, A. Hernandez-Mendoza, D. Guzman-de-Peῆa and H.S. Garcia, 2009). Physical methods include segregation of contaminated seeds from good seeds, which is effective but tedious when performing manually. The existing chemical methods of detoxification of aflatoxin are based on ammoniation of contaminated materials. Recently, researchers have focused on the biological detoxification method to degrade AFB_1 (Alberts J.F., Gelderblom A., Botha W.H., 2009, Cepeda, Franco C.M., Fente C.A., Vázqueza B1, Rodrígueza J.L., Prognonb P., Mahuzierb G., 1996, Celyk K., Denly M., Savas T., 2003).

Some strains of lactic acid bacteria and fungus have been reported to be effective in binding or degrading AFB_1 from contaminated liquid media and cereals (Zinedine, Mohamed Faid, Mohamed Benlemlih, 2005, Detroy R.W., Hesseltine C.W., 1968, Shantha T., 1999). It has been suggested that such AFB_1 degradation proceeds enzymatically with mainly laccase produced by fungus reacting with AFB_1[21–26] (Peltonen L., El-Nezami H., Haskard C., et al, 2001, Shantha T., 1999, Doyle M.P., Marth E.H., 1979, Kothe E., 2001, Shetty P.H., Hald B., Jespersen L., 2007, Alberts J.F., Engelbrecht Y., Steyn P.S., et al, 2006).

In this study, we tested the ability of three strains of edible fungus (*Pleurotus ostreatus* No. 2, *Pleurotus citrinopileatus* No. 1, and *Lentinus edodes* No. 54) to degrade AFB_1 by producing laccase, and investigated the effect of factors affecting the degradation of aflatoxin.

2 MATERIALS AND METHODS

2.1 *Materials*

AFB_1 (>99%) was purchased from Amresco-Aldrich, Inc. (St. Louis, MO, USA). Methanol and acetonitrile (HPLC-grade) were supplied by Merck & Co. Inc. (SinopharmGroup Chemical Reagent Co. Ltd., Shanghai, China). *Pleurotus ostreatus* No. 2, *Pleurotus citrinopileatus* No. 1, and *Lentinus edodes* No. 54 were provided by Linyi City, Shandong Province flavor of the mushroom industry.

2.2 *Preparation of strains tested for culture filtrate*

The strains tested were cultivated in 100 mL MSM broth at 30°C for 4d. The fermentation broth was centrifuged at 8000 rpm for 10 min, and then the supernatant was filtered to obtain the culture filtrate.

2.3 *Degradation of AFB_1 by culture*

An 800 μL culture filtrate was added to 0.05 mL AFB_1 (200 ng/mL) and incubated at 30°C for 6 d in the shaker incubator at 200 rpm. Then, the cells were centrifuged at 8000 rpm for 10 min, and the supernatant containing the residual AFB_1 was collected for further analysis.

2.4 *Determination of AFB_1 by high-performance liquid chromatography–mass spectrometry*

The acetonitrile–water (V:V = 84:16) was added to the supernatant containing the residual AFB_1. An aliquot of this solution was taken and acetonitrile was evaporated with nitrogen. Then, methanol–water (V:V = 1:9) was dissolved into the solution and analyzed by the high-performance liquid chromatography–mass spectrometry method [6] (Adel Hamidi, Reza Mirnejad, Emad Yahaghi, 2013). The optimal conditions were as follows: sample injection temperature 33°C; mobile phase acetonitrile and 0.1% formic acid solution; flow rate 0.3 mL/min; capillary voltage 3.0 kV; and extraction voltage 2 V. The ion source temperature was 120°C, and the desolvation temperature was 380°C.

2.5 *Factors affecting the degradation of AF*

In order to detect the effect of factors affecting the degradation of AFB_1, the culture filtrate was incubated with AFB_1 at different incubation conditions. The incubation was performed at temperatures of 20°C, 25°C, 30°C, 35, and 40°C, respectively. The pH of the solution was 5.0, 5.5, 5.8, 6.0, 6.5, and 7.0, respectively. The volume of the culture filtrate was 600 uL, 700 uL, 800 uL, 900 uL, and 1000 uL, respectively. The concentration of AFB_1 was 5 ng/mL, 50 ng/mL, 100 ng/mL, 150 ng/mL, 200 ng/mL, and 250 ng/mL, respectively. The incubation time was 1d, 2d, 4d, 6d, and 8d, respectively.

2.6 *Statistical analysis*

All experiments were carried out in triplicate. Statistical analysis was performed by the SPSS 17.0 statistical software to determine significant differences ($p < 0.05$).

3 RESULTS AND DISCUSSION

The results indicated that the AFB_1 degradation rate was strain specific, with the percent degradation ranging from 69.7% to 80.4% ($p < 0.05$). The amount of AFB_1 degradation of *Pleurotus citrinopileatus* No. 1 was notably higher than the other strains tested.

The effects of temperature on the reduction of AFB_1 are shown in Fig. 1. With the increasing incubation temperature, the ability of the three strains of edible fungus to degrade AFB_1

was first increased and then decreased ($p < 0.05$). The AFB_1 removal was higher at a temperature of 25°C (vs. 20, 30, 35, or 40°C) for *Pleurotus ostreatus* No. 2 and *Lentinusedodes* No. 54 and at a temperature of 35°C (vs. 20, 25, 30, or 40°C) for *Pleurotus citrinopileatus* No. 1. The AFB_1-degradation rate for the three strains of edible fungus was lowest at 40°C.

The effects of pH on the reduction of AFB_1 are shown in Fig. 2. With the increasing pH of the solution, the ability of the three strains of edible fungus to degrade AFB_1 was first increased and then decreased ($p < 0.05$). The AFB_1 degradation rate of *Pleurotus citrinopileatus* No. 1 was higher at the pH value of 5.5 to 6.5 (vs. 5 and 7). The AFB_1 degradation

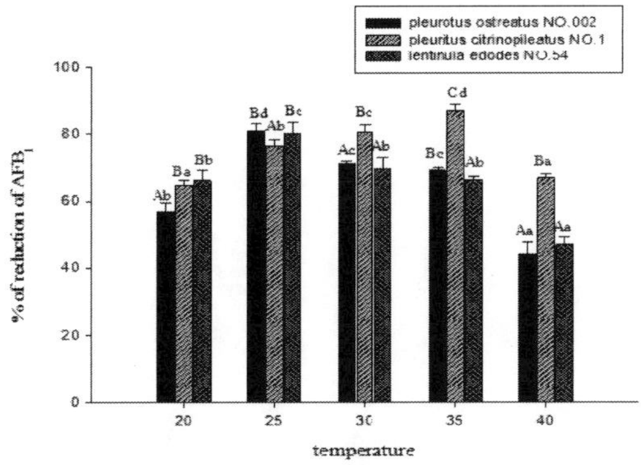

Figure 1. Effect of different temperatures on degradation rate of aflatoxins.

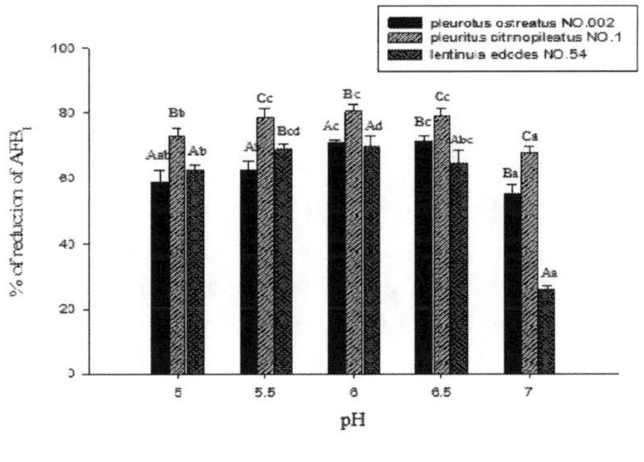

Figure 2. Effect of different pH values on the degradation rate of aflatoxins.

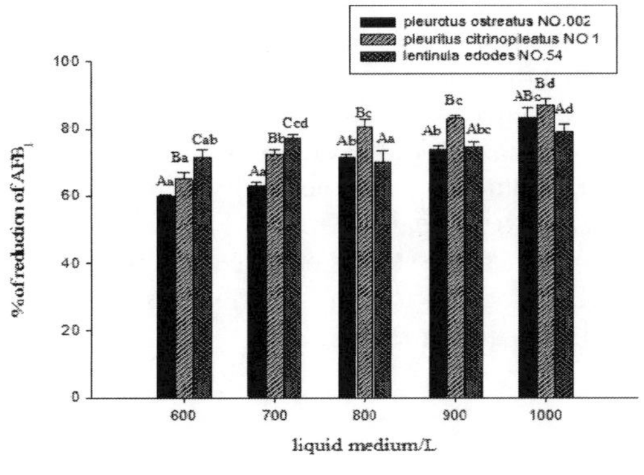

Figure 3. Effect of different loading fluid volumes on the degradation rate of aflatoxins.

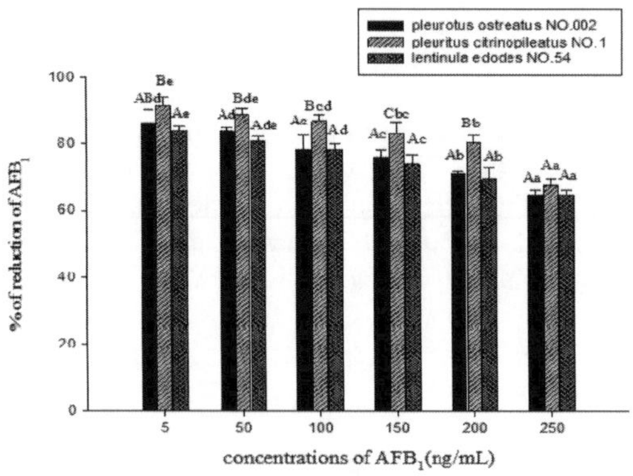

Figure 4. Effect of different concentrations on the degradation rate of aflatoxins.

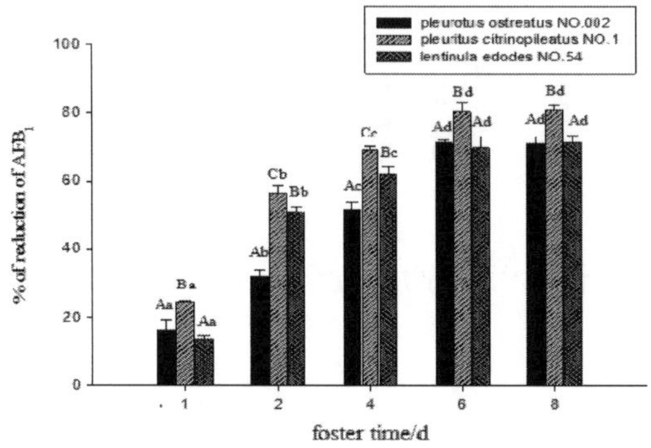

Figure 5. Effect of different incubation times on the degradation rate of aflatoxins.

rate of *Pleurotus ostreatus* No. 2 was higher at the pH value of 6.0 to 6.5 (vs. 5, 5.5, and 7). The AFB_1-degradation rate of *Lentinus edodes* No. 54 was the highest at pH 6.0. The AFB_1 degradation rate of the three strains of edible fungus was the lowest at pH 7.0.

The effects of the volume of culture filtrate on the reduction of AFB_1 are shown in Fig. 3. With the increasing volume of the culture filtrate, the ability of the three strains of edible fungus to degrade AFB_1 was increased ($p < 0.05$). When the volume of the culture filtrate was 1000 mL, the AFB_1 degradation rate of *Pleurotus ostreatus* No. 2 was highest (83.10%), followed by *Pleurotus citrinopileatus* No. 1 (86.97%) and *Lentinus edodes* No. 54 (79.05%).

The effects of concentration of AFB_1 on the reduction of AFB_1 are shown in Fig. 4. With the increasing concentration of AFB_1, the ability of the three strains of edible fungus to degrade AFB_1 was decreased ($p < 0.05$). The rate of degradation for the three strains was the highest (83.9%–91.7%) when the concentration of AFB_1 was 5 ng/mL. Their rate of degradation was the lowest (65.0%–68.0%) when the concentration of AFB_1 was 250 ng/mL.

The effects of incubation time on the reduction of AFB_1 are shown in Fig. 5. With the incubation time increasing from 1d to 6d, the ability of the three strains of edible fungus to reduce AFB_1 was increased significantly ($p < 0.05$). With the incubation time increasing from 6d to 8d, the ability of the three strains of edible fungus to reduce AFB_1 was increased slightly ($p > 0.05$). The AFB_1 degradation rate was obviously inhibited due to the decreasing enzyme activity with the increasing time.

4 CONCLUSION

Our study demonstrated that the three strains of edible fungus (*Pleurotus ostreatus* No. 2, *Pleurotus citrinopileatus* No. 1, and *Lentinus edodes* No. 54) examined could degrade AFB_1

significantly with strain specificity. The change in the incubation conditions (incubation time, incubation temperature, pH of solution, volume of culture filter, and AFB_1 concentration) could improve their degradation rate of AFB_1. Our study also showed that *Pleurotus citrinopileatus* No. 1 has a significant potential in degrading AFB_1 compared with the other strains.

ACKNOWLEDGMENT

This work was supported by the Funds of "Natural Science Foundation of Heilongjiang Province" (590024) and the "National Science and Technology Program" (2013 AA102208). The authors gratefully acknowledge the support.

REFERENCES

Adel Hamidi, Reza Mirnejad, Emad Yahaghi. The aflatoxin B1 isolating potential of two lactic acid bacteria [J]. Asian Pacific Journal of Tropical Biomedicine, 2013, 3(9): 732–736.

Alberts JF, Engelbrecht Y, Steyn PS, et al. Biological degradation of aflatoxin Bl by Rhodococcus erythropolis cultures [J]. International Journal of Food Microbiology, 2006, 109(1/2): 121–126.

Alberts JF, Gelderblom A, Botha WH. Degradation of aflatoxin B1 by fungal laccase enzymes [J]. International Journal of Food Microbiology, 2009, 135: 47–52.

Carolyn Haskard, Charlotte Binnion, Jorma Ahokas. Factors affecting the sequestration of aflatoxin by Lactobacillus rhamnosusstrain GG [J]. Chemico—Biological Interactions, 2000, 128: 39–49.

Celyk K, Denly M, Savas T. Reduction of toxic effects of aflatoxin B1 by using baker yeast (Saccharomyces cerevisiae) in growing broiler chicks diets [J]. Revista Brasileira de Zootecnia, 2003, 32(3): 615–619.

Cepeda, Franco CM, Fente CA, Vázqueza B1, Rodrígueza JL, Prognonb P, Mahuzierb G. Postcolumn excitation of aflatoxins using cyclodextrins in Liquid cartography for food analysis [J]. Journal of Chromatography A, 1996, 721(1): 69–74.

Detroy RW, Hesseltine CW. Isolation and biological activity of a microbial conversion product of aflatoxin Bl [J]. Nature, 1968, 219(51): 957–967.

Doyle MP, Applebaum RS, Brackett RE, Marth EHL. Physical, chemical and biological degradation of mycotoxins in foods and agricultural commodities [J]. Journal of Food Protection, 1982, 45(10): 964–971.

Hadar Y, Kerem Z, Gorodecki B. et al. Utilization of lignocellulosic waste by the edible mushroom Pleurotus [J]. Biodegradation, 1992, 3: 189–205.

Haskard CA, El-Nezami HS. Surface binding of aflatoxin B1 by lactic acid bacteria [J]. Appl Environ Microbiol, 2001, 67(7): 3086–3090.

Hernandez-Mendoza, A, Guzman-de-Peña D. and Garcia HS. Key role of teichoic acids on aflatoxin B1 binding by probiotic bacteria [J]. Journal of Applied Microbiology, ISSN 2009(107): 395–403.

Hernandez-Mendoza, A, Garcia, HS, Steele, JL. Screening of Lactobacillus casei strains for their ability to bind aflatoxin B1 [J]. Food and Chemical Toxicology, 2009, 47: 1064–1068.

Kothe E. Mating-type genes for basidiomycete strain improvement in mushroom farming [J]. Applied Microbiology and Biotechnology, 2001, 56(5/6): 589–601.

Peltonen L, El-Nezami H, Haskard C, et al. Aflatoxin B1 Binding by Dairy Strains of Lactic Acid Bacteria and Bifidobacteria [J]. Journal of Dairy Science, 2001, 84(10): 2152–2156.

Prudente AD, King JM. Efficacy and safety evaluation of ozonation to degrade aflatoxin in corn [J]. Journal of Food Science, 2002, 67(8): 2866–2872.

Shantha T. Fungal degradation of aflatoxin Bl [J]. Natural Toxins. 1999, 7(5): 175–178.

Shetty PH, Hald B, Jespersen L. Surface binding of aflatoxin B1 by Saccharomyces cerevisiae strains with potential decontaminating abilities in indigenous fermented foods [J]. International Journal of Food Microbiology, 2007, 113(1): 41–46.

Shetty PH, Jespersen L. Saccharomyces cerevisiae and lactic acid bacteria as potential mycotoxin decontaminating agents[J]. Trends in Food Science & Technology, 2006, 17(2):48–55.

Zinedine, Mohamed Faid, Mohamed Benlemlih, *In Vitro* Reduction of Aflatoxin B1 by Strains of Lactic Acid Bacteria Isolated from Moroccan Sourdough Bread [J]. International Journal of Agriculture Biology, 2005, 7(1): 67–70.

Zjatic S, Reverbefi M, Ricelli A, et al. Trametes versicolor: a possible tool for aflatoxin control [J]. Journal of Food Microbiology, 2006, 107(3): 243–249.

Advanced Engineering and Technology III – Xie (Ed.)
 ISBN 978-1-138-03275-0

Effect of different storage temperatures on respiration and marketable quality of sweet corn

Y. Xie, S. Liu, L. Jia & E. Gao
Beijing Vegetable Research Center, Beijing Academy of Agriculture and Forestry Sciences, National Engineering Research Center for Vegetables, Beijing, China

H. Song
College of Food Science, Fujian Agriculture and Forestry University, Fujian, China

ABSTRACT: The effects of different storage temperatures (0, 5, 10, 15, 20, 25 and 30°C) on respiration rate, sensory evaluation, weight loss, soluble sugar, vitamin C and soluble protein contents of sweet corn were studied within 10d. The results indicated that the higher the storage temperature, the higher the respiration rate. The respiration rate at 30°C was about 10 times higher than that at 0°C. With the increasing storage temperature, sensory evaluation, soluble sugar, vitamin C, and soluble protein contents decreased gradually, but the weight loss of sweet corn increased. The salable limits of sweet corn were 10, 6, 5, 4, 3, 2, and 3/8d at 0, 5, 10, 15, 20, 25 and 30°C, respectively. The quality of sweet corn stored at 0°C was the best.

1 INTRODUCTION

Sweet corn (*Zea mays L. Saccharata* Sturt) is a perishable food product with a high respiration rate, and loss of sweetness is the main quality degradation during storage. It has been reported that about 60% and 6% of the sugar in sweet corn was lost after 1d at 30 and 0°C, respectively (Brecht, 2004). Therefore, postharvest techniques, such as precooling for removal of field heat after harvest and refrigeration were applied to maintain its quality (Liu et al. 2012, Vigneault et al. 2007).

Storage temperature has a significant influence on the quality of fruits and vegetables after harvest. The higher the storage temperature, the higher the respiration rate (Bachmann & Earles, 2000). Currently, both domestic and foreign researchers have focused on the study of low-temperature storage of sweet corn. Sweet corn is mainly harvested in summer; the harvest and pile-up temperature is generally above 20°C. Meanwhile, the precooling equipment is scanty in field, the cold-chain logistics system is not perfect in China and other developing countries, and the circulation of sweet corn is usually under high temperature, resulting in quality degradation of sweet corn. However, the effect of different storage temperatures on respiration rate, salable limits and marketable quality of sweet corn has not reported so far.

In this study, our strategies focused on the storage temperature above 0°C, and we investigated the effect of different storage temperatures (0, 5, 10, 15, 20, 25, and 30°C) on respiration rate and sugar reduction of sweet corn. In addition, nutritional attributes, weight loss and sensory evaluation were investigated.

2 MATERIALS AND METHOD

2.1 *Plant materials and treatments*

'Wan tian' varieties of super sweet corn (in husks) were harvested at Gu An City of He Bei province, China, precooled and transported to the laboratory. Husks and external silks of

Table 1. Sensory evaluation standard of sweet corn.

Score	Flag leaves Color	Wilt	Decay	Kernel denting	Odor
9	fresh, deep green	none	none	none	aromatic odor
7	fresh, green	none	none	none	slight aromatic odor
5	slight yellow	slight	slight	slight	slight peculiar smell
3	partial gray	moderately severe	moderately severe	moderately severe	peculiar smell
1	large area white	severe	severe	severe	peculiar smell

sweet corn were removed. Sweet corn cobs were cut in to 1–2 cm pieces. Flag leaves were pulled from the cobs to leave a 3 cm-wide 'window' in which the kernels were exposed. Ears were collected for further experiment (Aharoni et al. 1996). All sweet corns were chosen without pest injury or diseases, and were packed with plastic cases in seven separate batches. The whole experiment included seven temperature treatments: sweet corns were unpacked and stored in seven cold storage temperatures at 0, 5, 10, 15, 20, 25, and 30°C, respectively. All experiments were performed in three replicates.

2.2 *Weight loss*

The weight of individual sweet corn was recorded on the day of harvest and after the different storage periods. The total weight loss was expressed as percentages.

2.3 *Sensory evaluation*

Taking color, wilting and decay of flag leaves, kernel denting and odor of sweet corn into account, the visual quality of sweet corn was scored on a 9 to 1 scale (Table 1).

2.4 *Respiration rate, soluble sugar, vitamin C and soluble protein contents*

The respiration rate was determined by GXH-3051A infrared CO_2 analytical instrument, and the velocity was 0.5 L/min. These nutritional constituents were analyzed in the kernel samples. Soluble sugar content was determined by the anthrone sulfuric acid method (Cao et al. 2007). Vitamin C content was determined by the molybdenum blue colorimetric method. Soluble protein content was determined by the Coomassie brilliant blue method.

3 RESULTS AND DISCUSSION

3.1 *Respiration rate*

The respiration rate of fruit and vegetables is an excellent indicator of the metabolic activity of the tissue, which is a useful guide to the potential storage life of fruit and vegetables. The respiration rates of sweet corn at 0, 5, 10, 15, 20, 25 and 30°C were 49.57, 62.34, 150.34, 234.67, 290.51, 380.65 and 480.34 mg $CO_2 \cdot kg^{-1} \cdot h^{-1}$, respectively (Fig. 1). The respiration rate at 30°C was about 10 times higher than that at 0°C. The respiration rate of sweet corn was the lowest at 0°C so that it could maintain quality, delay ripening and senescence. The lower the temperature, the lower the respiration rate, and the better the quality of sweet corn.

3.2 *Sensory evaluation*

The fresh sweet corn kernels with green and moist flag leaves and typical odor were full and yellow. The salable time limits and sensory evaluation of sweet corn at 0, 5, 10, 15, 20, 25

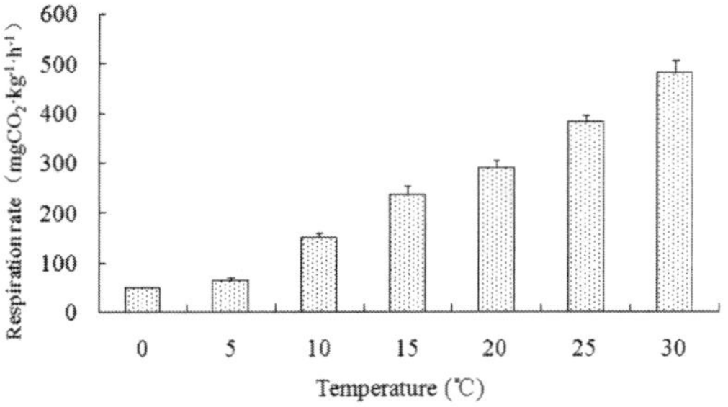

Figure 1. Effect of different temperatures on respiration rate of sweet corn during initial storage.

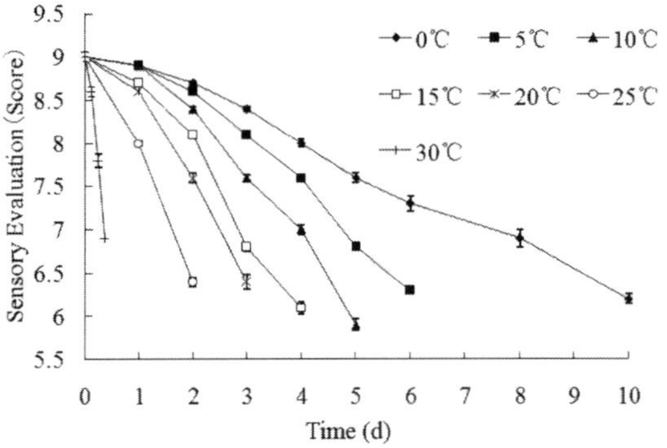

Figure 2. Effect of different temperatures on the sensory evaluation of sweet corn.

and 30°C were 10, 6, 5, 4, 3, 2, and 3/8d and 6.2, 6.3, 5.9, 6.1, 6.4, 6.5, and 6.9, respectively (Fig. 2). The low temperature delayed the wilting, the decay of flag leaves, and the denting of kernels, especially at 0°C. The lower the temperature, the higher the sensory evaluation, and the longer the shelf life of sweet corn. The grade of sweet corn at 30°C was approximately 6.9 after 3/8d. Although the grade of sweet corn was 6.2 at 0°C, the shelf-life had extended to 10d. Temperature management was critical for the sensory evaluation of sweet corn. So, sweet corn should be cooled within 1 hour of harvest. Precooling was complete when the core temperature of sweet corn dropped to approximately 0°C.

3.3 *Weight loss*

The weight loss of sweet corn increased gradually during cold storage. Generally, 5% weight loss will cause perishable vegetables to wilt or shrivel. However, the water loss of sweet corn is not only flag leaves but also kernels. Because the kernels of sweet corn were protected by flag leaves, the water loss of flag leaves was maintained. In addition, 10% weight loss of sweet corn is acceptable. The salable limits and weight losses of sweet corn at 0, 5, 10, 15, 20, 25, and 30°C were 10, 6, 5, 4, 3, 2, and 3/8d and 10.06, 8.64, 10.42, 10.07, 8.46, 7.25, and 3.34%, respectively (Fig. 3). With the storage time increasing, weight loss increased gradually. The higher the temperature, the more the weight loss. The weight loss of sweet corn was associated with water vapor that was inhibited effectively by the low temperature. So, it could prevent the weight loss of sweet corn significantly. The weight loss was rapid at 30°C but slow at 0°C.

3.4 *Soluble sugar*

Soluble sugar is an important nutritional indicator of sweet corn. The initial soluble sugar content of sweet corn was 13.67% after harvest. The salable limits and soluble sugar contents of sweet corn at 0, 5, 10, 15, 20, 25, and 30°C were 10, 6, 5, 4, 3, 2, and 3/8d and 11.72, 11.42, 10.77, 10.77, 9.53, 10.88, and 11.83%, respectively (Fig. 4). With the storage time increasing,

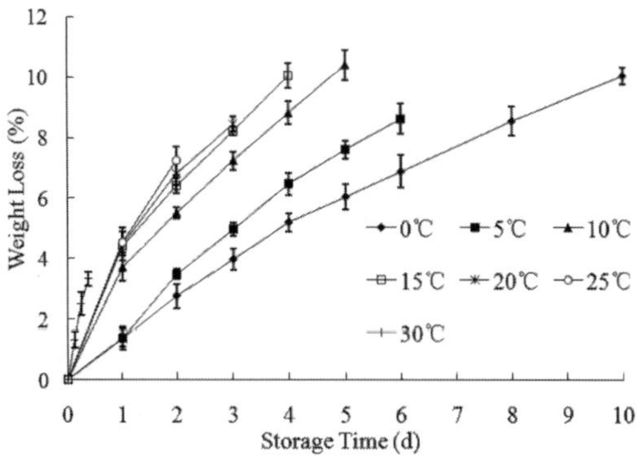

Figure 3. Effect of different temperatures on the weight loss of sweet corn.

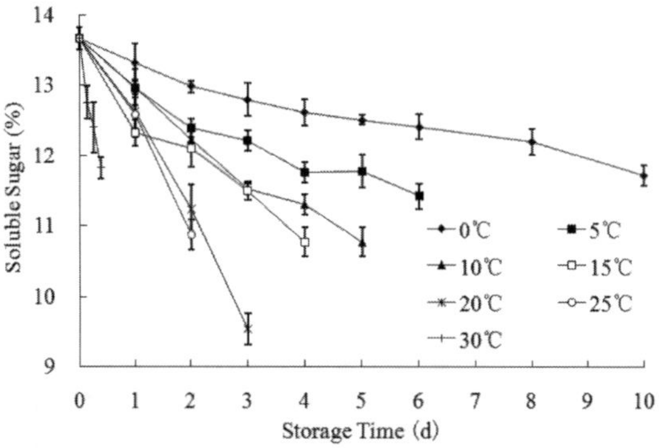

Figure 4. Effect of different temperatures on the soluble sugar of sweet corn.

the soluble sugar decreased gradually. The lower the temperature, the less the soluble sugar lost. The low temperature could extend the shelf life of sweet corn and keep its flavor well.

3.5 *Vitamin C*

The initial vitamin C content of sweet corn was 134.76 mg/100 g after harvest. The salable limits and Vitamin C contents of sweet corn at 0, 5, 10, 15, 20, 25, and 30°C were 10, 6, 5, 4, 3, 2, and 3/8d and 88.26, 79.25, 84.64, 82.50, 79.63, 81.96, and 108.58 mg/100 g, respectively (Fig. 5). The longer the storage time, the more the vitamin C lost. Similarly, the higher the temperature, the more the vitamin C lost. The vitamin C content of sweet corn decreased slowly at 0°C.

3.6 *Soluble protein*

Protein is one of the first nutrients recognized as essential for life. Proteins are usually made up of 20 or more amino acids (Fang et al. 2009). The initial soluble protein content of sweet corn was 5.98 mg/g after harvest. The salable limits and soluble protein contents of sweet corn at 0, 5, 10, 15, 20, 25 and 30°C were 10, 6, 5, 4, 3, 2, and 3/8d and 4.84, 4.12, 4.25, 4.76, 4.27, 4.30, and 4.66 mg/g, respectively (Fig. 6). With the temperature and storage time increasing, the soluble protein decreased gradually. The soluble protein content of sweet corn decreased slowly at 0°C.

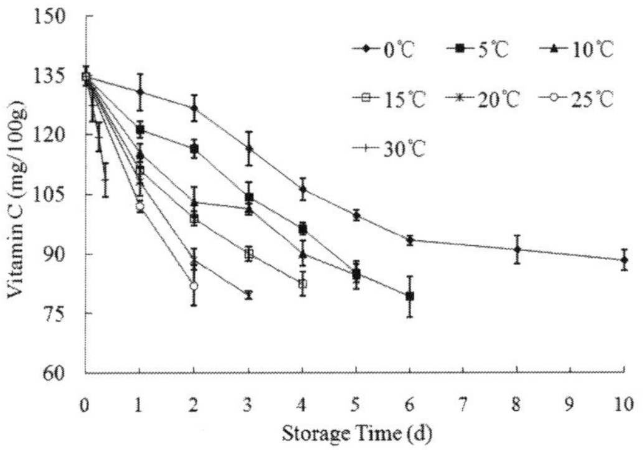

Figure 5. Effect of different temperatures on the vitamin C content of sweet corn.

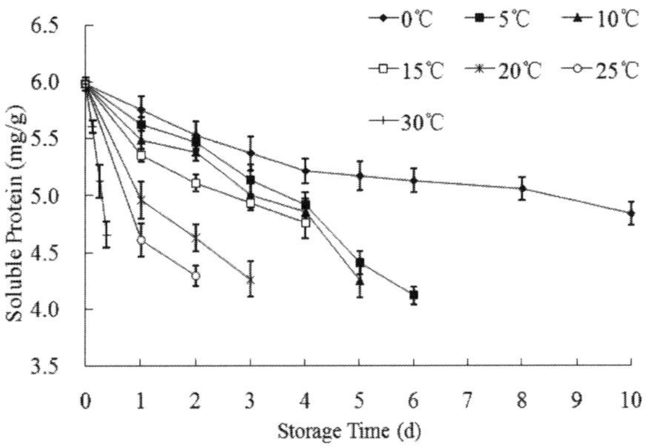

Figure 6. Effect of different temperatures on the soluble protein of sweet corn.

4 CONCLUSION

Differences in the respiration rate and quality of sweet corn at seven temperatures were significant. The temperature has an important effect on the quality of sweet corn. With the storage time increasing, the vitamin C, soluble sugar and soluble protein contents of sweet corn decreased gradually at seven temperature conditions. The lower the temperature, the lower the respiration rate. With the storage temperature increasing, the sensory evaluation, soluble sugar, vitamin C, and soluble protein contents of sweet corn decreased gradually, but the weight loss increased. The loss rates of the soluble sugar, vitamin C and soluble protein contents of sweet corn were 13.46, 19.43, and 19.73% at 30°C after 3/8 d, while at 0°C after 10d, the rates were only 14.24, 34.51, and 19.06%. Sweet corn could maintain good quality and extend the shelf life at 0°C. So the best storage method of sweet corn was at 0°C. In order to keep the weight of sweet corn, it could humidify with humidifiers or be packed with plastic films. Rapid precooling after harvest could remove the field heat and respiratory heat, reducing respiration intensity quickly. In addition, the salable limits and quality of sweet corn under different storage temperatures were determined, which contribute to providing the theoretical basis for the preservation and circulation of sweet corn, and promoting the development of reduced cold-chain logistics in China and other developing countries.

ACKNOWLEDGEMENTS

The research was funded by the National Science and Technology Support Program (No. 2015BAD19B02) and the Science and Technology Innovation Special Construction Funded Program of Beijing Academy of Agriculture and Forestry Sciences (No. KJCX20140205). Corresponding author: S. Liu.

REFERENCES

Aharoni, Y., Copel, A., Gil, M. & Fallik, E. 1996. Polyolefin stretch films maintain the quality of sweet corn during storage and shelf-life. *Postharvest Biology and Technology* 7(1): 171–176.

Bachmann, J. & Earles, R. 2000. Postharvest Handling of Fruits and Vegetables. ATTRA Horticulture Technical Note. http://www.attra.nvat.org.

Brecht, J.K. 2004. Postharvest quality maintenance guidelines. In: Gross K, Wang CY, Saltveit M, editors. The Commercial Storage of Fruits, Vegetables, and Florist & Nursery Crops. 3rd ed. Washington, D.C. USA.

Cao, J.K., Jiang, W.B. & Zhao, Y.M. 2007. Experiment Guidance of Postharvest Physiology and Biochemistry of Fruits and Vegetables. China Light Industry Press, Beijing.

Fang, Z.F., Luo, J., Qi, Z.L. Huang, F.R., Zhao, S.J., Liu, M.Y., Wang, S.W. & Peng, J. 2009. Effects of 2-hydroxy-4-methylthiobutyrate on portal plasma flow and net portal appearance of amino acids in piglets. *Amino Acids* 36(3): 501–509.

Liu, S., Zhang, H. & Zheng, S. 2012. Effect of Pressure Pre-cooling, Room Pre-cooling and cold storage on Quality of Sweet Corn. *Acta Horticulturae* 934: 1247–1254.

Vigneault, C., Goyette, B., Gariepy, Y., Cortbaoui, P., Charles, M.T. & Raghavan, V.G.S. 2007. Effect of ear orientations on hydrocooling performance and quality of sweet corn. *Postharvest Biology and Technology* 43(3): 351–357.

Advanced Engineering and Technology III – Xie (Ed.)

Antioxidant, antimutagenic and antitumor activities of flavonoids from the seed shells of *Juglans mandshurica*

H.Y. Xu
Department of Food Science and Engineering, Agricultural College, Yanbian University, Yanji, China

G.J. Xia
Department of Animal Science, Agricultural College, Yanbian University, Yanji, China

Y.H. Bao
Food Science and Engineering, Forestry College, Northeast Forestry University, Harbin, China

ABSTRACT: The bioactivities of flavonoids extracted and purified from the seed shells of *Juglans mandshurica* (FSSJM) were investigated for the first time. FSSJM exhibited significant antioxidant activity in 2,2′-Azino-bis-(3-ethylbenzothiazoline-6-sulfonic acid) (ABTS), 2,2-di(4-tert-octylphenol)-1-picrylhydrazyl (DPPH), hydroxyl radical-scavenging activity and lipid peroxidation inhibition assays with IC_{50} values of 9.52 μg/mL, 1.273 mg/mL, 1.479 mg/mL and 1.678 mg/mL, respectively. The inhibition of lipid peroxidation by FSSJM was better than that by VC, which exhibits an IC_{50} value of 2.753 mg/mL. FSSJM also showed significant antimutagenic activity against direct and indirect mutagen-induced mutations in *Salmonella typhimurium* strains TA98 and TA100 in a dose-dependent manner. MTT assays showed that FSSJM exhibited moderate cytotoxic activity against HepG-2, HepG-2B, N87 and MDA-MB-231 human tumor cells, with the most significant effects on HepG-2 cells. In conclusion, the results showed that FSSJM had significant antioxidant, antimutagenic and antitumor activities and might be further used in functional foods or in the development of anticancer agents.

1 INTRODUCTION

Reactive Oxygen Species (ROS), including the hydroxyl radical (OH·), alkoxyl radical (RO·), superoxide anion radical (O_2^-), singlet oxygen ($^1O_2$) and hydrogen peroxide (H_2O_2), are the products of normal cellular metabolism (Liu, Li, Sasaki, Asada & Koike, 2010). The "two-faced" character of ROS has been clearly substantiated (Mirian Valko, Rhodes, Moncol, Izakovic & Mazur, 2006). The beneficial effects of ROS occur at low/moderate concentrations and involve their physiological roles, for example, in cellular responses to anoxia, in the defense against infectious agents, and in a number of cellular signaling pathways, as well as in the induction of a mitogenic response (Forman, Maiorino & Ursini, 2010). However, ROS overproduction results in oxidative stress, a deleterious process that can be an important cause of damage to cellular structures, including lipids, membranes, proteins and DNA. ROS-induced damage can cause DNA mutation, which is a critical step in carcinogenesis. Elevated levels of oxidative DNA lesions have been noted in various tumors, strongly implicating similar damage in the etiology of cancer (Marian Valko, Leibfritz, Moncol, Cronin, Mazur & Telser, 2007).

Juglans mandshurica Maxim. belongs to the Juglandaceae, and it is a fast-growing walnut tree native to many Asian countries, especially China. This plant has been used in folk and traditional medicines to treat diseases in some countries (Graham, Quinn, Fabricant & Farnsworth, 2000; Machida, Yogiashi, Matsuda, Suzuki & Kikuchi, 2009). Although there have been some reports about flavonoids from the roots, leaves, and stem-barks of *Juglans mandshurica* (Min, Lee, Kim, Lee, Kim, Kim, Kim, Joung, Lee & Nakamura, 2003. Min, Lee, Lee, Kim, Bae, Otake, Nakamura & Hattori, 2002; Wang J.L., 2009), flavonoids from the seed shell of *Juglans mandshurica* have not been characterized. Flavonoids are polyphenolic compounds that are widely distributed in many plants, such as fruits, vegetables, grains, teas and some medicinal herbs (Li, Zhang, Wu, Wang, Liu & Wang, 2015). Some studies have shown that flavonoids can scavenge ROS, and they have various biological activities, such as antioxidant, anti-inflammatory, antivirus, and anticancer properties (Hoensch, H.P. & Oertel, R. 2015). Therefore, in recent years, interest in flavonoids has greatly increased due to the possible beneficial implications of these compounds for human health, including the treatment and prevention of cancer (Lirdprapamongkol, Sakurai, Abdelhamed, Yokoyama, Athikomkulchai, Viriyaroj, Awale, Ruchirawat, Svasti & Saiki, 2013; Ravishankar, Rajora, Greco & Osborn, 2013).

The *Juglans mandshurica* seed is a byproduct of the forestry industry, and its kernel has recently received a great deal of attention because of its rich nutritional value. Many seed shells are produced during kernel processing. Currently, most of these seed shells are discarded, although some are made into handicraft products or used as fuel. Thus, the seed shells are currently a source of waste and environmental pollution. Because the extraction of flavonoids from the seed shells of *Juglans mandshurica* does not result in the destruction of *Juglans mandshurica* trees, and seed shells are a continuously available resource, the development prospects for this material are wide open. Hence, the aim of the present study is to investigate the antioxidant, antimutagenic and antitumor activity of flavonoids from *Juglans mandshurica* seed shells. This finding would promote the use of flavonoids from *Juglans mandshurica* seed shells in functional foods. In addition, there would also be an increase in the added value of *Juglans mandshurica* seed, and the waste of this resource would be prevented.

2 MATERIALS AND METHODS

2.1 *Drugs and chemicals*

2,2′-Azino-bis-(3-ethylbenzothiazoline-6-sulfonic acid) (ABTS), 2,2-di(4-tert-octylphenol)-1-picrylhydrazyl (DPPH), phosphatidylcholine, 2,6-di-tert-butyl-4-methylphenol (BHT), 4-nitroquinoline-1-oxide (4 NQO), N-methyl-N-nitro-N-nitrosoguanidine (MNNG), benzo(α)pyrene [B(α)P], and 3-(4,5-dimethylthiazol-2-yl)-2,5-diphenyltetrazolium bromide (MTT) were purchased from Sigma Chemical Co. (St. Louis, MO, USA). VC was purchased from Tianjin Kermel Chemical Reagent Co. (Tianjin, China). Trolox was purchased from Merck Chemicals Co. (Germany). Rutin was purchased from the National Institute for the Control of Pharmaceutical and Biological Products (Beijing, China). 3-amino-1,4-dimethyl-5H-pyrido-(4,3-b)indole (Trp-P-1) was purchased from Hwa-Kwang Chemical (Japan). Dulbecco's modified Eagle's medium was obtained from Hyclone (Utah, America), and fetal bovine serum was obtained from Hangzhou Sijiqing Biotechnology Co. (Hangzhou, China).

2.2 *Flavonoid preparation*

Dried seed shells from *Juglans mandshurica* were gently triturated and sieved through a 60-mesh screen. The powder was extracted with EtOH-H_2O (60:40 v/v) with ultrasonic assistance. The extracts were filtered and evaporated, subjected to two rounds of purification using macroporous resin, the eluent were evaporated, and then freeze-dried under a vacuum. The resulting product was designated FSSJM.

2.3 *Determination of total flavonoids*

The total flavonoid content was determined using a rutin standard curve as previously described (Xie, Dong, Nie, Li, Wang, Shen & Xie, 2015). A 0.5 mL quantity of FSSJM solution was placed in a 10 mL volumetric flask, followed by 2 mL of EtOH-H_2O; then, 5% $NaNO_2$ (0.4 mL) was added. After 6 min, 10% $AlCl_3$ (0.4 mL) was added. After an additional 6 min, the reaction mixture was treated with 4 mL of 1 mol/L NaOH. Finally, EtOH-H_2O was added to bring the total volume of the reaction mixture to 10 mL, and the solution was mixed. After 15 min, the absorbance was measured against a blank at 510 nm. The calibration curve ($Y = 13.536X - 0.0005$, where Y is the absorbance value of the sample and X is the sample concentration) was measured within the range of 0–300 μg/mL ($R^2 = 0.9994$). The total FSSJM flavonoid content was 54.4%.

2.4 *Antioxidant activity assay*

FSSJM antioxidant activity was investigated by testing its ABTS, DPPH and hydroxyl radical-scavenging activity and its inhibition of lipid peroxidation.

2.4.1 *ABTS radical-scavenging activity*

The ability of FSSJM to scavenge $ABTS^{\cdot+}$ was measured according to a published method (Liang, Wu, Zhao, Zhao, Li, Zou, et al., 2012) with slight modifications. $ABTS^{\cdot+}$ was prepared by mixing an $ABTS^{\cdot+}$ stock solution (7.4 mmol/L in water) with 2.6 mmol/L potassium persulfate. This mixture was stored at room temperature in the dark for 16 h. $ABTS^{\cdot+}$ stock solution was diluted before use to an absorbance of 0.700 ± 0.020 at 734 nm with methanol. A 10 μL aliquot of FSSJM solution was added to 290 μL of $ABTS^{\cdot+}$ working solution and mixed thoroughly. The reaction mixture was allowed to stand at room temperature in the dark for 10 min, and the absorbance was recorded at 734 nm. Trolox was used to generate antioxidant standards for activity comparisons.

The radical-scavenging activity was calculated using the following formula (1):

$$Radical - scavenging\ activity\ \% = 1 - (A_1 - A_2) / A_{control} \times 100 \tag{1}$$

where A_1 is the absorbance of the reaction mixture, A_2 is the absorbance of the reaction mixture without ABTS solution, and A_{control} is the absorbance of the reaction mixture without FSSJM solution.

2.4.2 *DPPH radical-scavenging activity*

The FSSJM DPPH radical-scavenging assay was carried out according to a previously published method (Sharma, Ko, Assefa, Ha, Nile, Lee, Park, 2015), with slight modifications. In brief, 100 μL of FSSJM solution was added to 100 μL of DPPH working solution and mixed thoroughly. The reaction mixture was left at room temperature in the dark for 30 min, and the absorbance was recorded at 517 nm. The radical-scavenging activity was calculated as indicated for the ABTS method. Trolox was used to generate antioxidant standards for activity comparisons.

2.4.3 *Hydroxyl radical-scavenging activity*

The hydroxyl radical-scavenging activity was evaluated using the hydroxyl radical system generated by the Fenton reaction, according to a previously described method (Liu, Jia, Kan & Jin, 2012), with slight modifications. A 1 mL aliquot of FSSJM solution was added to the $FeSO_4$ (1 mL, 6 mmol/L) and H_2O_2 (1 mL, 6 mmol/L) mixture, mixed and then left for 10 min. Sodium salicylate (1 mL, 6 mmol/L) was added, and the mixture was placed in a 37°C water bath for 30 min and then centrifuged at 4000 rpm for 10 min. The supernatant absorbance was measured at 510 nm. The radical-scavenging activity was calculated as described for the ABTS method. Trolox and VC were used as antioxidant standards for activity comparisons.

2.4.4 *Lipid peroxidation inhibition*

The inhibition of lipid peroxidation by FSSJM was estimated by measuring the thiobarbituric acid-reacting substances (TBARS). The sample inhibition of lipid peroxidation was expressed as the inhibition of the sample of malondialdehyde (MDA) generation in the reaction system. The MDA concentration was determined using a previously described method (Li & Wang, 2011), with slight modifications. A 500 μL sample of 1% phosphatidylcholine in PBS was mixed with 200 μL of $FeSO_4$ (6 mmol/L) and 1 mL of FSSJM solution; then, 100 μL H_2O_2 (6 mmol/L) was added to the mixture. The reaction mixture was incubated at 37°C for 1 h. The reaction was stopped by adding 1 mL of trichloroacetic acid (TCA, 15%). Afterwards, 1 mL of thiobarbituric acid (TBA, 0.8%) was added, and the mixture was boiled for 15 min, cooled, and then centrifuged at 5000 rpm for 10 min. The supernatant absorbance was measured at 532 nm. The percent inhibition of lipid peroxidation was calculated as indicated for the ABTS method. Trolox and VC were used as antioxidant standards for activity comparisons.

2.5 *Antimutagenic activity assay*

The anti-mutagenic activity of FSSJM was assayed as described using *S. typhimurium* TA98 and TA100 tester strains (Ham, Kim, Moon, Chung, Cui, Han, Chung & Choe, 2009). The direct-acting mutagens MNNG and 4 NQO and the indirectly acting mutagens Trp-P-1 and B(α)P, which require an Aroclor 1254-induced S9 mixture for metabolic activation, were tested. In brief, 50 μL of FSSJM at different concentrations was added to 50 μL of direct or indirect mutagen (MNNG, 4 NQO, Trp-P-1 and B(α)P were used at 0.4 μg/plate, 0.15 μg/plate, 0.5 μg/plate and 10 μg/plate, respectively.), if bioactivation was necessary, 250 μL of S9 mixture was also added. Afterwards, 100 μL of preincubated *S. typhimurium* was added to the mixture, 0.2 M sodium phosphate buffer was added to bring the final volume to 700 μL, and the mixture was incubated at 37°C for 20 min. Two mL of molten top agar was added to the mixture, which was then gently mixed and poured onto minimal glucose agar plates. The plates were inverted and incubated at 37°C for 48 h. Each sample was assayed on triplicate plates for each run. The data are presented as the means of three experiments using different batches of FSSJM. The inhibition ratio was calculated using the following formula (2):

$$\text{Inhibition ratio (\%)} = \frac{N_1 - N_3}{N_1 - N_2} \times 100\% \qquad (2)$$

where N_1 is the number of histidine revertants induced by the mutagen alone, N_2 is the number of revertants induced in the presence of sample solution and solvent (negative control), and N_3 is the number of histidine revertants induced by the mutagen in the presence of FSSJM.

2.6 *Antitumor activity assay*

2.6.1 *Cell culture*

Human tumor HepG-2, HepG-2B, N87 and MDA-MB-231 cells were incubated at 37°C in a humidified atmosphere containing 5% CO_2/95% air in DMEM or both DMEM and RPMI-1640 supplemented with 10% fetal bovine serum, penicillin (100 IU/mL) and streptomycin (100 μg/mL). The cells were harvested by trypsinization until reaching 80% confluence and plated 18 h before treatment with the test drugs.

2.6.2 *MTT assay*

Cell proliferation was determined using an MTT assay to evaluate the cytotoxicity of FSSJM in four species of tumor cells (Tong, Yao, Zeng, Zhou, Chen, Ma & Wang, 2015). Cells (200 μl, 1×10^4 cells/mL) were seeded into each well of a 96-well plate and treated with various concentrations (100–1000 μg/mL) of FSSJM for 48 h. Then, 20 μL of MTT solution (5 mg/mL) was added to each well, and the plates were further incubated for 4 h at 37°C. At the end of the 4 h incubation, the solution was removed, and 150 μL of DMSO was added to each well to dissolve the formazan crystals. Finally, the absorbance was read at 490 nm using an ELIASA microplate reader. The percentage of cell proliferation inhibition was calculated using the following formula (3):

$$\text{Inhibition percentage }\% = \left(1 - \frac{A_1}{A_0}\right) \times 100\% \quad (3)$$

where A_1 is the absorbance of the reaction mixture and A_0 is the absorbance of the reaction mixture without FSSJM solution.

2.7 *Statistical analysis*

CurveExpert 1.3 was used for curve fitting. SPSS 11.5 was used for data analysis. An analysis of variance (ANOVA) was performed, and means comparisons were carried out by the Student-Newman-Keuls test. A value of $p < 0.05$ was considered significant.

3 RESULTS AND DISCUSSION

3.1 *FSSJM antioxidant activity assay*

3.1.1 *The ABTS radical scavenging activity of FSSJM*

The ABTS radical scavenging model is a widely used method for evaluating the free radical scavenging activities of antioxidants (Gavin & Durako, 2012; Katalinic, Milos, Kulisic & Jukic, 2006). The ABTS radical-scavenging ability of FSSJM is shown in Fig. 1 and compared with Trolox standards.

As shown in Fig. 1, the ABTS radical-scavenging ability increased from 12.85 ± 1.42% to 76.37 ± 0.21% when the FSSJM concentration increased from 2 to 24 μg/mL. The IC_{50} values for FSSJM and Trolox were 9.52 μg/mL (r = 0.9909) and 4.14 μg/mL (r = 0.9999), respectively. In spite of the lower ABTS radical-scavenging activity of FSSJM relative to Trolox, the results indicated that FSSJM has significant ABTS radical-scavenging activity.

3.1.2 *The DPPH radical scavenging activity of FSSJM*

The DPPH assay method is based on the reduction of alcoholic DPPH solution (which is dark blue in color) in the presence of a hydrogen-donating antioxidant when converted to the yellow colored non-radical diphenylpicrylhydrazine. Lower absorbance indicates higher the free radical scavenging activity (Yin, Yuan, Wu, Li, Shao, Xu, Hao & Wang, 2015).

The results of the DPPH radical-scavenging test for FSSJM are shown in Fig. 2 and compared with Trolox control standards. The effects of scavenging hydroxyl radicals were concentration dependent. Additionally, the DPPH radical-scavenging abilities of FSSJM and Trolox were 62.64 ± 7.91% and 81.57 ± 4.34%, respectively, at a concentration of 2.5 mg/mL. The IC_{50} values were 1.273 mg/mL (r = 0.9943) and 0.998 mg/mL (r = 0.9997), respectively. These results indicate that FSSJM has a strong DPPH radical scavenging activity.

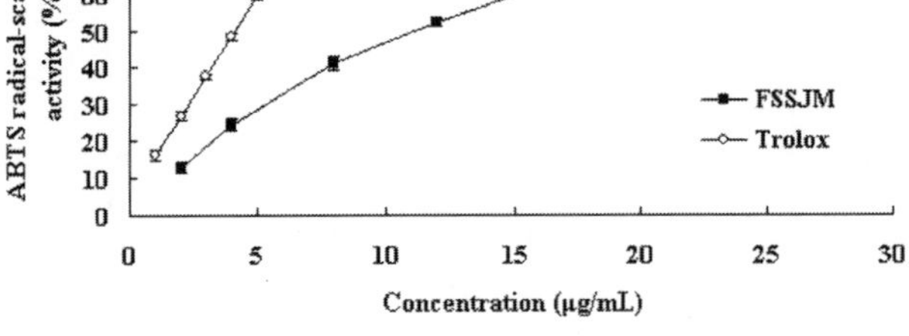

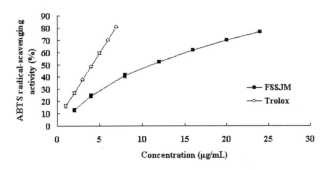

Figure 1. ABTS radical scavenging activity of FSSJM.

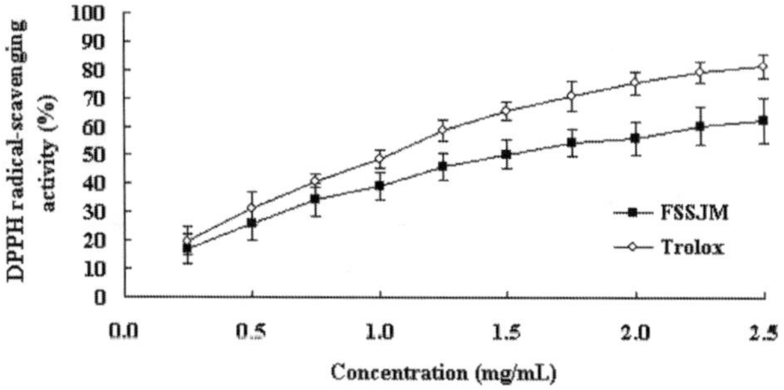

Figure 2. The DPPH radical scavenging activity of FSSJM.

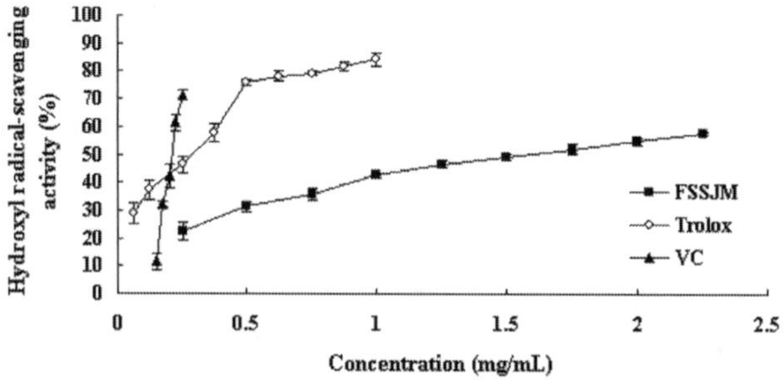

Figure 3. Hydroxyl radical scavenging activity of FSSJM.

3.1.3 *FSSJM scavenging activity against hydroxyl radicals*

Hydroxyl radicals are the most reactive known radicals, and they can attack and damage almost any bio-macromolecule in living cells. The best characterized biologically damaging property of the hydroxyl radical is its capacity to stimulate lipid peroxidation, which occurs when a hydroxyl radical is generated close to membranes and attacks the fatty acid side chains of the membrane phospholipids (Liu, Jia, Kan & Jin, 2012).

As shown in Fig. 3, the scavenging activity of FSSJM increased from 22.55 ± 3.06% to 57.86 ± 0.69% with increasing concentrations from 0.25 to 2.25 mg/mL. Trolox and VC control standards showed higher hydroxyl radical scavenging activity than FSSJM. The IC_{50} values for Trolox, VC and FSSJM were 0.218 mg/mL (r = 0.9906), 0.208 mg/mL (r = 0.9951) and 1.479 mg/mL (r = 0.9910), respectively. The results indicated that FSSJM had positive hydroxyl radical-scavenging activity.

3.1.4 *Lipid peroxidation inhibition activity*

Lipid peroxides are a mixture of extremely reactive lipid peroxidation products, and lipid peroxidation is a common process in all biological systems that has deleterious effects on the cell membrane and DNA (Kubow, 1992). Malondialdehyde (MDA) is a major oxidation product of peroxidized polyunsaturated fatty acids, and decreased MDA content has been employed as an important indicator of anti-lipid peroxidation (Huang, Xue, Niu, Jia & Wang, 2009).

As shown in Fig. 4, the inhibition of lipid peroxidation by FSSJM increased from 26.3 ± 2.31% to 57.1 ± 0.98% with increasing concentrations from 0.13 to 3.16 mg/mL. Trolox and VC were used as control standards. The IC_{50} values for Trolox, VC and FSSJM were 0.696 mg/mL (r = 0.9906), 2.753 mg/mL (r = 0.9951) and 1.678 mg/mL (r = 0.9910), respectively. The data showed that the lipid peroxidation inhibition of FSSJM was lower than that of Trolox but greater than that of VC. These results indicate that FSSJM has notable lipid peroxidation inhibition activity.

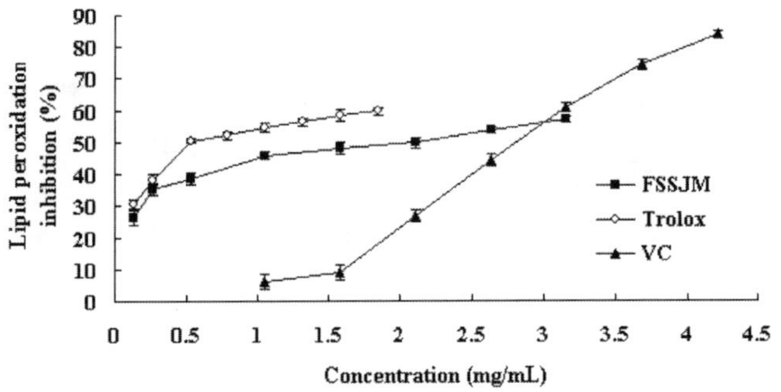

Figure 4. FSSJM lipid peroxidation inhibition.

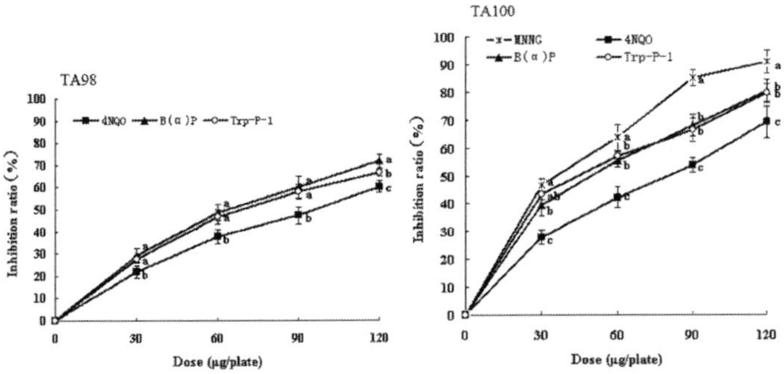

Figure 5. The FSSJM inhibition effect on mutagen-induced mutations in *S. typhimurium* TA98 and TA100. (a–c) Means with different letters at the same concentration are significantly different ($p < 0.05$) as determined by Student-Newman-Keuls tests. The results are presented as the mean ± SD (n = 3).

3.2 *FSSJM anti-mutagenicity assay*

In general, natural compounds that possess strong antioxidant activity often also have other bioactivities, such as anti-inflammatory, antiviral, anticancer and antimutagenic properties (Ham, et al., 2009; Kuete, Mbaveng, Tsaffack, Beng, Etoa, Nkengfack, Meyer & Lall, 2008). According to an Ames test on TA98 and TA100 *Salmonella typhimurium* strains, there were 18 ± 4 and 178 ± 6 revertants per plate, respectively, in the absence of FSSJM. These spontaneous mutation frequencies did not change in the presence of FSSJM, even at the highest concentration of 120 μg/plate. Thus, FSSJM is not mutagenic. Subsequently, we determined the anti-mutagenicity of FSSJM using the Ames test.

As shown in Fig. 5, FSSJM had significant anti-mutagenic activity against mutagen-induced direct and indirect mutations in *Salmonella typhimurium* TA98 and TA100. Anti-mutagenicity increased with increasing FSSJM concentration in a dose-dependent manner.

FSSJM at 120 μg/plate prevented the direct mutations induced by 4 NQO (0.15 μg/plate), inhibiting the mutation rate by 60.2% in *Salmonella typhimurium* TA98. Similarly, and the inhibition ratios for indirect mutations induced by B(α)P (10 μg/plate) or Trp-P-1 (0.5 μg/plate) were 71.7% and 66.6%, respectively. When the S9 mix was used, there was a greater increase in this inhibitory effect (Fig. 5). In *Salmonella typhimurium* TA100, 120 μg/plate FSSJM prevented the direct mutations induced by MNNG (0.4 μg/plate) or 4 NQO (0.15 μg/plate), with 91% and 69.1% inhibition, respectively. FSSJM inhibited indirect mutations induced by B(α)P (10 μg/plate) or Trp-P-1 (0.15 μg/plate), with 80.3% and 79.5% inhibition, respectively (Fig. 5).

Some investigators have used the Ames test to show that plant flavonoids have outstanding anti-mutagenic effects (Borges de Melo, Ataide Martins, Marinho Jorge, Friozi & Castro

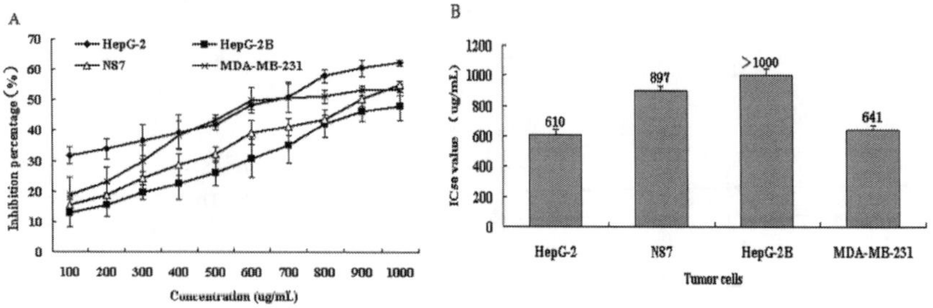

Figure 6. FSSJM inhibited proliferation in four tumor cell lines. (A) Comparison of the percent inhibition by FSSJM in the four tumor cell lines. (B) Comparison of the IC_{50} values for FSSJM in the four tumor cell lines. The results are represented as the mean ± SE (n = 6 per group).

Ferreira, 2010). Here, our data provide the first evidence that the flavonoids isolated from *Juglans mandshurica* seed shells could relatively strongly inhibit the mutagenic activity of both directly and indirectly acting mutagens in a dose-dependent manner. This finding suggested that FSSJM could protect the DNA or RNA in cells from mutagens. Thus, FSSJM was a candidate for further development as a functional food.

3.3 *FSSJM antitumor activity assay*

Previous studies also showed that flavonoids could efficiently inhibit the proliferation of tumor cells, and the inhibition was different in various tumor cells (Kim, Kim, Bak, Kang, Lee, Kim, Park, Kim, Jung & Lim, 2012; Kobayakawa, Sato-Nishimori, Moriyasu & Matsukawa, 2004). Therefore, HepG-2, HepG-2B, N87 and MDA-MB-231 tumor cells were selected to determine the antiproliferative effect of FSSJM using an MTT assay. A dose-dependent loss of cell viability was observed after tumor cells were treated with FSSJM for 48 h (Fig. 6 A). The inhibition effects of FSSJM on four species of tumor cells were all increased with the increase in FSSJM concentration. The IC_{50} values obtained by curve fitting were 610 μg/mL for HepG-2 cells, >1000 μg/mL for HepG-2B cells, 897 μg/mL for N87 cells, and 641 μg/mL for MDA-MB-231 cells (Fig. 6B). These results showed that FSSJM could significantly inhibit the proliferation of tumor cells and that the HepG-2 cells were the most sensitive to FSSJM.

Flavonoids such as taxifolin, kaempferol and quercetin were found in the roots, leaves, and stem bark of the plant, and they also had efficient antitumor activities (Chen & Chen, 2013; Russo, Spagnuolo, Volpe, Mupo, Tedesco & Russo, 2010; Zhang, 2010). These observations are relatively consistent with our own observations. These results suggest that FSSJM may be useful as an ingredient in functional food or for developing anticancer medicine.

4 CONCLUSIONS

This study is the first to extract and purify FSSJM and to confirm that FSSJM has diverse bioactivities, including antioxidant, antimutagenic and antitumor activities. These results support the further use of FSSJM in the development of functional foods or anticancer agents. The seed shells of *Juglans mandshurica* could serve as a new resource for obtaining flavonoid products.

ACKNOWLEDGEMENTS

This research was supported by Grant from the key scientific and technological project of science and technology department of Jilin Province (20150204063NY).

REFERENCES

Borges de Melo, E., Ataide Martins, J.P., Marinho Jorge, T.C., Friozi, M.C. & Castro Ferreira, M.M. 2010. Multivariate QSAR study on the antimutagenic activity of flavonoids against 3-NFA on *Salmonella typhimurium* TA98. *European Journal of Medicinal Chemistry* 45(10), 4562–4569.

Chen, A.Y. & Chen, Y.C. 2013. A review of the dietary flavonoid, kaempferol on human health and cancer chemoprevention. *Food Chemistry* 138(4), 2099–2107.

Forman, H.J., Maiorino, M. & Ursini, F. 2010. Signaling functions of reactive oxygen species. *Biochemistry* 49(5), 835–842.

Gavin, N.M. & Durako, M.J. 2012. Localization and antioxidant capacity of flavonoids in *Halophila johnsonii* in response to experimental light and salinity variation. *Journal of Experimental Marine Biology and Ecology* 416, 32–40.

Graham, J.G., Quinn, M.L., Fabricant, D.S. & Farnsworth, N.R. 2000. Plants used against cancer—an extension of the work of Jonathan Hartwell. *Journal of Ethnopharmacology* 73(3), 347–377.

Ham, S.S., Kim, S.H., Moon, S.Y., Chung, M.J., Cui, C.B., Han, E.K., Chung, C.K. & Choe, M. 2009. Antimutagenic effects of subfractions of Chaga mushroom (*Inonotus obliquus*) extract. *Mutation Research/Genetic Toxicology and Environmental Mutagenesis* 672(1), 55–59.

Hoensch, H.P. & Oertel, R. 2015. The value of flavonoids for the human nutrition: Short review and perspectives. *Clinical Nutrition Experimental* 3 (11), 8–14.

Huang, W., Xue, A., Niu, H., Jia, Z. & Wang, J. 2009. Optimised ultrasonic-assisted extraction of flavonoids from *Folium eucommiae* and evaluation of antioxidant activity in multi-test systems *in vitro*. *Food Chemistry* 114(3), 1147–1154.

Katalinic, V., Milos, M., Kulisic, T. & Jukic, M. 2006. Screening of 70 medicinal plant extracts for antioxidant capacity and total phenols. *Food Chemistry* 94(4), 550–557.

Kim, J.H., Kim, H., Bak, Y., Kang, J.W., Lee, D.H., Kim, M.S., Park, Y.S., Kim, E.J., Jung, K.Y. & Lim, Y. 2012. Naringenin derivative diethyl (5, 4′-dihydroxy flavanone-7-yl) phosphate inhibits cell growth and induces apoptosis in A549 human lung cancer cells. *Journal of the Korean Society for Applied Biological Chemistry* 55(1), 75–82.

Kobayakawa, J., Sato-Nishimori, F., Moriyasu, M. & Matsukawa, Y. 2004. G2-M arrest and antimitotic activity mediated by casticin, a flavonoid isolated from Viticis Fructus (*Vitex rotundifolia* Linne fil.). *Cancer Letters* 208(1), 59–64.

Kubow, S. 1992. Routes of formation and toxic consequences of lipid oxidation products in foods. *Free Radical Biology and Medicine* 12(1), 63–81.

Kuete, V., Mbaveng, A.T., Tsaffack, M., Beng, V.P., Etoa, F.X., Nkengfack, A.E., Meyer, J. & Lall, N. 2008. Antitumor, antioxidant and antimicrobial activities of *Bersama engleriana* (Melianthaceae). *Journal of Ethnopharmacology* 115(3), 494–501.

Li, C. & Wang, M.H. 2011. Antioxidant activity of peach blossom extracts. *Journal of the Korean Society for Applied Biological Chemistry* 54(1), 46–53.

Li, N., Zhang, P., Wu, H.G., Wang, J., Liu, F. & Wang, W.L. 2015. Natural flavonoids function as chemopreventive agents from Gancao (*Glycyrrhiza inflata* Batal). *Journal of Functional Foods* 19 (12), 563–574.

Liang, L., Wu, X., Zhao, T., Zhao, J., Li, F., Zou,Y., Mao, G. & Yang, L. 2012. *In vitro* bioaccessibility and antioxidant activity of anthocyanins from mulberry (*Morus atropurpurea* Roxb.) following simulated gastro-intestinal digestion. *Food Research International* 46(1), 76–82.

Lirdprapamongkol, K., Sakurai, H., Abdelhamed, S., Yokoyama, S., Athikomkulchai, S., Viriyaroj, A., Awale, S., Ruchirawat, S., Svasti, J. & Saiki, I. 2013. Chrysin overcomes TRAIL resistance of cancer cells through Mcl-1 downregulation by inhibiting STAT3 phosphorylation. *International Journal of Oncology*.

Liu, J., Jia, L., Kan, J. & Jin, C.H. 2012. *In vitro* and *in vivo* antioxidant activity of ethanolic extract of white button mushroom (Agaricus bisporus). *Food and Chemical Toxicology* 51, 310–316.

Liu, L., Li, W., Sasaki, T., Asada, Y. & Koike, K. 2010. Juglanone, a novel α-tetralonyl derivative with potent antioxidant activity from Juglans mandshurica. *Journal of natural medicines* 64(4), 496–499.

Machida, K., Yogiashi, Y., Matsuda, S., Suzuki, A. & Kikuchi, M. 2009. A new phenolic glycoside syringate from the bark of Juglans mandshurica MAXIM. var. sieboldiana MAKINO. *Journal of natural medicines* 63(2), 220–222.

Min, B.S., Lee, H.K., Lee, S.M., Kim, Y.H., Bae, K.H., Otake, T., Nakamura, N. & Hattori, M. 2002. Anti-human immunodeficiency virus-type 1 activity of constituents from Juglans mandshurica. *Archives of pharmacal research* 25(4), 441–445.

Min, B.S., Lee, S.Y., Kim, J.H., Lee, J.K., Kim, T.J., Kim, D.H., Kim, Y.H., Joung, H., Lee, H.K. & Nakamura, N. 2003. Anti-complement activity of constituents from the stem-bark of Juglans mandshurica. *Biological and Pharmaceutical Bulletin* 26(7), 1042–1044.

Ravishankar, D., Rajora, A.K., Greco, F. & Osborn, H.M.I. 2013. Flavonoids as prospective compounds for anti-cancer therapy. *The International Journal of Biochemistry & Cell Biology* 45(12), 2821–2831.
Russo, M., Spagnuolo, C., Volpe, S., Mupo, A., Tedesco, I. & Russo, G. 2010. Quercetin induced apoptosis in association with death receptors and fludarabine in cells isolated from chronic lymphocytic leukaemia patients. *British Journal of Cancer* 103(5), 642–648.
Sharma, K., Ko, E.Y., Assefa, A.D., Ha, S., Nile, S.H., Lee, E.T. & Park, S.W. 2015. Temperature-dependent studies on the total phenolics, flavonoids, antioxidant activities, and sugar content in six onion varieties. *Journal of food and drug analysis* 23(2), 243–252.
Tong, J., Yao, X.C., Zeng, H., Zhou, G., Chen, Y.X., Ma, B.X. & Wang, Y.W. 2015. Hepatoprotective activity of flavonoids from *Cichorium glandulosum* seeds *in vitro* and *in vivo* carbon tetrachloride-induced hepatotoxicity *Journal of Ethnopharmacology* 174(11), 355–363.
Valko, M., Leibfritz, D., Moncol, J., Cronin, M.T., Mazur, M. & Telser, J. 2007. Free radicals and antioxidants in normal physiological functions and human disease. *International Journal of Biochemistry and Cell Biology* 39(1), 44–84.
Valko, M., Rhodes, C., Moncol, J., Izakovic, M. & Mazur, M. 2006. Free radicals, metals and antioxidants in oxidative stress-induced cancer. *Chemico-biological interactions* 160(1), 1–40.
Wang J.L., Z.S.X., Li T.J., Zhang F.Q., Wang J.J., Zhang S.J. 2009. Chemical constituents from bark of Juglans mandshurica. *Chinese Traditional and Herbal Drugs* 39(4), 490–493.
Xie, J.H., Dong, C.J., Nie, S.P., Li, F., Wang, Z.J., Shen, M.Y. & Xie, M.Y. 2015. Extraction, chemical composition and antioxidant activity of flavonoids from *Cyclocarya paliurus* (Batal.) Iljinskaja leaves. *Food Chemistry* 186 (11), 97–105.
Yin, D.D., Yuan, R.Y., Wu, Q., Li, S.S., Shao, S., Xu, Y.J., Hao, X.H. & Wang, L.S. 2015.Assessment of flavonoids and volatile compounds in tea infusions of water lily flowers and their antioxidant activities. *Food Chemistry* 187(11), 20–28.
Zhang, X.X., Xiao, W.F., Qi, S., Hou, W.B. 2010. Recent advance on plant sources, bioactivities, pharmacological effects and pharmacokinetic studies of taxifolin. *Asian Journal of Pharmacodynamics and Pharmacokinetics* 10(1), 35–43.

Advanced Engineering and Technology III – Xie (Ed.)
© 2017 Taylor & Francis Group, London, ISBN 978-1-138-03275-0

Effects of ultra-high pressure on the properties and structural characteristics of chitosan

Xing-ke Li, Li Zhang & Hua Zhang
Department of Food and Biological Engineering, Zhengzhou University of Light Industry, Zhengzhou, Henan Province, China

ABSTRACT: Chitosan were processed by different pressure (20, 200, 400, 800, and 1000 MPa) and effects of ultra-high pressure on thickening, emulsifying properties and structure of chitosan were studied by laser light scattering instrument, X-Ray Diffraction (XRD), Thermogravimetric Analysis (TGA), FT-IR spectroscopy, UV–vis spectroscopy and Scanning Electron Microscopy (SEM). Our results showed that the viscosity of chitosan dropped from 9.81 pa · s to 9.36 pa and emulsifying properties were changed with increasing pressure. Ultra-high pressure effectively reduced the Molecular Weight (MW) of chitosan. Thermal stability of chitosan declined with maximum degradation temperature decreasing from 300°C to 278°C. The crystalline structure of chitosan was destroyed by ultra-high pressure since diffraction peaks disappeared. FT-IR and UV–vis spectra showed that ultra-high pressure did not cause significant changes in the chemical structure of chitosan. SEM observed that particle morphology of chitosan tended to flaking under ultra-high pressure.

1 INTRODUCTION

Chitosan [β-(1,4)-2-amino-2-deoxy-D-glucopyranose] is a modified, natural carbohydrate polymer derived by deacetylation of chitin, which is the second most abundant biopolymer in nature next to cellulose and is found in the exoskeleton of crustaceans, fungal cell walls and other biological materials (No et al., 2007; Shahidi et al., 1999). Among abundant naturally occurring polysaccharides, which are neutral or acidic, chitosan is distinguished for its polycationic nature. As a natural, nontoxic and biodegradable biopolymer, chitosan has received considerable attention in the food industry due to its physiological properties, nutritional and biochemical activities. Thus, chitosan has been approved as a food additive in Korea and Japan since 1995 and 1983 respectively (KFDA, 1995; Weiner, 1992), whereas in China, chitosan has been approved as a food thickener since 2007 (GB2760, 2014).

But its solubility only in dilute acid and high viscosity restrict its application to a certain extent. Many different modification methods were adopted to improve the solubility of chitosan, such as chemical, enzymatic and other methods (Xia et al., 2008; Feng et al., 2011). Every method had its advantages and disadvantages. In addition, there is a growing interest in physical modification of chitosan using innovative processing techniques. In our previous study, we found the viscosity of chitosan was decreased when it is experienced sterilization under high pressure and temperature.

High pressure processing is a food processing method which has shown great potential in the food industry. High-pressure treatment alters the structure of biopolymers such as proteins and starch, providing the possibility to produce other bio with novel textures. High-pressure processing has been identified as a potential area for food texture engineering (Norton et al., 2008).

Therefore, the objective of this study was to study the effects of high pressure on the properties and structural characteristics of chitosan.

2 MATERIALS AND METHODS

2.1 *Preparation of chitosan*

Chitosan was purchased from Golden-Shell Pharmaceutical Co. Ltd and processed under different pressures (20, 200, 400, 800 and 1000 MPa) by UDS-III typed diamond press (China).

2.2 *Determination of thickening properties*

The viscosity of chitosan was assayed by DV2T Viscosity Instruments (BROOKFIELD, USA).

2.3 *Determination of emulsifying properties*

Emulsifying Activity Index (EAI) and Emulsifying Stability Index (ESI) were determined using the turbidimetric method. Turbidity was calculated by T = 2.303 A/L. Emulsifying activity index was defined as the emulsion droplet surface area formed per gram emulsifiers. ESI was defined as the time when the turbidity of the emulsion measured at 500 nm reached half of the initial (Li et al., 2011).

2.4 *Characterization of chitosan*

The average-weight molecular weight (Mw) and molecular weight distribution (Mw/Mn) of all kinds of chitosan were measured by laser light scattering instrument. The potentiometry method was used to determine the degree of deacetylation of these chitosan (Lin et al., 1992).

2.5 *Thermogravimetric Analysis (TGA)*

TGA was performed on a Mettler Toledo TGA/SDTA851 Thermo gravimeter (Mettler Toledo Corp., Zurich, Switzerland) with STARe software (version 9.01) was used to analyze the thermal stability of the samples. Samples were heated from 30 to 500°C at a heating rate of 10°C/min during the analysis.

2.6 *FT-IR and UV–vis spectroscopy*

Fourier Transform Infrared (FT-IR) spectrum was recorded on a Nicolet Nexus 470 instrument (Nicolet Instrument, Thermo Company, Germany). Samples were prepared as KBr pellet and scanned against a blank KBr pellet background at wave number range 4000–400 cm^{-1} with resolution of 4.0 cm^{-1} (Hou et al., 2013).

0.1 g of chitosan was dissolved in 50 ml 1% (w/v) HCl solution. After complete dissolution, the solution was centrifuged at 3000 × g for 10 min to remove the insoluble material (Zhang, 2012). UV–vis absorption spectra were obtained using a UV1100 spectrophotometer (Techcomp Ltd., China) in the range of 200–500 nm.

2.7 *X-Ray Diffraction (XRD)*

The X-ray diffraction patterns of the chitosan samples were measured using a Bruker AXS D8 Advance diffract meter (Germany) under the following conditions: Cu-Kα radiation, 40 kV, 40 mA and measurement range 2θ 4–40° (Zhang, 2012).

2.8 *Electron microscopy*

Morphological characterization of particles in original chitosan powder was performed using a scanning electron microscope (JSM-6490 LV SEM, FEI, Japan). The sample was coated with spraying gold powder to make it conductive.

2.9 *Statistical analysis*

All experiments were repeated 3 times each. Least significant differences ($P < 0.05$) were used to determine differences among treatments.

3 RESULTS AND DISCUSSION

3.1 *Effects of ultra-high pressure on viscosity of chitosan*

Thickening properties is the basic properties of polysaccharides. The viscosity of chitosan by different pressure treatment was showed in Figure 1. With increasing pressure, the viscosity of the chitosan was slightly reduced which might be due to the breaks of molecular chain of chitosan. This result was accord to the study of Li Feng which showed that the viscosity of cellulose decreased slightly after ultra-high pressure (Li, 2008).

3.2 *Effects of ultra-high pressure on emulsifying properties of chitosan*

Chitosan, a deacetylated derivative of chitin, which contains a large number of hydrophobic groups (acetyl group) and hydrophilic group ($-NH_2$) (Li et al., 2011). Figure 2 shows the emulsifying properties of chitosan. Emulsifying activity of chitosan increased with the increase of pressure, reached the peak at 400 MPa. This result indicated that molecular

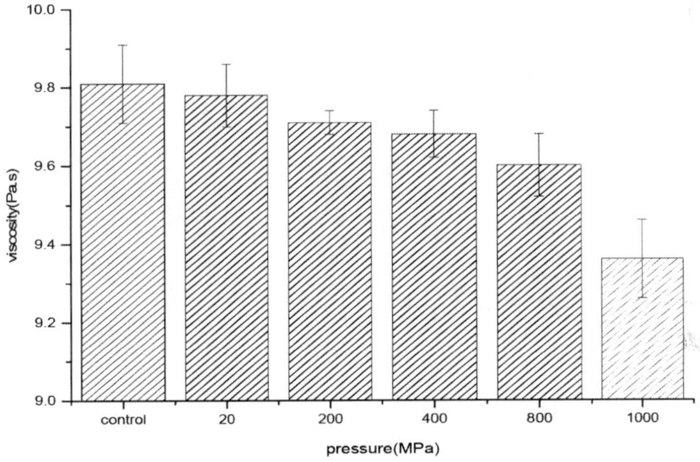

Figure 1. Effects of ultra-high pressure on viscosity of chitosan.

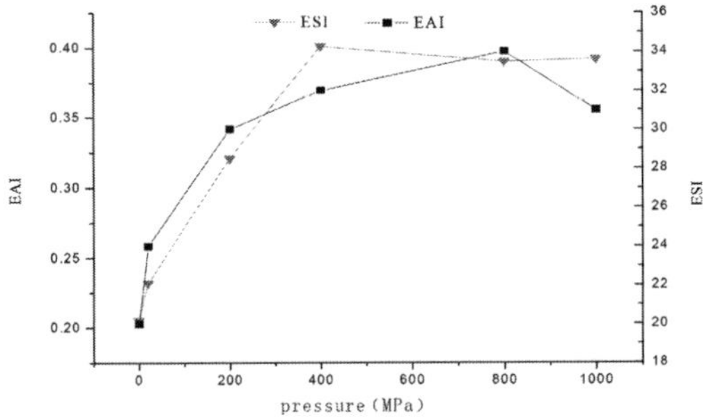

Figure 2. Effects of ultra-high pressure on emulsifying properties of Chitosan.

structure of chitosan was destroyed by ultra-high pressure strengthening its polar groups hydrated and making internal hydrophobic groups exposed.

3.3 *Effects of ultra-high pressure on characterization of chitosan*

Table 1 shows the Mw, Mn, Mw/Mn and degree of deacetylation of chitosan. Compare to the control, Mw and Mw/Mn decreased with the increasing pressure. It was suggested that UHP treatment resulted in damaging of the internal structure of chitosan and the degraded of molecules, while high pressure has no influence on degree of deacetylation of chitosan.

3.4 *Effects of ultra-high pressure on characterization of chitosan*

TG and DTG curves of chitosan under different pressures are shown in Figure 3. TG curves of chitosan showed two thermal events. The first thermal event is observed at 54°C, which is attributed to the loss of water. The second thermal event occurs at 250–400°C. The maximum degradation temperature (Tmax) of original chitosan was at about 300°C, but the Tmax of chitosan treated at 800,1000 MPa, especially 1000 MPa decreased to 278°C, which might be attributed to the decrease in molecular weight and the main chain was broken. The findings shows that the structure of chitosan was destroyed treated at 1000 MPa. Meanwhile the DTG curves appear multi-peak phenomenon under 800, 1000 MPa and the event occurred exothermic reaction at about 450°C, which might be attributed to the damage of the structure of chitosan under ultra-high pressure.

Table 1. Characterization of chitosan.

	Mw ($\times 10^5$ Da)	Mn ($\times 10^5$ Da)	Mw/Mn	DD
Control	1.623	1.251	1.297	85.1%
20 mpa	1.603	1.253	1.279	84.7%
200 mpa	1.597	1.260	1.267	84.9%
400 mpa	1.561	1.251	1.248	85.2%
800 mpa	1.551	1.251	1.239	85.0%
1000 mpa	1.424	1.216	1.171	84.8%

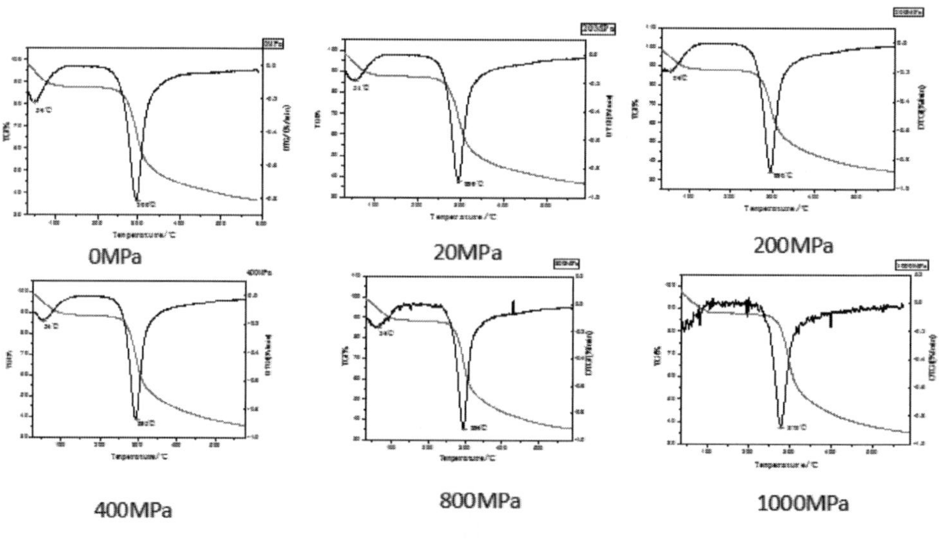

Figure 3. TG and DTG curves of chitosan under different pressures.

3.5 *FI-IR and UV-vis spectra analysis*

Infrared spectroscopy is a technology to determine functional groups and chemical bonds. FI-IR spectra of chitosan are shown in Figure 4. For the FT-IR spectrum of chitosan, the broad band at around 3436 cm^{-1} was attributed to —NH and —OH stretching vibration, as well as inter- and extra-molecular hydrogen bonding of chitosan molecules. The characteristic absorption bands of HMWC appeared at 1644 cm^{-1} (Amide I), 1594 cm^{-1} (–NH_2 bending) and 1320 cm^{-1} (Amide III) (Feng, 2011). The original chitosan and chitosan treated by ultra-high pressure did not show great differences in FT-IR spectra, which suggested that the ultra-pressure had no obvious effect on chemical structure of chitosan.

Figure 5 shows the UV-vis spectrum of chitosan. Based on the data reported in the published literature, the broad absorption band might be ascribed to C = O group (Cai et al.,

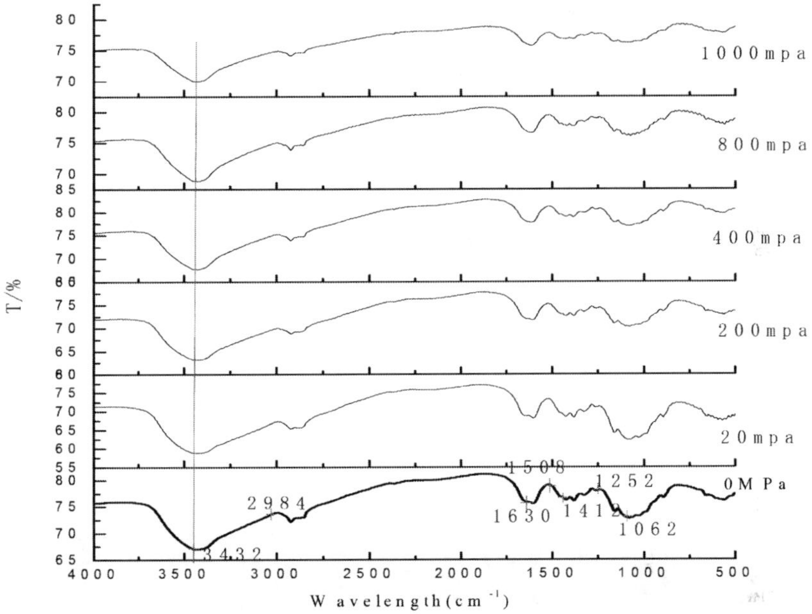

Figure 4. FT-IR of chitosan treated by different pressure.

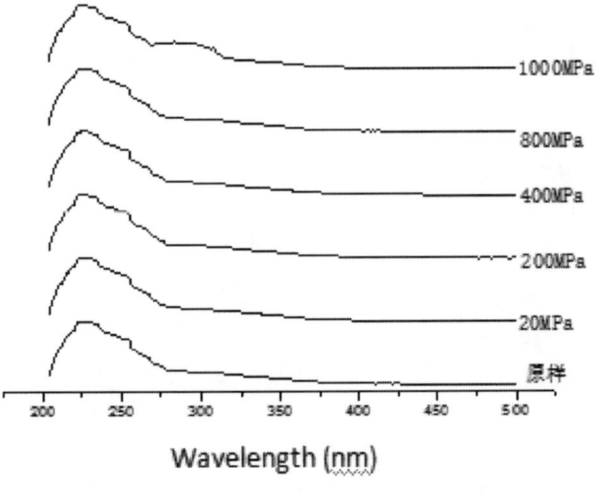

Figure 5. UV-vis of chitosan treated at different pressure.

2010). The UV-vis spectrum of chitosan were not changed with the increase of ultra-high pressure. Chitosan treated at 1000 MPa has a small absorption in about 300 nm, which might be the molecular weight of chitosan lower and the increase of C = O groups.

3.6 *Crystalline structure*

The structure in the crystalline region of chitosan particles was determined by X-ray diffraction. Figure 6 shows that there are two different crystalline morphology of original chitosan which belong to single crystal structure. The XRD patterns of chitosan exhibited characteristic peaks at 2θ of about 12° and 20° were called Form I, Form II respectively (Qin et al., 2002). The changes of XRD patterns of under 20,200,400 pressure was slight while the characteristic peaks at 2θ of about 12° was disappeared when treated at 800, 1000 MPa. This result indicated that the crystal structure of chitosan were destroyed and the crystalline were decreased when treated under 800, 1000 MPa pressure.

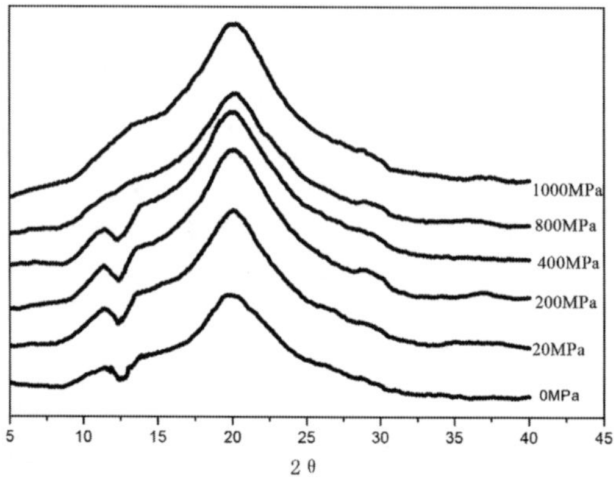

Figure 6. X-Ray Diffraction pattern of chitosan.

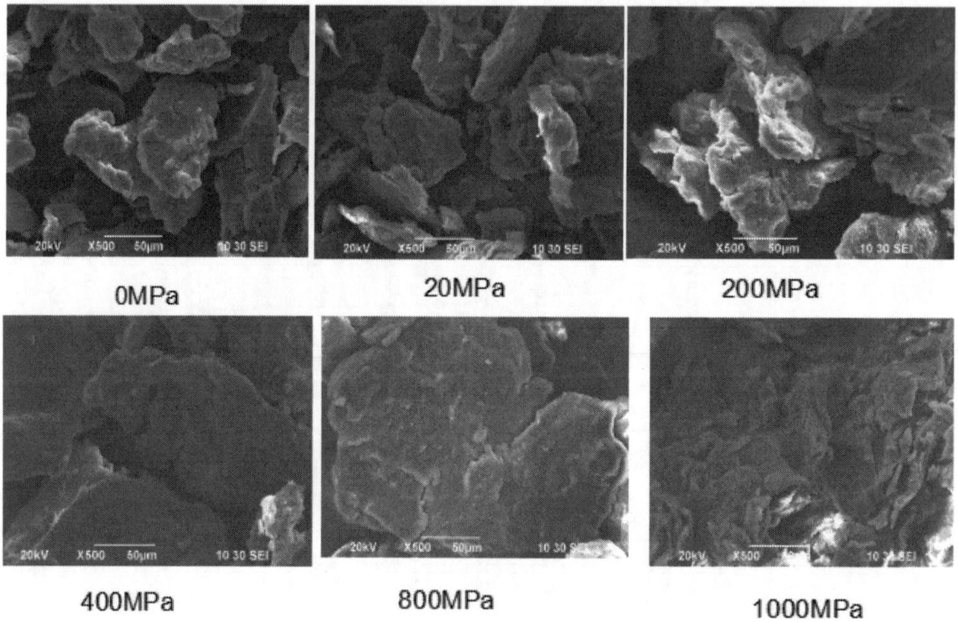

Figure 7. Scanning electron micrographs of chitosan treated at different pressure (500 ×).

3.7 *Morphology*

SEM micrographs of different pressure chitosan are shown in Figure 7. The degrees of change in the granule morphology of chitosan after 20,200,400 MPa pressure was slight, while the degrees of change in the granule morphology of chitosan after 800, 1000 MPa pressure was obvious. The particles of chitosan were observed to be in the form of flakes and the structure of chitosan become porous when treated at 800, 1000 MPa pressure. The findings suggested that the process of ultra-high pressure, crystalline structure of chitosan was destroyed. These findings were similar with some previous studies concerning ultra-high pressure of soybean dietary fibers (Li, 2008).

4 CONCLUSION

With increasing pressure, the viscosity of chitosan dropped from 9.81 pa · s to 9.36 pa · s and emulsifying properties of chitosan changed. Mw of chitosan decreased from 1.623 × 105 Da to 1.424 × 105 Da and Mw/Mn decreased from 1.297 to 1.171. The thermal stability of chitosan declined with maximum degradation temperature of chitosan decreasing from 300°C to 278°C. The infrared spectra obtained that chitosan treated at UHP O-H bond were changed. The crystalline structure of chitosan was destroyed by ultra-high pressure since diffraction peaks disappeared. FT-IR and UV-vis spectra showed that ultra-high pressure did not cause significant changes in the chemical structure of chitosan. SEM observed that particle morphology of chitosan tended to flaking under ultra-high pressure. These findings suggest that ultra-high pressure is an effective method to modify chitosan and some physico-chemical properties of chitosan will change during the preparation process, which may affect its application and deserve to be further researched.

ACKNOWLEDGEMENTS

This research was financially supported by the funds of Education Department of Henan province (NOS. 14 A550008) and of Key Teacher of Zhengzhou University of Light Industry.

REFERENCES

Cai, Q.Y., Gu, Z.M., Chen, Y., Han, W.Q., Fu, T.M., & Song, H.C., et al. (2010). Degradation of chitosan by an electrochemical process. Carbohydrate Polymers, 79(3), 783–785.

Feng, Y.W., & Xia, W.S. (2011). Preparation, characterization and antibacterial activity of water-soluble O-fumaryl-chitosan. Carbohydrate Polymers, 83(3), 1169–1173.

[GB2760-2014] Hygienic standards for uses of food additives, Published 2014, Implemented 2015 May 24.

Hou W.L., Yang T., & Yang Y.D. (2013). Infrared spectrometric and thermogravimetric analysis of maleoyl chitosan. Physical and Chemical testing. 49(10): 1163–1165.

[KFDA] Korea Food and Drug Administration. (1995). Food additives code. Seoul, Korea.

Li Feng. (2008). Modifying Soybean Dietary Fiber by Ultra-High Pressure Treatment. Soybean Science, 27, 141–144.

Li, X.K., & Xia, W.S. (2011). Effects of concentration, degree of deacetylation and molecular weight on emulsifying properties of chitosan. International journal of biological macromolecules, 48, 768–772.

Lin, R.X., Jiang, S.H., & Zhang, M.S. (1992). The determination of degree of deacetylation. Chemistry Bulletin, 3, 39–42.

No, H.K., Meyers, S.P., Prinyawtwatkul, W., & Xu, Z. (2007). Application of chitosan for improvement of quality and shelf life of foods: a review. Journal of Food Science, 72, 87–99.

Shahidi, F., Arachchi, J.K.V., & Jeon, Y.J. (1999). Food applications of chitin and chitosans. Trends in Food Science and Technology, 10, 37–51.

Qin, C., Du, Y., & Zong, L., et al. (2003). Effect of hemicellulase on the moleculai weight and structure of chitosan. Polymer Degradation and Stability. 80, 435–441.

Tomás Norton & Sun, D.W. (2008). Recent Advances in the Use of High Pressure as an Effective Processing Technique in the Food Industry. Food Bioprocess Technology, 1, 2–34.
Xia, W.S., Liu, P., & Liu, J. (2008). Advance in chitosan hydrolysis by non-specific cellulases. Bioresource Technology, 99, 6751–6762.
Zhang, W., Zhang, J.L., & Xia, W.S. (2012). Physicochemical and structural characteristics of chitosan nanopowders prepared by ultrafine milling. Carbohydrate Polymers. 87(1): 309–313.

Advanced Engineering and Technology III – Xie (Ed.)
© 2017 Taylor & Francis Group, London, ISBN 978-1-138-03275-0

Optimization of extraction parameters of flavonoids from the seed shell of *Juglans mandshurica* and evaluation of antioxidant activity *in vitro*

H.Y. Xu
Department of Food Science and Engineering, Agricultural College, Yanbian University, Yanji, China

Y.H. Bao
Food Science and Engineering, Forestry College, Northeast Forestry University, Harbin, China

ABSTRACT: The seed shell of *Juglans mandshurica* is a waste product from seed kernel processing in the food industry. For better utilization of the resource, the Box-Behnken Design (BBD) combined with Response Surface Methodology (RSM) was used to optimize the extraction conditions of total flavonoids from the seed shell of *Juglans Mandshurica* (TFSSJM). The optimal parameters obtained were as follows: solid/liquid ratio 1:24 (g/mL); ethanol concentration 57%; and extraction time 131 min. The yield of flavonoids was 9.27 ± 0.04 mg/g under these optimal conditions. TFFSJM were purified by macroporous resin and purified flavonoids (FSSJM-1) were obtained. The antioxidant activity of FSSJM-1 was analyzed. The data demonstrated that FSSJM-1 could scavenge ABTS, DPPH, and hydroxyl radicals, and found a dose–effect relationship. The antioxidant capability was better than BHT. The IC_{50} values were 26 μg/mL, 1.273 mg/mL, and 0.622 mg/mL, respectively. The results indicated that the optimal extraction parameters could be used as an effective and reliable method, and flavonoids extracted from the seed shell of *Juglans mandshurica* had significant antioxidant activity, which can be used or further developed as a potential source of natural antioxidant.

1 INTRODUCTION

Plant-derived flavonoids are a group of polyphenolic compounds ubiquitously found in fruits, vegetable grains, and some medicinal plants and herbs (Peterson and Dwyer, 1998). They have attracted more and more attention due to their beneficial effects on many diseases, including cancer, cardiovascular disease, inflammation, and neurodegenerative disorders. (Hodek et al., 2002; Tripoli et al., 2007). Some reports have also shown that biological activity of flavonoids is mainly attributed to their antioxidant properties (Duffy et al., 2008; Hughes, 2008; Pham-Huy et al., 2008).

Juglans mandshurica, is a fast-growing walnut tree native to many Asian countries, especially in China. This plant has been used as folk and traditional medicines for the treatment of cancer in some countries (Graham et al., 2000; Machida et al., 2009). Some active substances such as flavonoids, quinones, and polyphenols have been found in roots, leaves, stem-barks, and fresh unripe fruits of this plant (Li et al., 2003; Min et al., 2002; Park et al., 2012). Furthermore, they show remarkable bioactivity effects in many aspects. However, the study on flavonoids in the seed shell of *Juglans mandshurica* has not been reported. The seed of *Juglans mandshurica* seed is the by-product of forestry. Moreover, its kernel has recently received a great deal of attention because of its rich nutrition. During the processing of the kernel, a lot of seed shells are produced as garbage. Currently, in addition to some seed shells being made into handicraft products or used as fuel woods, many of them were discarded as waste. For the better utilization of the seed shell of *Juglans mandshurica,* more research should be done on its composition and bioactivity. Moreover, agricultural and industrial residues have been considered as materials to obtain antioxidants (Bagchi et al., 2000; Graham et al., 2000; Guangyan Pan, 2012; Lafka et al., 2007).

In order to achieve a higher yield of extraction, techniques and methods for natural extraction have been developed successively. One of such methods is the Response Surface Methodology (RSM). The RSM is a collection of mathematical and statistical techniques that are useful for the modeling and analysis of problems in which a response of interest is influenced by several variables and the objective is to optimize this response (Baş and Boyacı, 2007). In addition, it has been widely applicable for different purposes in chemical, biochemical, and engineering processes, and industrial research. The main advantage of RSM is the reduced number of experimental trials needed to evaluate multiple parameters and their interaction (Khuri and Mukhopadhyay, 2010; Yin et al., 2011). Therefore, it is less laborious and time-consuming.

The aim of the present study was to select and validate the optimal conditions of extraction for flavonoids from the seed shell of *Juglans mandshurica* using the Box-Behnken Design (BBD) with RSM. Moreover antioxidant activities of purified flavonoids, including ABTS, DPPH, and hydroxyl radical-scavenging effects, were evaluated *in vitro*. We hope that this study will be helpful in providing further bioactivity information about flavonoids extracted from the seed shell of *Juglans mandshurica* for the development and application of the resource.

2 MATERIALS AND METHODS

2.1 *Materials and instruments*

Seeds of *Juglans mandshurica* were provided by Forest Enterprise of Mudanjiang in Heilongjiang Province (China). Rutin was purchased from the National Institute for the Control of Pharmaceutical and Biological Products (Beijing, China). HPD-600 macroporous resin was purchased from Tianjin Orui Biotechnology Co. (Tianjin, China). ABTS (2,2′-Azino-bis-(3-ethylbenzoth-iazoline-6-sulfonic acid)), DPPH (2,2-Di(4-tert-octylphenl)-1-picrylhydrazyl), and BHT (2,6-Di-tert-butyl-4-methylphenol) were purchased from Sigma Chemicals Co. (Shanghai, China). VC was purchased from Tianjin Kermel Chemical Reagent Co. (Tianjin, China). All other chemical reagents used in the experiments were of analytical grade.

RE-52A rotavapor (Shanghai Yarong Biochemistry Instrument Factory, Shanghai, China) was used for determining the concentration of the sample, and TU-1810 spectrophotometer (Persee Universal Instrument Co., Beijing, China) was used for analyzing total flavonoid content and antioxidant activity.

2.2 *Extraction of flavonoids*

The kernels of seeds were taken out, and its shells were pulverized in a knife mill and passed through a 60-mesh sieve. The powder (2.5 g) of the seed shell was weighed accurately, and total flavonoids were extracted with ethanol solvent (solid/liquid ratio 1:10 to 1:50, g/mL and ethanol concentration 30% to 90%) in an Erlenmeyer flask heated by a water bath. The extraction time ranged from 0.5 h to 3.5 h, and the extraction temperature varied from 35°C to 60°C. Then, the extract was filtered and the solution was collected and added to a defined volume in a volumetric flask (50 mL) for further analysis. Total flavonoid yield and reducing power were chosen as the indices of the single-factor experiment.

2.3 *Box-Behnken Design (BBD)*

On the basis of the single-factor experimental results, the BBD was applied to evaluate the main and interaction effects of the factors in the experimental region: solid/liquid ratio (X_1), ethanol concentration (X_2), and extraction time (X_3) on total flavonoid yield (Y). The independent variables were coded at three levels (−1, 0, 1), and the complete design

Table 1. Independent variables and coded levels of the BBD.

	Independent variables		
Coded levels	Solid/liquid ratio (X_1) (g/mL)	Ethanol concentration (X_2) (%)	Extraction time (X_3) (h)
−1	1:10	70%	2.5
0	1:20	50%	3
1	1:30	30%	3.5

consisted of 15 experimental points including three replications of the center points (all variables were coded as zero). The actual and coded levels of the independent variables used in the experimental design are presented in Table 1.

In the study, a second-order polynomial Eq. (1) was used to describe the effects of variables on the response:

$$Y = \beta_0 + \sum_{j=1}^{3}\beta_j X_j + \sum_{j=1}^{3}\beta_{jj} X_j^2 + \sum\sum_{t<j}\beta_{ij} X_i X_j \tag{1}$$

where Y is the predicted response, β_0, β_j, β_{jj} and β_{ij} are the regression coefficients for intercept, linearity, square, and interaction, respectively, while X_i and X_j are the coded levels of the independent variables. All the experiments were repeated three times, and the flavonoid yields were expressed as average values.

2.4 *Determination of total flavonoid content*

Total flavonoid content was determined using a sodium nitrite and aluminum chloride colorimetric method, as described previously (Gong et al., 2012) with slight modifications. Briefly, a 1 mL sample was placed in a 10 mL volumetric flask, and 1 mL ethanol/water (60:40 v/v) and 0.4 mL of sodium nitrite (5 g/100 mL) were added and mixed with the sample. A 0.4 mL aluminum chloride solution (10 g/100 mL) was added after 6 min. At 6 min later, 4 mL sodium hydroxide solution (1 mol/L) was added, and the total volume was made up to 10 mL with ethanol/water (60:40 v/v) and mixed with the solution. After 15 min, absorbance was measured against a blank at 510 nm. The total flavonoid content was calculated from a calibration curve of rutin and expressed as rutin equivalents (mg/g dry basis). The calibration curve (Y = 13.536X − 0.0005, where Y is the absorbance value of the sample and X is the concentration of the sample) ranged from 0 to 300 μg/mL ($R^2 = 0.9994$).

The flavonoid yield was calculated using the following formula:

$$\text{Flavonoid yield (mg/g)} = CNV/m \tag{2}$$

where *C* is the mass concentration of the sample fluid calculated using the calibration curve of rutin (mg/mL), *N* is the dilution, *V* is the extract volume (mL), and *m* is the quantity of powder (g).

2.5 *Determination of reducing power*

The determination of reducing power was performed as described previously (Zhang et al., 2011) with slight modifications. A 2 mL extract sample (varying concentration) was mixed with 2.5 mL of 0.2 mol/L phosphate buffer (pH 6.6) and 2.5 mL of 1% potassium ferricyanide. The mixture was incubated at 50°C for 20 min, and then quickly chilled in an ice bath. Thereafter, 2.5 mL of 10% trichloroacetic acid was added, and the mixture was centrifuged at 4000 rpm for 10 min (TGL-16G Centrifuge, Shanghai Anting scientific

Instrument Factory, Shanghai, China). The supernatant (2.5 mL) was mixed with distilled water (2.5 mL) and ferric chloride (0.5 mL, 0.1%) for 10 min. Absorbance was measured at 700 nm against a blank. The increasing absorbance of the reaction mixture indicated the increasing reducing power.

2.6 *Purification of crude flavonoids*

The crude flavonoid sample obtained under the optimized parameters was purified using a column (8 × 96 cm) packed with HPD-600 macroporous resin. The concentration of the liquid sample of 1.2 mg/mL was used, and macroporous resin was eluted using 4BV distilled water and 2BV 90% ethanol. The flow rate was maintained at 2 mL/min. The purified flavonoids were collected and evaporated at a thermal decompression of 45°C, which was labeled as FSSJM-1.

2.7 *Determination of the antioxidant activity of flavonoids*

2.7.1 *ABTS radical-scavenging activity*

The activity of FSSJM-1 to scavenge $ABTS^{\cdot+}$ was carried out according to the method described previously (Gülsen Tel, 2012) with slight modifications. $ABTS^{\cdot+}$ was prepared by mixing an ABTS stock solution (7.4 mmol/L in water) with 2.6 mmol/L potassium persulfate. This mixture was stored at room temperature in the dark for 16 h. Before analysis, the $ABTS^{\cdot+}$ solution was diluted with methanol (MeOH) to obtain an absorbance of 0.700 ± 0.020 at 734 nm. Then, 0.15 mL of the appropriately diluted sample was added to 2.85 mL $ABTS^{\cdot+}$ working solution and mixed thoroughly. The reaction mixture was kept at room temperature in the dark for 10 min. Then, absorbance was recorded at 734 nm. The absorbance of the reagent control was measured as done previously.

Radical scavenging activity was calculated using the following formula:

$$\text{Radical scavenging activity } \% = (A_0 - A_1)/A_0 \times 100 \quad (3)$$

where A_1 is the absorbance of the reaction mixture and A_0 is the absorbance of the reaction mixture without the sample solution. VC and BHA were used as antioxidant standards for comparison of the activity.

2.7.2 *DPPH radical scavenging activity*

The activity of FSSJM-1 to scavenge DPPH radicals was performed according to the method described previously (Sharma et al., 2015) with slight modifications. DPPH stock solution (0.15 mmol/L) was prepared with ethanol and stored in the dark. The appropriately diluted sample (2 mL) was added to the DPPH working solution (2 mL) and mixed thoroughly. The reaction mixture was kept at room temperature in the dark for 30 min. Absorbance was recorded at 517 nm. The absorbance of the reagent control and the sample solution was measured as done previously.

Radical scavenging activity was calculated using the following formula:

$$\text{Radical scavenging activity } \% = 1 - (A_1 - A_2)/A_0 \times 100 \quad (4)$$

where A_1 is the absorbance of the reaction mixture, A_2 is the absorbance of the reaction mixture without the DPPH solution, and A_0 is the absorbance of the reaction mixture without the sample solution. VC and BHA were used as antioxidant standards for comparison of the activity.

2.7.3 *Hydroxyl radical-scavenging activity*

The hydroxyl radical-scavenging activity was evaluated according to the method (Zhao et al., 2012) with slight modifications. The appropriately diluted sample (1 mL) was added to the mixture of $FeSO_4 \cdot 7H_2O$ (2 mL, 6 mmol/L) and H_2O_2 (2 mL, 6 mmol/L), mixed, and placed for 10 min. Salicylic acid (2 mL, 6 mmol/L) was added and mixed 37°C for 30 min. Then, the mixture was centrifuged at 4000 rpm for 10 min. The absorbance of the supernatant was measured at 510 nm. Radical scavenging activity was calculated as mentioned for the ABTS method.

2.8 *Statistical analysis*

All analyses were performed in triplicate. The experimental results obtained are expressed as means ± SE. Statistical analysis was performed using the software Minitab 15.0. Data were analyzed by using analysis of variance (ANOVA). $P < 0.05$ was considered as statistically significant.

3 RESULTS AND DISCUSSION

3.1 *Single-factor experiments*

Extraction parameters including solid/liquid ratio, ethanol concentration, extraction time, and extraction temperature were investigated. The results are shown in Fig. 1.

The effect of the solid/liquid ratio on flavonoid yield and reducing power was analyzed, and the results are shown in Fig. 1A. It can be seen from Fig. 1A that there was a significant increase in both flavonoid yield and reducing power with increasing solid/liquid ratios from 1:10 to 1:30. Thereafter, the further increase in the solid/liquid ratio led to a slight increase in both flavonoid yield and reducing power. This is because that the increase in the solvent volume would increase the interfacial area between tiny bubbles and samples, and thus the cavitation effect of bubble collapse would be more intense. While considering the cost of experimental and further work, 1:30 was determined as the best solid/liquid ratio in the single-factor experiment.

The effect of ethanol concentration on flavonoid yield and reducing power was analyzed, and the results are shown in Fig. 1B. It can be observed from Fig. 1B that flavonoid yield and reducing power started to increase with increasing ethanol concentration. Furthermore, flavonoid yield was similar at ethanol concentrations of 40%, 50%, and 60%, but reducing power reached a maximum at 50%, followed by the quick decrease in both flavonoid yield and reducing power with a further increase in the ethanol solution concentration. This may be because different ethanol concentrations have different polarities. The lower ethanol concentration yielded stronger polarity of the solution, which favored the extraction of polar

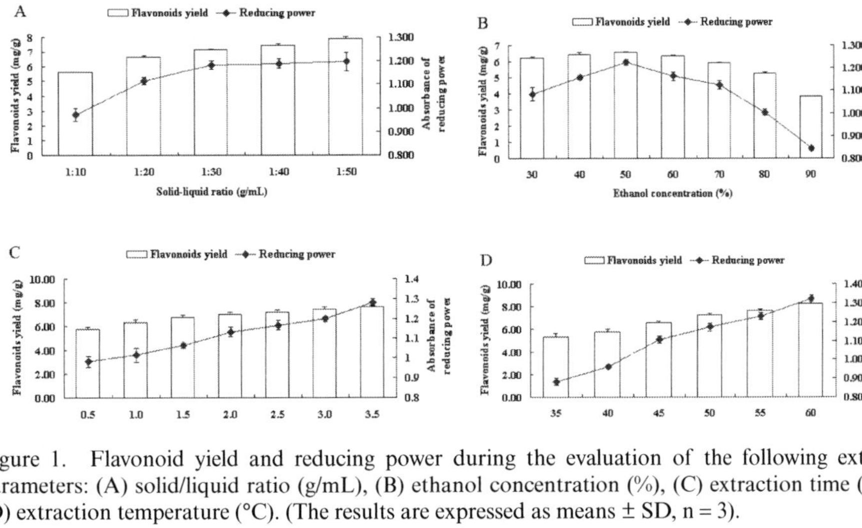

Figure 1. Flavonoid yield and reducing power during the evaluation of the following extraction parameters: (A) solid/liquid ratio (g/mL), (B) ethanol concentration (%), (C) extraction time (h), and (D) extraction temperature (°C). (The results are expressed as means ± SD, n = 3).

compounds. On the contrary, higher ethanol concentration was suitable for slightly polar compounds. Hence, the ethanol concentration of 50% was determined as the best solvent in the single-factor experiment.

The effect of extraction time on flavonoid yield and reducing power was analyzed, and the results are shown in Fig. 1C. It can be seen from Fig. 1C that flavonoid yield and reducing power remarkably increased with increasing extraction time. Then, the further increase in extraction time led to a slight increase in both flavonoid yield and reducing power. This is because of limited flavonoids in a quantitative powder. Considering the cost of time, 2 h was determined as the best extraction time in the single-factor experiment.

The effect of extraction temperature on flavonoid yield and reducing power was analyzed, and the results are shown in Fig. 1D. It can be seen from Fig. 1D that flavonoid yield and reducing power increased with the increasing extraction temperature. This may be due to higher extraction temperature that results in the increasing speed and quantity of flavonoid dissolution. However, an exorbitant temperature may result in the degradation or isomerization of bioactivity components. Therefore, 55°C was determined as the extraction temperature in the BBD test.

On the whole, in the single-factor experiment, different factors had different effects on flavonoid yield and reducing power. The solid/liquid ratio, ethanol concentration, and extraction time were chosen for further optimization in the BBD test.

3.2 *Analysis of BBD results*

3.2.1 *Model fitting*

According to the values obtained in the single-factor experiment, three factors were subjected to the BBD with RSM. The experiments were randomized and the results are summarized in Table 2.

The experimental data were analyzed using the Minitab 15.0 software for statistical analysis of variance (ANOVA), regression coefficients, and regression equation. The significance of each coefficient was determined using the *t* test and p-value and the result are given in Table 3. The following second-order polynomial stepwise equation is obtained as follows:

$$\begin{aligned} Y = 9.14642 + 0.385542X_1 + 0.350966X_2 + 0.377379X_3 - 0.572011X_1^2 \\ - 1.06189X_2^2 - 0.132747X_3^2 - 0.308851X_1X_2 + 0.0140868X_1X_3 \\ - 0.0575388X_2X_3 \end{aligned} \quad (5)$$

Table 2. Arrangement and experimental results of the Box-Behnken Design.

Run	X_1 Solid/liquid ratio (g/mL)	X_2 Ethanol concentration (%)	X_3 Extraction time (h)	Flavonoid yield (mg/g)
1	−1	−1	0	6.213
2	1	−1	0	7.849
3	−1	1	0	7.794
4	1	1	0	8.195
5	−1	0	−1	8.015
6	1	0	−1	8.510
7	−1	0	1	8.345
8	1	0	1	8.897
9	0	−1	−1	7.098
10	0	1	−1	7.654
11	0	−1	1	8.365
12	0	1	1	8.690
13	0	0	0	9.459
14	0	0	0	9.020
15	0	0	0	8.960

Table 3. Regression coefficients of the quadratic polynomial model for the experimental results.

Term	Coefficient	St. dev.	T	P
Constant	9.14642	0.2193	41.706	0.000
X_1	0.385542	0.1343	2.871	0.035
X_2	0.350966	0.1343	2.613	0.047
X_3	0.377379	0.1343	2.810	0.038
X_1^2	−0.572011	0.1977	−2.894	0.034
X_2^2	−1.06189	0.1977	−5.372	0.003
X_3^2	−0.132747	0.1977	−0.672	0.532
$X_1 \cdot X_2$	−0.308851	0.1899	−1.626	0.165
$X_1 \cdot X_3$	0.0140868	0.1899	0.074	0.944
$X_2 \cdot X_3$	−0.0575388	0.1899	−0.303	0.774

$R^2 = 92.4\%$.

Table 4. Analysis of variance (ANOVA) for the quadratic polynomial model.

Source	DF	Seq. SS	Adj. SS	F	P
Regression	9	8.7669	0.97410	6.75	0.024
Linear	3	3.3139	1.10462	7.66	0.026
Square	3	5.0575	1.68582	11.68	0.011
Interaction	3	0.3956	0.13186	0.91	0.497
Residual error	5	0.7215	0.14429		
Lack-of-fit	3	0.5732	0.19105	2.58	0.292
Pure error	2	0.1483	0.07414		
Total	14				

As shown in Table 3, the coefficient of determination (R^2) was the proportion of variability in the data explained or accounted for by the model. Therefore, the R^2 value of 92.4% was desirable (Baş and Boyacı, 2007).

The linear variables X_1, X_2, X_3 and the quadratic variables X_1^2 had significant influences ($P < 0.05$) on flavonoid yield. The quadratic variables X_2^2 were statistically highly significant ($P < 0.01$). The quadratic variables X_3^2 and two-variable interaction X_1X_2, X_2X_3 and X_1X_3 had no significant influence ($P > 0.05$) on flavonoid yield. By observing the linear and quadratic coefficients, the order of factors influencing the response value of the flavonoid yield was determined as follows: solid/liquid ratio > extraction time > ethanol concentration.

ANOVA was used to analyze the model for significance and suitability, and the results are given in Table 4.

As shown in Table 4, the P-value of the model was 0.024 ($p < 0.05$), indicating the significance of the model. The P-value of the lack-of-fit was 0.722 ($p > 0.05$), indicating no significance. The lack-of-fit was bad, so the model was significant (Khuri and Mukhopadhyay, 2010). These results indicated that the build model could describe the relationship between the independent coded variables and the response value, which was desirable.

3.2.2 *Response surface analysis*

Three-Dimensional (3D) response surfaces and Two-Dimensional (2D) isograms are the graphical representations of regression function. The response kyrtograph and isograms are shown in Fig. 2 and Fig. 3 for independent variables (solid/liquid ratio, ethanol concentration and extraction time). They were obtained by keeping two of the variables constant, which indicated the changes in flavonoid yield under different conditions.

The combined effect of the solid/liquid ratio and ethanol concentration on flavonoid yield is shown in Fig. 2A and Fig. 3A. It can be seen from Fig. 2A and Fig. 3A that at low and high levels of the ultrasonic power and ultrasonic time the flavonoid yield was minimum. At a certain solid/liquid ratio, the flavonoid yield increased with the increasing of ethanol

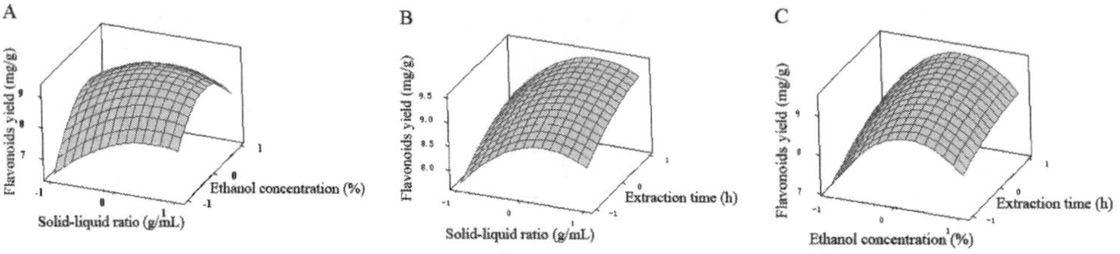

Figure 2. Response surface plot showing interactive effects of independent variables on flavonoid yield: (A) at varying solid/liquid ratio and ethanol concentration, (B) varying solid/liquid ratio and extraction time, and (C) varying ethanol concentration and extraction time.

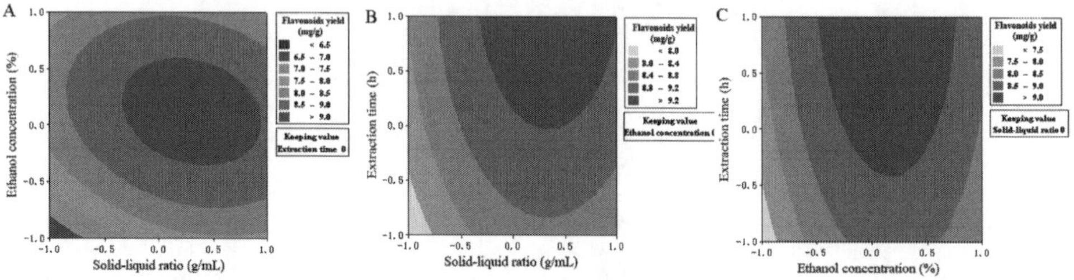

Figure 3. Contour plot showing interactive effects of independent variables on flavonoid yield: (A) at varying solid/liquid ratio and ethanol concentration, (B) varying solid/liquid ratio and extraction time, and (C) varying ethanol concentration and extraction time.

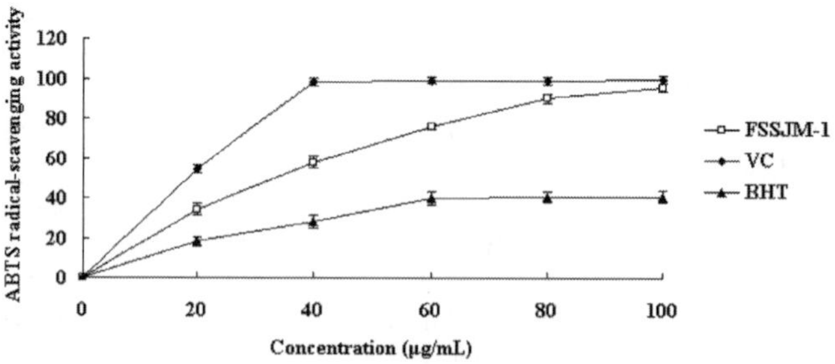

Figure 4. ABTS radical-scavenging activity of FSSJM-1.

concentration. However, the increase in ethanol concentration did not significantly affect the flavonoid yield at a certain solid/liquid ratio. The combined effect of the solid/liquid ratio and extraction time was not significant, as shown in Fig. 2B and Fig. 3B, as well as the combined effect of ethanol concentration and extraction time, as shown in Fig. 3C and Fig. 4C.

The response surface indicated that the flavonoid yield showed remarkable increases depending on the solid/liquid ratio, ethanol concentration, and extraction time. Hence, this was in good agreement with our findings using the ANOVA.

3.2.3 *Optimization and verification*

To obtain the highest flavonoid yield from the seed shell of *Juglans mandshurica*, the following optimal conditions were determined by computation: solid/liquid ratio 1:23.86 (g/mL); ethanol concentration 57.02%; and extraction time 2.19 h. The flavonoid yield of the model was predicted as 9.41 mg/g. For the practical application, the extraction conditions were adjusted as follows: solid/liquid ratio 1:24 (g/mL); ethanol concentration 57%; and extraction time 131 min. To further test the reliability of the experimental method, verification experiments were carried out three times under these optimal conditions. The resulting mean flavonoid yield was 9.27 ± 0.04 mg/g with a Relative Standard Deviation (RSD) of 1.83%. The result indicated that the experimental values were consistent with the predictive values. Therefore, the extraction conditions obtained by RSM were effective, reliable, and feasible.

3.3 *Antioxidant activity analysis of FSSJM-1*

Some reports have shown that flavonoids exhibited extensive free radical-scavenging activities through their reactivity as hydrogen or electron-donating agents, and metal ion chelating properties (Erlund, 2004; Fraga et al., 1987; Heim et al., 2002; Pham-Huy et al., 2008). The crude flavonoids obtained under the optimized parameters were purified using HPD-600 macroporous resin, which was labeled as FSSJM-1. The antioxidant activity of FSSJM-1 was analyzed by the ABTS and DPPH radical-scavenging test.

3.4 *Antioxidant activity analysis of FSSJM-1*

3.4.1 *ABTS radical-scavenging activity*

ABTS radical is often used to evaluate the total antioxidant activity of natural products (Kim and Son, 2011). The scavenging activity of FSSJM-1 on $ABTS^{\cdot +}$ was investigated and the results are shown in Fig. 4.

As shown in Fig. 4, the scavenging activity increased with the increasing FSSJM-1 concentration increasing. However, FSSJM-1 exhibited lower ABTS radical-scavenging activity compared with VC. The scavenging percentage of VC reached 98.22% at the concentration of 40 μg/mL. However, FSSJM-1 exhibited much higher ABTS radical-scavenging activity than BHT. At the concentration of 100 μg/mL, the scavenging percentage of FSSJM-1 on $ABTS^{\cdot +}$ was 96.26%, but that of BHT was only 40.97%. The result indicated that TFSSJM had a significant ABTS radical-scavenging activity.

3.4.2 *DPPH radical-scavenging activity*

DPPH radical scavenging is considered a good *in vitro* model that is widely used to assess antioxidant efficacy (the antioxidant from Njavara rice bran). The scavenging activity of FSSJM-1 on the DPPH radical was investigated and the results are shown in Fig. 5.

As shown in Fig. 5, the scavenging capability increased with the increasing FSSJM-1 concentration. However, FSSJM-1 exhibited lower DPPH radical-scavenging activity compared with VC. The scavenging percentage of VC reached 80% at the concentration of 1.5 mg/mL. However, FSSJM-1 exhibited higher DPPH radical-scavenging activity compared with BHT. At the concentration of 2.5 mg/mL, the scavenging percentage of FSSJM-1 on the DPPH radical was 67.82%, but that of BHT was only 14.11%. These results indicated that FSSJM-1 had a significant DPPH radical-scavenging activity.

3.4.3 *Hydroxyl radical-scavenging activity*

Hydroxyl radicals are considered responsible for causing the aging of the human body and some diseases (Siddhuraju and Becker, 2007), interacting with the purine and pyrimidine bases of DNA, as well as leading to the formation of sulfur radicals that are able to combine

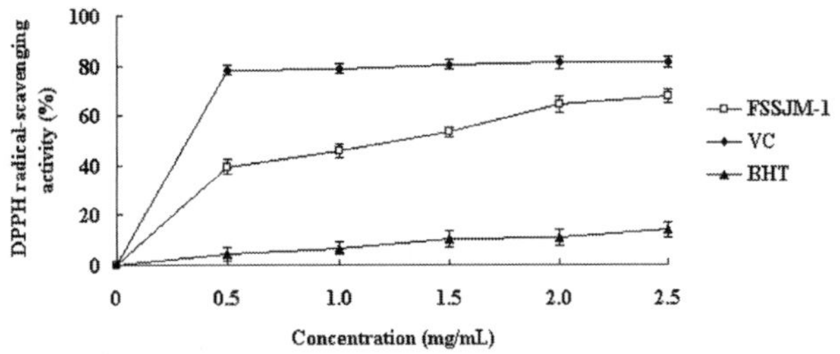

Figure 5. DPPH radical-scavenging activity of TFSSJM-1.

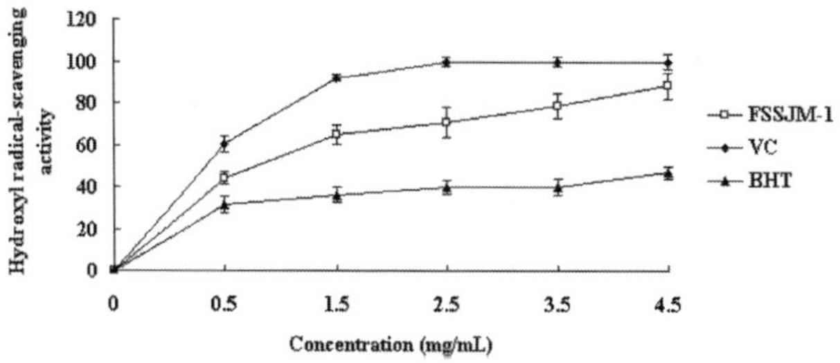

Figure 6. Hydroxyl radical-scavenging activity of TFSSJM.

with oxygen to generate oxysulfur radicals, which can damage biological molecules (Huang et al., 2009). The scavenging activity of FSSJM-1 on the hydroxyl radical was investigated, and the results are shown in Fig. 6.

As shown in Fig. 6, the scavenging capability increased with the increasing TFSSJM concentration. Its IC_{50} value was 0.622 mg/mL. However, TFSSJM exhibited lower hydroxyl radical-scavenging activity compared with VC. The scavenging percentage of VC reached 99.87% at the concentration of 2.5 mg/mL. However, TFSSJM exhibited higher hydroxyl radical-scavenging activity compared with BHT. At the concentration of 4.5 mg/mL, the scavenging percentage of TFSSJM on the hydroxyl radical was 88.21%, but that of BHT was only 46.97%. These results indicated that TFSSJM had a significant hydroxyl radical-scavenging activity.

The above data demonstrated that FSSJM-1 could scavenge ABTS, DPPH, and hydroxyl radicals. Compared with VC and BHT, the antioxidant activity of FSSJM-1 fell behind VC, but was better than BHT.

The results studied will possibly arouse the interest in using flavonoids from the seed shell of *Juglans mandshurica* as an inexpensive source of health-promoting additives, e.g. natural antioxidants. Thus, flavonoids from the seed shell of *Juglans mandshurica* may be beneficial to the antioxidant protection system in the food industry, even in the human body against oxidative damage. Further study is in progress to identify the components of flavonoids from the seed shell of *Juglans mandshurica* and elucidate their antioxidant mechanisms.

4 CONCLUSIONS

In this study, extraction conditions of flavonoids from the seed shell of *Juglans mandshurica* was investigated with a three-variable, three-level experiment Box-Behnken Design based on response surface methodology to enhance the flavonoid yield. The optimal parameters obtained were as follows: solid/liquid ratio 1:24 (g/mL); ethanol concentration 57%; and extraction time 131 min. In addition, the antioxidant activity of flavonoids from the seed shell of *Juglans mandshurica* was evaluated. The results indicated that it could scavenge ABTS, DPPH, and hydroxyl radicals, and showed significant antioxidant activity.

ACKNOWLEDGMENTS

This research was supported by the grant from the Key Scientific and Technological Project of Science and Technology Department of Jilin Province (20150204063NY).

REFERENCES

Bagchi, D., Bagchi, M., Stohs, S.J., Das, D.K., Ray, S.D., Kuszynski, C.A., Joshi, S.S., Pruess, H.G. 2000. Free radicals and grape seed proanthocyanidin extract: Importance in human health and disease prevention. *Toxicology* 148, 187–197.

Baş, D., Boyacı, İ.H. 2007. Modeling and optimization I: Usability of response surface methodology. *J. Food Eng.* 78, 836–845.
Duffy, K.B., Spangler, E.L., Devan, B.D., Guo, Z., Bowker, J.L., Janas, A.M., Hagepanos, A., Minor, R.K., DeCabo, R., Mouton, P.R. 2008. A blueberry-enriched diet provides cellular protection against oxidative stress and reduces a kainate-induced learning impairment in rats. *Neurobiol. Aging* 29, 1680–1689.
Erlund, I. 2004. Review of the flavonoids quercetin, hesperetin, and naringenin. Dietary sources, bioactivities, bioavailability, and epidemiology. *Nutr. Res.* 24, 851–874.
Fraga, C.G., Martino, V.S., Ferraro, G.E., Coussio, J.D., Boveris, A. 1987. Flavonoids as antioxidants evaluated by *in vitro* and *in situ* liver chemiluminescence. *Biochem. Pharmacol.* 36, 717–720.
Gong, Y., Hou, Z., Gao, Y., Xue, Y., Liu, X., Liu, G. 2012. Optimization of extraction parameters of bioactive components from defatted marigold (*Tagetes erecta* L.) residue using response surface methodology. *Food Bioprod. Process.* 90, 9–16.
Graham, J.G., Quinn, M.L., Fabricant, D.S., Farnsworth, N.R. 2000. Plants used against cancer—an extension of the work of Jonathan Hartwell. *J. Ethnopharmacol.* 73, 347–377.
Guangyan Pan, G.Y., Chuanhe Zhu, Julin Qiao 2012. Optimization of ultrasound-assisted extraction (UAE) of flavonoids compounds (FC) from hawthorn seed (HS). *Ultrason. Sonochem.* 19, 486–490.
Gülsen Tel, M.A., Mehmet E. Duru, Mehmet Öztürk 2012. Antioxidant and Cholinesterase Inhibition Activities of Three Tricholoma Species with Total Phenolic and Flavonoid Contents: The Edible Mushrooms from Anatolia. *Food Anal. Methods* 5, 495–504.
Heim, K.E., Tagliaferro, A.R., Bobilya, D.J. 2002. Flavonoid antioxidants: chemistry, metabolism and structure-activity relationships. *J. Nutr. Biochem.* 13, 572–584.
Hodek, P., Trefil, P., Stiborová, M. 2002. Flavonoids-potent and versatile biologically active compounds interacting with cytochromes P450. *Chem. Biol. Interact.* 139, 1–21.
Huang, W., Xue, A., Niu, H., Jia, Z., Wang, J. 2009. Optimised ultrasonic-assisted extraction of flavonoids from Folium eucommiae and evaluation of antioxidant activity in multi-test systems *in vitro*. *Food Chem.* 114, 1147–1154.
Hughes, D.A. 2008. Dietary antioxidants and human immune function. *Nutrition Bulletin* 25, 35–41.
Khuri, A.I., Mukhopadhyay, S. 2010. Response surface methodology. Wiley Interdiscip. Rev: *Comput. Stat.* 2, 128–149.
Kim, Y.-J., Son, D.-Y. 2011. Antioxidant effects of solvent extracts from the dried jujube (Zizyphus jujube) sarcocarp, seed, and leaf via sonication. *Food Sci. Biotechnol.* 20, 167–173.
Lafka, T.-I., Sinanoglou, V., Lazos, E.S. 2007. On the extraction and antioxidant activity of phenolic compounds from winery wastes. *Food Chem.* 104, 1206–1214.
Li, G., Lee, S.Y., Lee, K.S., Lee, S.W., Kim, S.H., Lee, S.H., Lee, C.S., Woo, M.H., Son, J.K. 2003. DNA topoisomerases I and II inhibitory activity of constituents isolated from *Juglans mandshurica. Arch. Pharm. Res.* 26, 466–470.
Machida, K., Yogiashi, Y., Matsuda, S., Suzuki, A., Kikuchi, M. 2009. A new phenolic glycoside syringate from the bark of *Juglans mandshurica* MAXIM. var. sieboldiana MAKINO. *J. Nat. Med.* 63, 220–222.
Min, B.S., Lee, H.K., Lee, S.M., Kim, Y.H., Bae, K.H., Otake, T., Nakamura, N., Hattori, M. 2002. Anti-human immunodeficiency virus-type 1 activity of constituents from *Juglans mandshurica. Arch. Pharm. Res.* 25, 441–445.
Park, G., Jang, D.S., Oh, M.S. 2012. *Juglans mandshurica* leaf extract protects skin fibroblasts from damage by regulating the oxidative defense system. *Biochem. Biophys. Res. Commun.* 421, 343–348.
Peterson, J., Dwyer, J., 1998. Flavonoids: Dietary occurrence and biochemical activity. *Nutr. Res.* 18, 1995–2018.
Pham-Huy, L.A., He, H., Pham-Huy, C. 2008. Free radicals, antioxidants in disease and health. *Int. J. Biomed. Sci.* 4, 89–96.
Sharma, K., Ko, E.Y., Assefa, A.D., Ha, S., Nile, S.H., Lee, E.T., & Park, S.W. 2015. Temperature-dependent studies on the total phenolics, flavonoids, antioxidant activities, and sugar content in six onion varieties. *Journal of Food and Drug Analysis* 23(2), 243–252.
Siddhuraju, P., Becker, K. 2007. The antioxidant and free radical scavenging activities of processed cowpea (*Vigna unguiculata* (L.) *Walp.*) seed extracts. *Food Chem.* 101, 10–19.
Tripoli, E., Guardia, M.L., Giammanco, S., Majo, D.D., Giammanco, M. 2007. Citrus flavonoids: Molecular structure, biological activity and nutritional properties: A review. *Food Chem.* 104, 466–479.
Yin, X., You, Q., Jiang, Z. 2011. Optimization of enzyme assisted extraction of polysaccharides from Tricholoma matsutake by response surface methodology. *Carbohydr. Polym.* 86, 1358–1364.
Zhang, G., He, L., Hu, M. 2011. Optimized ultrasonic-assisted extraction of flavonoids from *Prunella vulgaris* L. and evaluation of antioxidant activities *in vitro*. In nov. *Food. Sci. Emerg. Technol.* 12, 18–25.

Miscellaneous

Advanced Engineering and Technology III – Xie (Ed.)
© 2017 Taylor & Francis Group, London, ISBN 978-1-138-03275-0

Measurement of control indicators for municipality- and county-level marine functional zoning—with Putian as an example

Fa-ming Huang
State Oceanic Administration, Third Institute of Oceanography, Xiamen, Fujian Province, China

Ju-ying Li
College of Environment and Ecology, Xiamen University, Xiamen, Fujian Province, China

ABSTRACT: As an important content of the new round of marine functional zoning, control indicators play an important role in managing the use of regional maritime space. In the present work, Putian was taken as a research example to analyze the use of maritime space over the past decade and its trends. The linear trend prediction method was used to predict the demand up to 2020 and discuss the determination method for target values of reclamation scale, mariculture area, and other main relevant control indicators, with a view to providing the reference for the research on the theories and methods of marine functional zoning.

1 INTRODUCTION

With the rapid development of marine economic industry, the contradiction between human development needs and marine space resources or marine environments has increasingly intensified, and thus the research on marine zoning or planning has become an important issue of marine management.

With a long history of overseas marine spatial zoning, many achievements have been made in this field. Traditional maritime zoning management is only aimed at a single marine species or a small range of uniquely valuable marine waters, resulting in great limitations (Agardy 2000). Since 2006, ecosystem-based marine spatial zoning has been increasingly considered to be an important way for achieving the sustainable development of ocean and coastal zones (Taussik 2007). Pikitch, *et al.* believed that marine zoning was necessary for the management of marine fisheries, and might effectively protect fish habitats and environmental factors by specifying the type and extent of human activity in time and space, so that the abundance and diversity of fishery resources could be improved (Pikitch *et al.* 2004, Roberts *et al.* 2001). Douvere, *et al.* elaborated the advantages and practical significance of marine spatial zoning for marine management through comparing experiences from Germany, Belgium, the Netherlands, etc., and recommended inter-country marine spatial zoning from the concept of large area (Douvere & Ehler 2009). Gilliland believed that marine spatial zoning need to focus on management objectives, marine usage priorities, stakeholder involvement, and other key factors (Gilliland & Dan 2008). Based on the difference in ecological factors, Day used the GIS method to divide the South Australian waters into four levels and set the corresponding management objectives (Day *et al.* 2008). Overall, overseas marine spatial zoning mainly takes into account the nature of the marine environment, and thus constitutes a constraint of marine usage for humans through the formation of management tools.

Furthermore, marine functional zoning, a national and staged proposition presented by China by drawing on Western spatial zoning thought, needs to take into account the natural and social properties of maritime space in the implementation process, in order to play the role of macro-control in marine management with regard to the reasonable arrangements for various types of marine usage in various sectors. In the process of continuous practice, the study of the marine functional zoning system has gradually deepened. For zoning system and theory, LUO Mei-xue, *et al.* made recommendations on optimizing the indicator system, balancing protection, and development, etc. through province-level zoning preparation (Luo 2010). LIU Shu-fen believed that increasing controlling indicators will play the role of enhanced regulation after analyzing all aspects of the control system (Liu *et al.* 2014). GUO Pei-fang and YANG Shun-liang proposed the conflicts between marine functional zoning and marine versatility or social needs multiplicity as well as the differences between the rigid boundary of marine functional zoning and the arrangement of actual marine usage, and made recommendations (Guo 2009, Yang 2008). As confirmatory research, LI Jin, *et al.* expressed the spatial compliance of province- and municipality-level marine functional zoning using a functional equation and modeled for quantitative evaluation (Li *et al.* 2009). Taking Guangxi waters as an example and using the average synthetic model, LIU Yang carried out research and verification upon the implementation of marine functional zoning, and thus concluded that marine functional zoning was well implemented (Liu *et al.* 2009). For the research on suitability, reclamation as a functional zoning type that is most destructive for marine natural attributes is the main object of study. After comprehensive consideration of marine ecology, exploitation and utilization situation, natural conditions, social economy, and other factors, YU Yong-hai, *et al.* selected an indicator evaluation system to evaluate the suitability of reclamation in Liaoning Province (Yu *et al.* 2011). From three aspects including marine functional zoning control, total reclamation area trend forecasting, and total reclamation demand, WANG Jiang-tao probed into the method for determining the control indicators of total reclamation area (Wang *et al.* 2010). Based on statistical analysis, FU Yuan-bin proposed the methods for grading reclamation intensity and assessing reclamation potential (Fu *et al.* 2010). The methods related to reclamation control are of significance reference to determining the control indicators.

The preparation for the marine functional zoning currently focuses on delineating functional boundaries and determining the basic functions of maritime space and control requirements. There are no clear and uniform methods for determining the target values of controlling indicators. Great subjectivity usually uses the macro values obtained from relevant planning as the basis. The analysis of actual local marine usage situation and marine usage development trend is insufficient. A theoretical basis for determining target values is lacking. In the present work, Putian was taken as a research example to analyze the use of maritime space over the past decade and its trends, reasonably analyze and predict the future marine usage demand, and discuss the determination method for target values of reclamation scale, mariculture area, and other main relevant control indicators.

2 RESEARCH REGIONS AND METHODS

2.1 *Overview of research regions*

For marine development, Putian focuses on the development of marine economy by taking the 'three bays and two islands' as the main development pattern and using the main dominant resources in various maritime spaces. Port logistics and portside industries are mainly arranged in Xinghua Bay and on the north shore of Meizhou Bay. Mariculture and seed breeding are mainly arranged in Pinghai Bay and Nanri Island. Coastal tourism places are based on Meizhou Island, Nanri Island, and other islands. In addition, marine energy and other emerging marine industries are important development orientations.

According to the data analysis of marine usage with usufruct in Putian (Table 1), since the implementation of maritime space utilization hierarchy in 2002, the marine usage in Putian has achieved a total area of about 30,241.68 hm^2, as of the end of December 2014.

As an important pillar industry in Putian, marine fishery accounts for the most part, i.e., 84.94% of the total marine usage area; therefore, the reclamation aquaculture accounts for about 6.45% of the total marine fishery area. With the increasing emphasis on the development of agricultural product deep processing and ecological recreational fisheries

Table 1. Data analysis of marine usage with usufruct in Putian.

Usage type	Level	Number	Area (hm^2)	Total (hm^2)	Proportion
Fishery	Province level	3	700.61	25,686.68	84.94%
	Municipality level	10	1,934.49		
	County or district level	506	23,051.58		
Industry	Province level	43	1,263.87	2,967.46	9.81%
	County or district level	67	1,703.58		
Transport	National level	1	213.01	1,257.74	4.16%
	Province level	57	1,032.64		
	Municipality level	4	12.09		
Submarine engineering	Municipality level	8	329.81	329.81	1.09%
Total (hm^2)	30,241.68				

in Putian, the increasing trend of marine fishery is expected to be slow. This is followed by marine usage for industry and transport, accounting for 9.81% and 4.16%, respectively. A total of 1703.58 hm^2 of marine usage for industry is used for the salt industry, and the rest are mostly involved in the construction of portside industrial park in Meizhou Bay. With the construction and development of portside industry, the demand for reclamation will expand in the future.

2.2 *Research methods*

During this round of marine functional zoning preparation, the main protection goals for the exploitation of marine resources up to 2020 need to be scientifically and rationally made, based on the development orientation for economical and intensive utilization of marine resources in the region, the demand guarantee of fishermen's marine usage, and the sustainable development of marine areas. Mariculture functional zone and reclamation area are important controlling indicators, as well as important ways to use maritime spaces in Putian. Thus, the prediction and determination of their target values is of great significance to the future development and management of marine fisheries in Putian. The main goal of prediction is to presumably determine the future status of a thing based on its past and present characteristics. Generally, prediction methods include subjective prediction and objective prediction. The former relies mainly on the predictor's experience to judge, so the predicted result is subjective. The latter mainly depends on the aid of a mathematical model, which is independent of the predictor's judgment, so the predicted result is more reliable (Liu 2005). In the present work, the prediction on trend for aquaculture, reclamation and other maritime spaces may be estimated by employing a mathematical statistical method in accordance with the trend of change in the corresponding marine usage size during the past period of time, in order to meet the requirements of coordination with marine usage needs and realize the target control.

Many mathematical models and methods have been used for prediction. The linear trend prediction method was mainly employed for the present study.

Prediction model: Linear model:

$$Y_t = a + bx \tag{1}$$

Linear-quadratic model:

$$Y_t = a + bx + cx^2 \tag{2}$$

where x is the time series; Y_t is the predicted value; and a, b, c are parameters.

3 RESULTS

3.1 *Necessity of controlling indicators*

As an important tool to ensure effective implementation of planning and intensive management, indicator control plays an important role in the planning system (Wang & Liu 2011). The controlling indicators for this round of marine functional zoning are used to carry out the total quantity control and the minimum retention rate control on the different types of marine usage, which is of positive significance to the reasonable coordination of all types of marine usage, the implementation of the 'five marine usages', and the formation of healthy and coordinated marine spatial development pattern.

As a marine usage way is greatly changing the natural properties of the sea and affecting marine ecosystems, reclamation may cause irreversible damage to hydrodynamic conditions of waters, coastal topography, coastal vegetation, aquatic habitat, etc. Statistical data show that the accumulated coastal wetland loss caused by national large-scale land reclamation has reached 50% in about 40 years (Guan 2012). Nevertheless, as the main way to ease the contradiction between supply and demand of land in coastal areas, the reclamation demand will continue to be strong under the marine economic trend. Thus, the implementation of indicator control for construction reclamation scale helps to slow the growth rate of reclamation, effectively balance marine economic development and marine environmental protection, avoid blind reclamation, waste and over-exploitation of resources, etc., in order to achieve sustainable exploitation of marine resources.

Food is one of the world's three major issues, and obtaining the protein from the ocean is an important way to solve the current human food sources. However, because of over-exploitation, the world's fish resources have almost been exhausted; therefore, mariculture has become an essential part of marine economic structure (Wang 2007). In recent years, in pursuit of rapid development, some areas have tended to develop the industry, so mariculture has been neglected to varying degrees, impeding the healthy development of marine fishery. Therefore, the implementation of indicator management for the functional zone of mariculture helps to ensure the spatial occupancy ratio of marine usage for fishery and reasonable marine usage demand, leading to the healthy development of marine fishery.

There are two main types of MPAs (Marine Protected Areas): marine nature reserves and special marine reserves. The former is to protect rare, endangered, and economic species and their habitats, as well as the marine natural objects that are of great scientific, cultural, scenic, or ecological value. The latter is to protect the maritime space with special geographic conditions, ecosystem resources, and special marine exploitation needs (State Oceanic Administration 2013). The implementation of indicator management for the area of protected marine functional zone helps to guarantee the systematic construction of protected zones, avoid the destruction and consumption of critical habitats and resources, and protect the integrity and biodiversity of ecosystems.

Marine development focuses on coastal maritime space and coastline resources. The utilization of marine resources has gradually become saturated. The marine resources reserve for the development is scarce. To protect the sustainable utilization of marine resources, the reasonable control of the minimum retention rate for reserved areas, natural coastline and regulated and repaired coastline length, and other indicators is necessary to appropriately retain marine spatial resources and natural coastline resources.

In the marine functional zoning preparation process, all the controlling indicators, except for reclamation controlling indicator, will be further shifted from the state-level marine functional zoning to the province-, municipality- and county-level marine functional zoning. The gradual shifting of controlling indicators helps to effectively intensify the efforts of marine functional zoning on the management and control of marine development and protection, and guide the actual operation of zoning management. Thus, it plays an important role in reasonably controlling the marine usage scale and intensity and adjusting the marine usage structure proportion and way.

3.2 *Prediction on the area of functional zone for mariculture*

From the analysis of 2003–2013 historical data for fishery development and aquaculture area in Putian (Putian Municipal Statistics Bureau 2004–2014), it can be seen that the mariculture area increased greatly from 2004 to 2006, declined significantly in 2007 and afterwards increased at lower growth rates, showing a trend of year-on-year growth. Mariculture production has also continued to grow, but specific yield has remained at a stable level, fluctuating at about 29t/hm^2.

The regression function model will be established according to the time-dependent trend of statistical data for mariculture area in the last decade. Let $x_{2003} = 1$, and the rest can be done in the same manner according to the time series. Thus, we obtain a linear regression equation as follows:

$$Y = 403x + 16241.182 \quad (3)$$

The coefficient of determination (R^2) is 0.877. The curve fits well the actual value. According to the above result, it can be predicted that the mariculture area will reach 23,495.182 hm^2 by 2020 ($x = 18$).

The agriculture and fishery zones determined within the scope of marine functional zoning in Putian mainly include Houhai agriculture and fishery zone, Shicheng agriculture and fishery zone, Pinghai Bay agriculture and fishery zone, and Nanri Island agriculture and fishery zone, with a total area of 36,937.06 hm^2. Agriculture and fishery zone means the maritime space adapted to expand agricultural development space and exploit marine biological resources and available for agricultural land reclamation, fishing port and hatchery and other fishery infrastructure construction, mariculture and fishing production, as well as important fishery species conservation (State Oceanic Administration 2013), which will include but not be limited to the maritime space for mariculture. The area of target retaining maritime space for mariculture will allow enough room for the future development of fishery. The upper limit of prediction will meet the required area under the policy of developing aquaculture, and the lower limit of prediction will also ensure the required mariculture area at the current production level. Based on the mariculture area determined through the quantitative prediction, the area of maritime space for mariculture in 2020 will be approximately 24,000 hm^2, accounting for 64.98% of the total area of agriculture and fishery zones in Putian.

3.3 *Prediction on total reclamation area*

Historical data analysis: In 2002, the statistical system for the use of maritime spaces was established in China, which provides a basis for the method of predicting the future trend of total reclamation area using a mathematical method. By analyzing the land reclamation area with usufruct in Putian during 2005 to 2014, we can see the growth trend and law of land reclamation. Fig. 3 shows the general trend of year-on-year growth for the total land reclamation area with usufruct in Putian. Before 2010, the land reclamation area with usufruct increased greatly every year, substantially by more than 30%. In 2010, the growth was particularly evident with an increase of up to about 55%, and during 2011 to 2014, the land reclamation area entered a relatively stable period of growth, basically at the growth rate of 10% to 15%.

The maximum number and area of land reclamation projects with usufruct were reached in the year 2010. The analysis shows the following main reasons: first, in May 2009, the State Council issued the *Opinions on Supporting Fujian Province to Speed up the Construction of the Economic Zone on the West Side of the Straits*, which has brought strategic development opportunities to the harbor shipping, port-side industry, and other aspects in Fujian Province; second, the new Meizhou Bay Port was established in August 2009, which integrated Xiaocuo Port and Douwei Port under the jurisdiction of Quanzhou Port Authority, as well as Xiuyu Port and Dongwu Port under the jurisdiction of Putian Port Authority, and was planned as one of the three-hundred-million-tonnage ports prioritized in Fujian Province. At that time, the state's efforts to control the scale of reclamation have not been intensified. In this situation, the efforts on the development and construction of coastal areas, especially the south bank of Meizhou Bay, were enhanced, showing strong demand for reclamation.

Reclamation potential assessment: Reclamation intensity generally refers to the ratio of reclamation area to total marine usage area of the Bay. Reclamation is substantially a coastal project, so the development intensity for land reclamation is reflected using the reclamation

Table 2. Intensity grades of reclamation (Fu et al. 2010).

Intensity level	R-value/($hm^2 \cdot km^{-1}$)	Connotation of indicator
I	0~10	Great potential for development, slight stress of reclamation
II	10~20	A certain potential for development, less stress of reclamation
III	20~50	A certain influence on the subsequent development, a certain stress of reclamation
IV	50~100	Intensive and economical use of reclamation, greater stress of reclamation
V	R≥100	Inappropriate addition of reclamation, but if necessary, backfilling on the basis of the existing reclamation, strong stress of reclamation

area (hm^2) per unit length of coastline (km), denoted by R, and a different intensity level is given for the corresponding connotation of indicator (Table 2). When the reclamation area per kilometer coastline is 100 hm^2, it will be deemed that he health of coastal resources and the environment is at a critical point. On the one hand, it is shown that the stress of reclamation on the ocean exceeds the limit of the ecosystem in the Bay in a good condition; on the other hand, it is shown that the effects of reclamation on the health of coastal ecosystems are not significant, and the stress is within the acceptable range.

Based on the concept of carrying capacity of the marine environment, reclamation potential can be understood as the greatest reclamation area that the maritime space can bear in a specific area within a period of time, under the condition that the ecosystem in the maritime space is in a virtuous circle and the marine functional structure is essentially unchanged (Liu & Feng 2011). According to the reclamation intensity grading standard, reclamation potential can be calculated as follows:

$$P = \begin{cases} S_{\text{III}} - S & \text{(when R value is between 0~20)} \\ S_N - S & \text{(when R value is between 20~100)} \\ 0 & \text{(when R value is greater than 100)} \end{cases} \tag{4}$$

where P is the reclamation potential, S_{III} is the maximum reclamation scale within the region at the critical level, S_N is the maximum reclamation scale within the region at the current intensity level, and S is the existing reclamation scale within the region.

The length of the mainland coastline in Putian is 336 km. The calculation for reclamation intensity during 2005 to 2014 is shown in Fig. 4. At the end of 2014, the reclamation area in Putian reached 1888.88 hm^2 and the reclamation intensity value R was 5.62, belonging to the intensity level of I. The reclamation development potential P was 14,911.12 hm^2. According to the current growth rate of reclamation in Putian, it is predicted that reclamation will reach a critical health limit (i.e. in case of giving full play to reclamation potential) within 29 years.

Reclamation prediction: Regression analysis is conducted according to the historical data for reclamation area with usufruct in Putian. Time series is taken as the independent variable. Let $x_{2005} = 1$, and the linear equation for the trend of reclamation area can be obtained. Comparison of the quadratic equation with the linear equation showed that the one with greater goodness-of-fit is the quadratic curve ($R^2 = 0.981$). Thus, the linear equation for trend can be calculated as follows:

$$Y = 1.843 + 137.711x + 5.786x^2 \tag{5}$$

It is predicted that the total reclamation area will reach 3,686.435 hm^2 by 2020 ($x = 16$). Therefore, it is recommended that the increase in the total reclamation area should not exceed 3,600 hm^2 by 2020.

4 DISCUSSION

Ocean space is limited, so the exploitation of the ocean cannot always keep the volume growth. The development of traditional aquaculture in Putian has gradually become dominant, but the growth has been mainly associated with the aquaculture area in the recent ten years. So, strategic restructuring needs to be carried out in fishery, in order to develop new models of sustainable fisheries. The control of reclamation will be combined with the reduction of occupied natural coastline, the optimization of the layout of artificial coastline, and other measures, in order to improve the efficiency in the use of maritime spaces and achieve the intensive and economical use of the sea.

In the present work, through the analysis of historical data, the mariculture area and the reclamation area in Putian up to 2020 was predicted, in order to reasonably determine target values of the corresponding control indicators for marine functional zoning in Putian. Reclamation control indicator need not be shifted to municipality- and county-level marine functional zoning, but a reasonable predictive value can be used as a reference for the implementation and management of zoning. In addition, the prediction method can also provide a reference for the determination of controlling indicators for province-level zoning.

The prediction on controlling indicators is mainly based on the trend and law of many years of marine usage data, but error still persists for the area with a large degree of annual variation. The target values of final controlling indicators mainly reflect the potential marine usage demand of economic and social development during the life of marine functional zoning, which needs to expand to some extent. The implementation of management of functional zoning and maritime space usage will take into account marine usage timing and necessity, in order to remove any unnecessary marine usage demand. For the approved reclamation project, group layout will be proposed, and artificial coastline tortuosity and other recommendations will be added, in order to minimize the occupation of natural coastline and the harm to the marine function and ecological environment.

REFERENCES

Agardy, T. 2000. Information needs for marine protected areas: scientific and societal. *Bulletin of Marine Science* 66 (3): 875–888.

Day, V. et al. 2014. The Marine Planning Framework for South Australia: A new ecosystem-based zoning policy for marine management. *Marine Policy* 32 (4): 535–543.

Douvere, F. & Ehler, C.N. 2009. New perspectives on marine usage management: Initial findings from European experience with marine spatial planning. *Journal of Environmental Management* 90 (90): 77–88.

Fu, Y.B. et al. 2010. Quantitative evaluation method for reclamation intensity and potential. *Ocean Development and Management* 27 (1): 27–30.

Gilliland, P.M. & Dan, L. 2008. Key elements and steps in the process of developing ecosystem-based marine spatial planning. *Marine Policy* 32 (5): 787–796.

Guan, D.M. (ed.) 2012. *China Coastal Wetlands*. Beijing: Ocean Press.

Guo, P.F. 2009. Contradictions and changes in marine functional zoning. *Ocean Development and Management* 26 (5): 26–30.

Li, J. et al. 2009. Analysis of compliance between municipality- and province-level marine functional zoning spaces. *Marine Science Bulletin* 28 (5): 1–6.

Liu, S.F. (ed.) 2005. *Forecasting Methods and Techniques*. Beijing: Higher Education Press.

Liu, S.F. et al. 2014. Marine functional zoning control system. *Marine Environmental Science* 33 (3): 455–458.

Liu, Y. et al. 2009. Evaluation method and empirical study for implementation of marine function zoning. *Ocean Development and Management* 26 (2): 12–17.

Liu, Y. & Feng, A.P. 2011. A method for prediction and analysis of regional reclamation area demand. *Chinese Fisheries Economics* 29 (6): 92–97.

Luo, M.X. 2010. Several technical methods for marine function zoning preparation in Fujian Pikitch, E. et al. 2004. Babcock E, et al. Ecosystem-based fishery management. *Science* 305 (5682): 346–347.

Province. *Journal of Oceanography* 29 (2): 290–294.
Putian Municipal Statistics Bureau 2004–2014. *Putian Statistical Yearbook*. http://www.stats-pt.gov.cn/tjnj/index.htm
Roberts, C.M. et al. 2001. Effects of marine reserves on adjacent fisheries. *Science* 294 (5548): 1920–1923.
State Oceanic Administration 2013. *Technical guidelines for municipality- and county-level marine functional zoning preparation*.
Taussik, J. 2007. The opportunities of spatial planning for integrated coastal management. *Marine Policy* 31 (5): 611–618.
Wang, J.T. et al. 2010. Determination method for controlling indicators of total reclamation area—with Tianjin for example. *Marine Technology* 29 (2): 97–100.
Wang, J.T. & Liu, B.Q. 2011. Control system for marine function zoning. *Marine Science Bulletin* 30 (4): 371–376.
Wang, Q.Y. (ed.) 2007. *Sustainable Development of Aquaculture: Challenges and Countermeasures*. Beijing: Ocean Press.
Yang, S.L. & Luo, M.X. 2014. Some problems in marine functional zoning preparation. *Ocean Development and Management* 25 (7): 12–18.
Yu, Y.H. et al. 2011. Evaluation method for suitability of reclamation. *Marine Science Bulletin* 30 (1): 81–87.

Advanced Engineering and Technology III – Xie (Ed.)

Synergetic development of film industry chain and its evaluation research

W.G. Xia
School of Management and Economics, Beijing Institute of Technology, Beijing, China
Beijing Film Academy, Beijing, China

X. Zhang
School of Management and Economics, Beijing Institute of Technology, Beijing, China

ABSTRACT: The synergetic development of film industry chain is an inevitable choice in response to external challenges such as environmental uncertainties, competition in the international market, as well as the inherent requirement to solve internal problems including conflicts of interest and information silo. Based on the analysis and discussion on the connotation of the synergetic development of film industry chain, this article put forward the double-layer network chain structure of film industry chain, built comprehensive evaluation index systems for the synergetic development of film industry chain from the core layer and the radiation layer, respectively, and carried out an evaluation research on the advantages and disadvantages of such synergetic development in China by means of the fuzzy analytic network process.

1 INTRODUCTION

Film industry is a forerunner for the development of cultural industries, and the governance of film industry chain is the core problem for the development of film industry. China's inland market grossed RMB 44.069 billion yuan at the box office in 2015 with a year-on-year growth of 48.7%, and became the second largest film market next to North America. However, many problems are exposed in the development of China's film industry. For example, there is low attendance of many films and differences in income sharing at the box office between issuing companies and theaters of some films, and barriers to capital and talent of the industry that are not yet completely broken. It is thus clear that how to integrate the elements in each link of the industry and effectively allocate resources to realize the synergetic development of film industry chain has become a theoretical and practical problem that needs to be solved urgently.

With respect to the research on the connotation of industrial chain synergy, previously, researchers put forward that industrial chain synergy is a process involving knowledge innovation and deepening and integration of labor division (Liu and Huang, 2007). Later on, domestic scholars took automobile industry and IoT as an example and analyzed the connotation and purpose of industrial chain synergy (Tan, 2010; Zhu, 2012). Research on the synergy of film industry chain has summarized the development pattern of the motion picture mega-industry value chain to externally seek industrial linkage and internally realize synergetic advantages (Qian, 2008). It has put forward the strategy for industrial upgrading from the perspective of vertical integration and horizontal integration (Mi and Yu, 2011). In addition, researchers have analyzed the approaches to industrial chain synergy from the perspective of industrial linkage such as contents, audiences, and products (Guo and Yan, 2012). Advanced research has been done on the synergetic evaluation of supply chain.

Scholars have established the synergetic frame system from different perspectives such as strategy, information, and knowledge synergy, and carried out evaluation research by using the analytic hierarchy process and the fuzzy evaluation method (Tang and Huang, 2005; Zhou and Zhao, 2008; Lu, Liu and Wang, 2013).

It is thus clear that numerous scholars have put forward the approaches to and strategies for it from different perspectives. However, there are still limitations in relevant researches: first, there is a lack of in-depth research on the connotation of the synergetic development of film industry chain and its necessity to promote the development of film industry; second, there is a lack of consideration on the synergetic development of film industry chain from the perspective of "motion picture mega-industry" and in combination with the structural evolution trend of film industry chain; third, there is a lack of evaluation research on the advantages and disadvantages of the synergetic development of China's film industry chain from the perspective of quantitative analysis. To address these shortcomings, this article carries out the research as follows.

2 SYNERGETIC DEVELOPMENT OF FILM INDUSTRY CHAIN

The scientific and reasonable development of film industry cannot be separated from the cooperation of each subject in the industrial chain. Synergetic development plays an important role in promoting the overall operation of film industry chain. From our viewpoint, the synergetic development of film industry chain comprises core parts such as production, distribution, allocation of elements, and communication on film products, and external connections of the chain, which are linked and integrated on the industrial interface around film derivatives to jointly promote the efficient, orderly, and open development of the whole film industry chain.

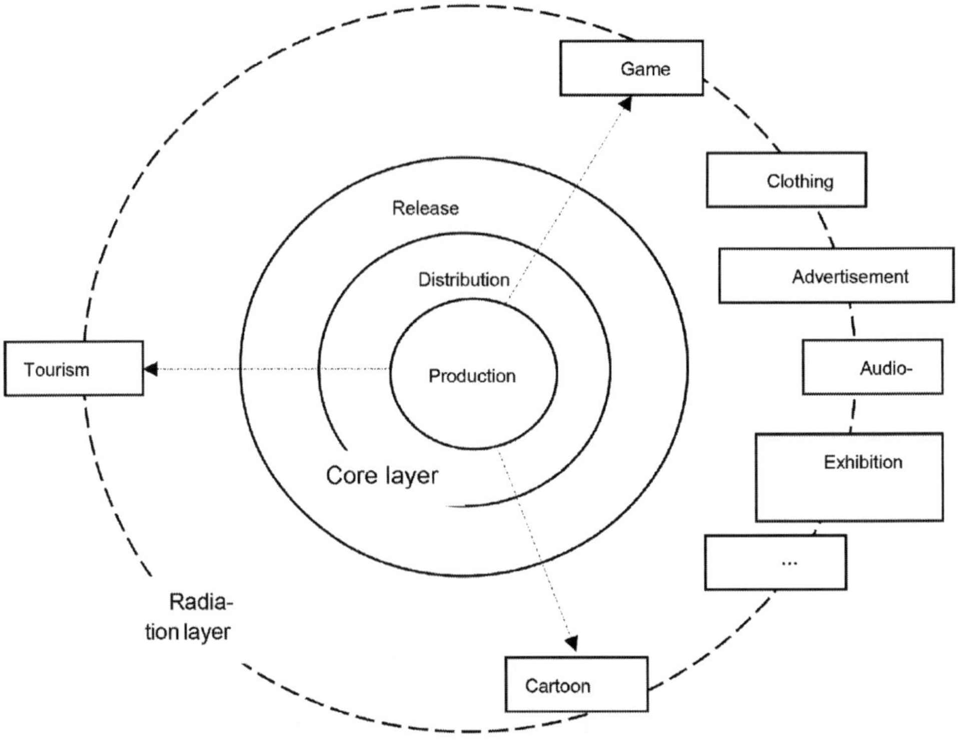

Figure 1. Double-layer network chain structure of film industry chain.

With the start of the new economic year, film industry keeps extending to various relevant industrial chains by the "window extension effect" and forms a "motion picture mega-industry" (Wang and Li, 2007; Barry, 2005). According to the product category in the industrial chain, this article divides film industry chain into core layer and radiation layer, and forms the double-layer network chain structure of film industry chain, as shown in Fig. 1.

3 ESTABLISHMENT OF EVALUATION INDEX SYSTEMS FOR THE SYNERGETIC DEVELOPMENT OF FILM INDUSTRY CHAIN

In order to realize the synergetic development of film industry chain, there should be a scientific and reasonable evaluation on its practical situation. In addition, the key to such evaluation is to establish systematic and reasonable evaluation index systems. The evaluation index systems established for the synergetic development of film industry chain are presented in Table 1 and Table 2.

Table 1. Evaluation index system for synergetic development of core layer of film industry chain.

Target layer	Criterion layer	Index layer
Synergetic development of the core layer of film industry chain	Supply-demand synergy (A_1)	Average time from production to release (a_{11})
		Construction of information management system (a_{12})
		Number of films released annually (a_{13})
		Audience satisfaction (a_{14})
	Organizational synergy (A_2)	Enterprise strategic alliances (a_{21})
		Vertical integration of the enterprise (a_{22})
		Industry-university-research cooperation (a_{23})
	Spatial synergy (A_3)	Industrial cluster scale (a_{31})
		Completeness of intermediaries (a_{32})
		International cooperation (a_{33})
	Value synergy (A_4)	Reasonableness of capital flows (a_{41})
		Construction of diversified financing channels (a_{42})
		Accumulated annual box office (a_{43})
		Profit distribution coordination (a_{44})

Table 2. Evaluation index system for synergetic development of radiation layer of film industry chain.

Target layer	Criterion layer	Index layer
Synergetic development of the radiation layer of film industry chain	Technology synergy (C_1)	Spread of film and TV technologies (c_{11})
		Application of high and new technologies in film and television (c_{12})
	Product synergy (C_2)	Development of film derivatives (c_{21})
		Film promotion of creative elements (c_{22})
	Process synergy (C_3)	Creative design stage (c_{31})
		Content production stage (c_{32})
		Marketing stage (c_{33})
	Market synergy (C_4)	Joint cultivation of consumer groups (c_{41})
		Joint development of regional markets (c_{42})

4 EVALUATION AND ANALYSIS OF THE SYNERGETIC DEVELOPMENT OF CHINA'S FILM INDUSTRY CHAIN BASED ON FUZZY-ANP

Fuzzy analytic network process is a comprehensive evaluation method that combines the Analytic Network Process (ANP) with the fuzzy comprehensive evaluation method. The overall train of thought is to first calculate the stable weight of the evaluation indices that have internal network correlation characteristics by means of ANP, and then carry out a quantitative evaluation of the qualitative indices in line with the fuzzy linear transformation principle and maximum membership principle by means of the fuzzy comprehensive evaluation method, and finally provide the comprehensive evaluation result. So far, fuzzy ANP has been widely applied in various research fields such as engineering management (Zhou and Gao, 2012) and business management (Yu, Zhou and Li, 2013).

Based on the comprehensive evaluation index system, we invited six experts and scholars from relevant universities and enterprises who are familiar with the film industry, to evaluate the related indices of the criterion layer and index layer with weighted scoring, and score the synergetic development of China's film industry chain. The specific comprehensive evaluation process is as follows.

4.1 *Determination of the evaluation index weight based on ANP*

First, we build the ANP models of the synergetic development of the core layer and the radiation layer separately according to the evaluation index systems established above and in line with the logical structure of ANP. The core layer is taken as an example, as shown in Fig. 2.

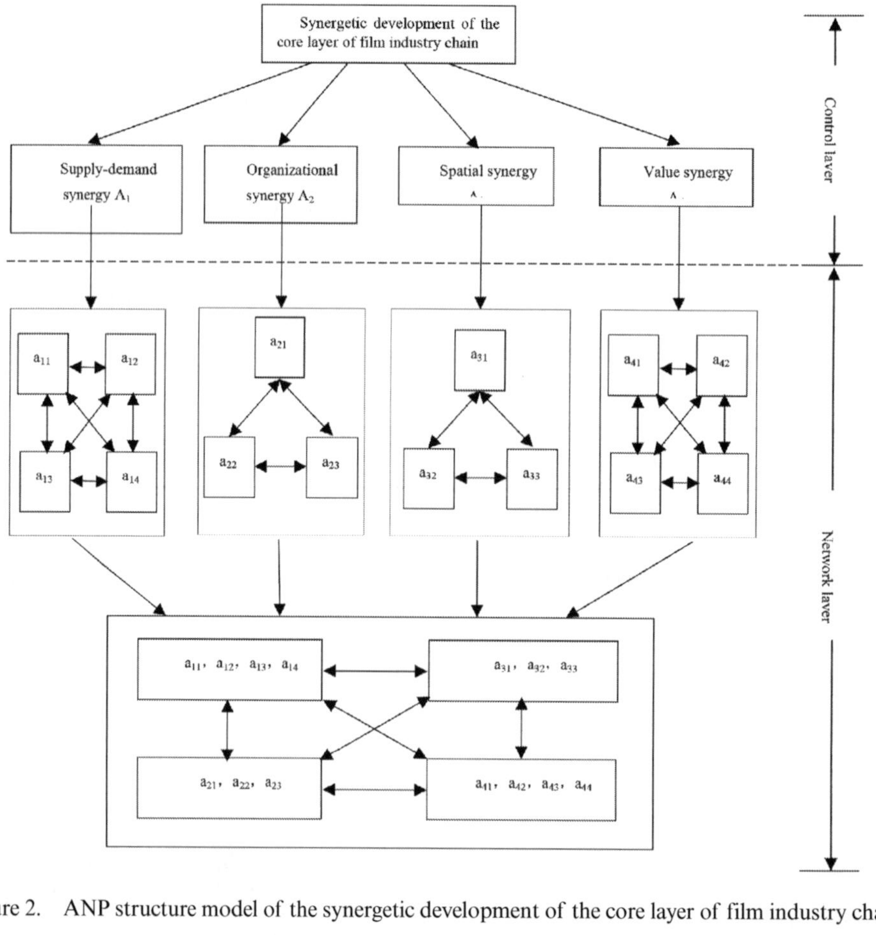

Figure 2. ANP structure model of the synergetic development of the core layer of film industry chain.

Next, based on the average weighted score calculated by experts and scholars, the Rule of Nine Score is adopted for the judgment of every two, to establish the judgment matrix of the criterion layer, as shown in Table 3, which takes the core layer as an example. By calculation, we obtain the consistency inspection result of this judgment matrix, i.e., 0.0539 (<0.1), which indicates that the judgment matrix is acceptable.

Then, according to the above process, the indices of the criterion layer are taken as the judging criteria for the judgment of every two, the construction of the judgment matrix of the network layer, and the consistency inspection, respectively.

Finally, after the consistency inspection, Super Decisions software is used to calculate and obtain the matrix W, the weighted hyper-matrix $\overline{W}$, and the limit matrix. As the limit matrix is convergent and single, we can obtain the stable weight of the evaluation indices of the core layer and the radiation layer of film industry chain, as shown in Table 4.

4.2 *Fuzzy comprehensive evaluation of the synergetic development of China's film industry chain*

First, we summed up and sorted out the scores evaluated by the six experts for the synergetic development of China's film industry chain. Frequencies of the experts' score on different evaluation values are listed in Table 5, taking the core layer as an example.

Table 3. An example of the judgment matrix of the evaluation index based on ANP.

Take V as the criterion	A_1	A_2	A_3	A_4
A_1	1	3	4	1/2
A_2	1/3	1	3	1/3
A_3	1/4	1/3	1	1/4
A_4	2	3	4	1

Table 4. Evaluation index weight of the synergetic development of film industry.

Layer	Index of criterion layer	Index weight of criterion layer	Index of network layer	Index weight of network layer
Core layer	A_1	0.3181	a_{11}	0.0203
			a_{12}	0.0516
			a_{13}	0.0956
			a_{14}	0.1507
	A_2	0.1552	a_{21}	0.0423
			a_{22}	0.0932
			a_{23}	0.0198
	A_3	0.0775	a_{31}	0.0278
			a_{32}	0.0104
			a_{33}	0.0392
	A_4	0.4491	a_{41}	0.0477
			a_{42}	0.0686
			a_{43}	0.2008
			a_{44}	0.1321
Radiation layer	C_1	0.0931	c_{11}	0.0208
			c_{12}	0.0723
	C_2	0.2929	c_{21}	0.1945
			c_{22}	0.0984
	C_3	0.0778	c_{31}	0.0010
			c_{32}	0.0422
			c_{33}	0.0257
	C_4	0.5361	c_{41}	0.3086
			c_{42}	0.2276

Table 5. Evaluation results of the synergetic development of the core layer of China's film industry.

Evaluation index	Evaluation results						
	Very good	Good	Relatively good	Average	Not good	Bad	Very bad
a_{11}	0	0	2	2	2	0	0
a_{12}	0	0	0	0	4	2	0
a_{13}	0	0	4	2	0	0	0
a_{14}	0	0	0	4	0	2	0
a_{21}	0	0	0	4	2	0	0
a_{22}	0	0	0	4	2	0	0
a_{23}	0	0	0	0	0	6	0
a_{31}	0	0	0	4	2	0	0
a_{32}	0	0	0	2	2	2	0
a_{33}	0	0	0	6	0	0	0
a_{41}	0	0	0	0	2	4	0
a_{42}	0	0	0	0	2	0	4
a_{43}	0	2	4	0	0	0	0
a_{44}	0	0	0	2	4	0	0

Then, we used the Matlab software for the calculation of fuzzy linear transformation. Taking the supply-demand synergy A1 of the core layer as an example, we obtain the evaluation matrix E_1^1 as follows:

$$E_1^1 = \begin{bmatrix} 0 & 0 & 0.333 & 0.333 & 0.333 & 0 & 0 \\ 0 & 0 & 0 & 0 & 0.667 & 0.333 & 0 \\ 0 & 0 & 0.667 & 0.333 & 0 & 0 & 0 \\ 0 & 0 & 0 & 0.667 & 0 & 0.333 & 0 \end{bmatrix}$$

With the weight set W_1^1 of the A1 index layer and the weighted evaluation operator M(·,+), we conduct the fuzzy operation of E_1^1, and obtain the evaluation vector of the supply-demand synergy of the core layer as follows:

$$B_1^1 = W_1^1 \cdot E_1^1 = (0, 0, 0.0705, 0.1391, 0.0411, 0.0674, 0)$$

Similarly, we obtain the evaluation vectors of the organizational synergy, spatial synergy and value synergy of the core layer of China's film industry chain: B_2^1 = (0, 0, 0, 0.0903, 0.0452, 0.0198, 0), B_3^1 = (0, 0, 0, 0.0613, 0.0128, 0.0035, 0), and B_4^1 = (0, 0.0669, 0.1338, 0.0440, 0.1268, 0.0318, 0.0457), respectively.

According to B_1^1, B_2^1, B_3^1, and B_4^1, we can build the comprehensive evaluation matrix B^1 of the core layer of China's film industry chain as follows:

$$B^1 = \begin{bmatrix} B_1^1 \\ B_2^1 \\ B_3^1 \\ B_4^1 \end{bmatrix} = \begin{bmatrix} 0 & 0 & 0.0705 & 0.1391 & 0.0411 & 0.0674 & 0 \\ 0 & 0 & 0 & 0.0903 & 0.0452 & 0.0198 & 0 \\ 0 & 0 & 0 & 0.0613 & 0.0128 & 0.0035 & 0 \\ 0 & 0.0669 & 0.1338 & 0.0440 & 0.1268 & 0.0318 & 0.0457 \end{bmatrix}$$

With the criterion layer index weight W^1 and the weighted evaluation operator M(·, +), we obtain the comprehensive evaluation vector U^1 of the core layer by calculation, i.e.:

$$U^1 = W^1 \cdot B^1 = (0, 0.0301, 0.0825, 0.0828, 0.0781, 0.0391, 0.0205)$$

Lastly, according to the above calculation, we obtain the index evaluation vectors (E_1^2, E_2^2, E_3^2 and E_4^2) and the comprehensive evaluation vector (U^2) of the criterion layer of the radiation layer, as shown in Table 6.

Table 6. Comprehensive evaluation results of synergetic development of China's film industry chain.

Layer	Evaluation criteria	Evaluation vector Very good	Good	Relatively good	Average	Not good	Bad	Very bad	Result
Core layer	Overall	0	0.030	0.0825	0.0828	0.078	0.039	0.021	Average
	A_1	0	0	0.070	0.139	0.041	0.067	0	Average
	A_2	0	0	0	0.090	0.045	0.020	0	Average
	A_3	0	0	0	0.061	0.013	0.003	0	Average
	A_4	0	0.067	0.134	0.044	0.127	0.032	0.046	Relatively good
Radiation layer	Overall	0	0	0.001	0.204	0.132	0.031	0.019	Average
	C_1	0	0	0	0.038	0.031	0.024	0	Average
	C_2	0	0	0	0.066	0.065	0.098	0.065	Not good
	C_3	0	0	0.017	0.054	0.003	0.003	0	Average
	C_4	0	0	0	0.330	0.206	0	0	Average

4.3 *Analysis of the results*

Table 6 summarizes the above calculation results, and based on the principle of maximum membership, we obtain the comprehensive evaluation results of China's film industry chain under various criteria, as shown in table 6:

1. Analysis of the evaluation index weight. From Table 4, we can see that for the synergetic development of the core layer of film industry chain, in the criterion layer, the value synergy is the most important evaluation factor; in the network layer, the accumulated annual box office is the most important evaluation index. For the synergetic development of the radiation layer of film industry chain, in the criterion layer, market synergy is a piece of evaluation criteria with the highest weight; in the network layer, the joint cultivation of consumer groups is the most important evaluation index.
2. Analysis of evaluation results under different evaluation criteria. From Table 6, we can see that, first, the overall synergetic development situation of the core layer of China's film industry chain is just average. Given the aspect of the value synergy, the core layer of China's film industry chain development is good. The development situations in terms of supply-demand synergy, organizational synergy, and spatial synergy are limited. This shows that although the China's film industry chain has achieved rapid growth in the whole output dominated by the box office in recent years, problems still exist in the product supply and demand relationship, the cooperative relationship between organizations, regional cooperation, and other issues. Secondly, the overall synergetic development situation of the radiation layer of China's film industry chain is not good. In the radiation layer, the evaluation of the development in technology synergy, process synergy, and market synergy is just average, especially the product synergy which is in bad development condition. This shows that under the trend of industrial convergence, a preliminary "motion picture mega-industry" development trend has been formed in China's film industry chain. However, China's film industry chain lacks technical exchanges, process cooperation, and market expansion with the related industries, especially the collaboration at the product level which is much lower, showing its limited exchanges and communication with other industries.

5 CONCLUSIONS

The synergetic development of film industry chain is an inevitably choice in response to external challenges such as environmental uncertainties and international market competition, as well as the inherent requirement to solve internal problems such as conflicts of interests and information silo. Based on the above research, this paper argues that the synergetic development of film industry chain should make the core links of film industry chain such as the production, distribution, and release center on film products for element configuration and communication, and realize supply-demand synergy, organizational synergy, spatial synergy, and value synergy. Meanwhile, it is required that the film industry chain realizes the linkage and integration with the related industrial chains at the industrial level, so as to realize technology synergy, product synergy, process synergy, and market synergy and promote the efficient, orderly, and open development of the overall film industry chain.

However, this research still has some limitations. For example, further in-depth research on the connotation and framework systems of the film industry chain is required, the evaluation index system needs to clearly reflect the characteristics of film industry chain, and the case research lacks the evaluation and research on the synergetic development of the regional film industry chain or industrial cluster. These problems will be considered for the future directions of research.

REFERENCES

Barry R.L. 2006. The Motion Picture Mega-Industry. *Tsinghua University Press.*

Guo X.W., Yan N. 2012. Reflect on the Film Industry Chain Upgrade in the Digital Times. *Journal of Northwest University (Philosophy and Social Sciences Edition)*, (5): 187–188.

Liu H.B., Huang Z.H. 2007. An Empirical Research on the Synergy Effects in the Integration of Industry Chains: The perspective of circular economy theory. *Technology Economics*, (9): 26–28.

Lu Z.N., Liu C.Q., Wang G.D. 2013. Research on the Synergetic Performance Evaluation Index System of the Photovoltaic Industry Chain. *Science & Technology and Economy*, (1): 106–110.

Mi L.Z., Yu H. 2011. Research of China's Film Industry Based on Value Chain. *Economic Forum*, (1): 101–104.

Qian Z.Z. 2008. Dynamic Progress of Value Chain and the Sustainable Competitive Advantages of the American Film Industry. *Forum of World Economics & Politics*, (6): 112–117.

Tan H.L. 2010. Research on the Automobile After-Sales Service Management System Facing to the Industry Chain synergy. *Kunming University of Science and Technology.*

Tang X.B., Huang Y.Y. 2005. Evaluation of SCM Strategy and Model. *Journal of Information*, (1): 89–91.

Wang H.L., Li G.P. 2007. From Ideas to Commodities Running Procedure and Growth of Creative Industries Based on Survey of One-Idea-Diversiform-Development. *China Industrial Economics*, (8): 60–67.

Yu S.K., Zhou L.S., Li C. 2013. Research on the Application of ANP-Fuzzy Method in the Performance Evaluation of Electric Power Enterprise. *Chinese Journal of Management Science*, (1): 167–175.

Zhou R.F., Zhao J.X. 2008. Construction of the Evaluation Index System for the Supply Chain Collaboration. *Statistics and Decision*, (13): 66–68.

Zhou X.G., Gao X.D. 2012. Method of Selection Construction Projects Based on FANP Model and Its Application. *System Engineering Theory and Practice*, (11): 111–118.

Zhu R. 2012. Research on Collaboration of the Internet of Things Industry Chain Based on Value Net Theory. *Nanjing University of Posts and Telecommunications.*

Advanced Engineering and Technology III – Xie (Ed.)
© 2017 Taylor & Francis Group, London, ISBN 978-1-138-03275-0

Author index